Automotive Powertrains

Autotech 95 Organizing Committee

C Bale
Tickford Limited

I Birchall
Prodrive Engineering Limited

J Chelsom
City University

M Dunn (Chair)
Consultant

P Extance
Lucas Advanced Engineering Centre

C Hickman
MIRA

R Johnson
SDRC Engineering Services

P Marshall
Consultant

I Noble
Consultant

S Ocock
Rover Group Limited

B Southall
Perkins Technology Limited

D Tidmarsh
University of Central England

J Turner
University of Southampton

M Williams
Jaguar Cars Limited

IMechE Seminar Publication

I MECH E
150th Anniversary
1847 - 1997

Automotive Powertrains

Selected papers from Autotech 95
7–9 November 1995

Organized jointly by the Automobile Division
of the Institution of Mechanical Engineers (IMechE)

and in association with the Institution of Electrical Engineers

IEE

IMechE Seminar Publication 1996 – 8

MEP

Published by Mechanical Engineering Publications Limited for
The Institution of Mechanical Engineers, London.

First Published 1996

ISSN 1357-9193
ISBN 1 86058 020 3

A CIP catalogue record for this book is available from the British Library
Printed by The Ipswich Book Company, Suffolk, UK

Contents

Hybrid Powertrains

Diesel Engine Developments

Engine Developments

Driveline Developments

Transmissions

Emissions

Mechanical Developments

Related Titles of Interest

Title	Author	ISBN
Automotive Manufacturing	IMechE Seminar 1996–4	1 86058 023 8
Automotive Systems, Sensors, and Signalling	IMechE Seminar 1996–5	1 86058 022 X
Automotive Passenger Safety	IMechE Seminar 1996–6	1 86058 042 4
Automotive Refinement	IMechE Seminar 1996–7	1 86058 021 1
Set of 5 Autotech Volumes (including the 4 titles listed above and this title)	IMechE Seminar 1996	1 86058 019 X
Handbook of Vehicle Design Analysis	J Fenton	0 85298 963 6
Application Of Power Train And Fuel Technologies To Meet Emission Standards	IMechE Conference 1996–5	0 85298 996 2
Diesel Fuel Injection Systems	IMechE Seminar 1995–3	0 85298 980 6
Using Natural Gas in Engines	IMechE Seminar 1996–2	1 86058 033 5

Hybrid Powertrains

Critical components for fuel cell hybrid powertrains

C C BOURNE MSc
Rover Advanced Technology Centre, University of Warwick, UK
R A DAMS BTech, CEng, MIChemE
CJB Developments Limited, Portsmouth, UK
C M SEYMOUR BEng, CEng
Intensys Limited, Ulverston, UK
J G SMITH BSc
Lucas AEC, Solihull, UK

SYNOPSIS

A fuel cell/battery hybrid drive system for a car is discussed in terms of the requirements of a Rover 200 Series vehicle. Components critical to further development are those associated with fuel processing, the fuel cell peripheral systems and the power buffer store.

1 INTRODUCTION

An increasing awareness of pollution issues and a desire to reduce both local and global emissions from industrial and transport activities has initiated further investigations of alternative power plants for road vehicles. These efforts have been given impetus by the California legislation requiring Zero Emission Vehicles (ZEVs) to be available in that state by 1998. The influence of the Californian action has produced similar proposals elsewhere in North America and the commercial implications have aroused considerable activity in Europe and Japan. Apart from the export trade to the USA, European interest centres on reducing urban pollution, but it is not so well focused as in the USA or Japan.

The immediate technical response to these pressures has been the re-development of vehicles powered by secondary batteries. These eliminate local pollution, though they may still produce high emission levels at the points of power generation. Moreover, the power generation, transmission and recharging processes introduce inefficiencies which reduce the overall advantages.

Possibly the largest obstacle to battery vehicles, at least in Europe, Japan and the USA, is the difference between the capabilities of such vehicles and those using the highly developed form of the internal combustion engine which is familiar in all industrialised countries and which is provided with a world-wide fuel infrastructure. The short range of electric vehicles together with the feature limitations due to restricted energy storage and the relatively long refuelling time have to be countered with financial concessions or special operational privileges which are politically difficult to control (1). Thus battery vehicles may only provide an interim solution except in those situations where the limited performance is acceptable.

It is therefore important to examine more advanced solutions which have the potential to satisfy the sophisticated customer whilst giving improved efficiency and minimal emissions. Fuel cells have the potential to be used in this way and their application to road vehicles has been proposed from the 1960s (2). The obstacles to such fuel cell applications have traditionally been low power per unit weight and volume and the preference for hydrogen as the basic fuel. However, the last decade has seen developments in fuel cells that render them considerably more attractive and which prompt a new examination of their role as vehicle power sources. In particular, the Ballard Company in Canada has produced trial quantities of high power density fuel cell stacks which are being used in demonstration buses (3).

In view of the promising prospects for fuel cell systems, Rover and Lucas (vehicle systems) together with Intensys and CJB Developments (fuel cell systems) have been examining the feasibility of installing a fuel cell power train in a medium size car. This paper reviews the important issues to be faced before such a system can become a practical reality.

2 FUEL POWER SYSTEMS FOR VEHICLES

There are several types of fuel cell but the one most appropriate for transport applications is the so-called Solid Polymer Fuel Cell (abbreviated to SPFC). The basic cell consists of porous catalysed electrodes coated on either side of a thin sheet of ionically conducting polymer. The essential elements are shown in figure 1. This particular cell system is attractive since-

a) High power densities (~10 kW/m^2) are available from modern electrode constructions. In addition unit cell thickness is of the order of millimetres.

b) The cells operate at normal temperatures (though optimum output is at about 90°C).

c) The supporting structures involved can be made from readily available materials.

d) The form of construction (stacks of thin laminate components) is amenable to mass production.

The SPFC also presents some challenges, notably-

a) Its optimum fuel is pure hydrogen. The hydrogen can be mixed with other gases, but some gases notably carbon monoxide poison the electrode catalysts and reduce the output. Air is normally used to provide the oxygen supply for the positive electrode.

b) The water content of the polymer electrolyte membrane needs to be kept within a proscribed range. Too much water reduces output by affecting electrode operation and too little dries out the electrolyte membrane, reducing its conductivity.

c) Maximum power is obtained at approximately half the cell open circuit voltage. This requires the power systems involved to tolerate a fairly large voltage range.

The size determining factor in fuel cell systems is the maximum power requirement. This is the main factor in determining the number of cells required and the

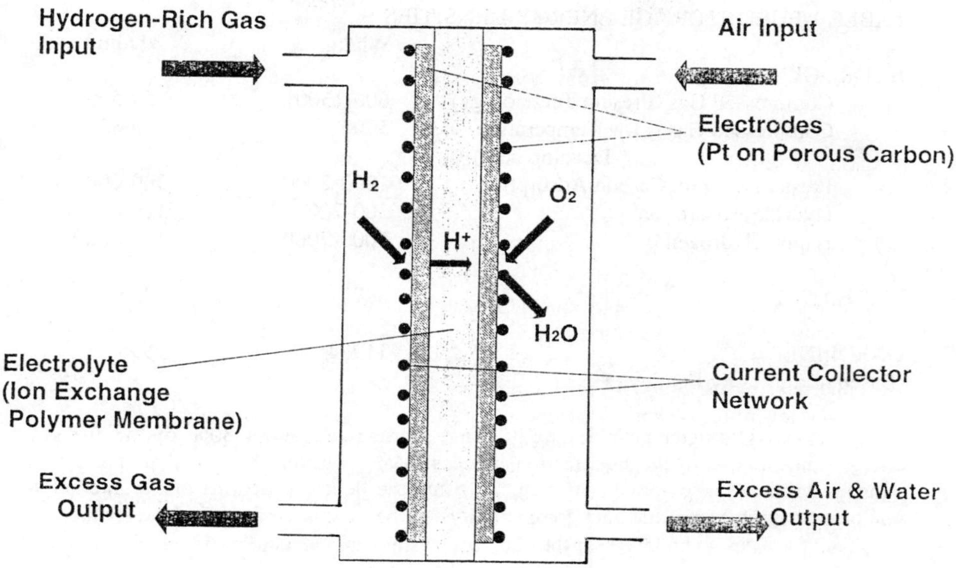

Figure 1 Essentials of a Solid Polymer Fuel Cell

electrode areas involved. It also determines the maximum hydrogen and air flow requirements. The fuel system's energy storage is determined only by the size of the fuel tank (in contrast to traction batteries where the energy requirement is usually the size determining factor for the system).

2.1 Ancillary Systems

Current SPFC systems require air and fuel gases at about 200 kPa. In order to remove the water produced at the positive (air) electrode, a flow equivalent to twice the stoichiometric value is required (i.e. about 2000 dm^3/h per kW at 200 kPa). The pumping power required is considerable and may amount to 20% of the system's gross output. Energy recovery from the exhaust gases may be considered to improve the overall system efficiency.

As noted earlier, pure hydrogen is the preferred fuel. However, the available means of hydrogen storage leave much to be desired when compared to the energy storage density of common liquid hydrocarbon fuels. A comparison is shown in Table 1

TABLE 1-FUEL STORAGE ENERGY DENSITIES

	Wh/kg	Wh/dm^3
HYDROGEN		
Compressed Gas (Present Technology)	600-1300	350-500
Compressed Gas (Low Temperature Development)	5000	500-850
Pressurised with Carbon Adsorption	1000-2000	500-600
Hydride Absorption	300-700	1000-3000
Liquid Hydrogen	5000-7000	1500-2000
METHANOL	5472	4323
GASOLINE	11994	9595

(Based in figures by Bentley (4))

Liquid hydrogen provides fairly dense storage but about 30% of the fuel's energy value is lost in the liquefaction process. Many workers believe that fuel cell vehicles will need to co-exist with vehicles using the developed forms of the internal combustion engine and that fuel compatibility will be necessary for initial acceptance. This then provides a challenge for the chemical system engineers in providing a suitable hydrogen generator.

Other ancillary systems are associated with cooling the fuel cell stack and other associated components. At full power, the fuel cell stack may only be 50-60% efficient and appropriate heat removal is required.

It should be noted that apart from the fuel tank itself, the sizes of all the ancillary components are determined by the maximum power requirements.

3 VEHICLE SYSTEM REQUIREMENTS

In order to define the required power characteristics for a fuel cell based system, studies were carried out by Rover on the basis of a Rover 200 series platform. The vehicle characteristics are given in Table 2.

TABLE 2-ROVER 200 CHARACTERISTICS

Vehicle type	4/5 seat Saloon
Kerb Weight	1100 kg
Payload	400 kg
Installed Power (Diesel)	75 kW
Acceleration (0-100 km/h)	16 s
Maximum Speed	125 km/h

This vehicle, with its existing engine replaced by an alternative electric power train weighing 400 kg would require cruising power (on the level) to its traction motor as shown below.

Speed (km/h)	Power (kW)
50	10
80	16.5
115 (UK limit)	29.0

Thus the fuel cell stack must be capable of producing 29 kW continuously over periods of several hours. However, the largest power requirements occur during acceleration. Tests with vehicles show that the worst case acceleration occurs when joining high speed traffic streams. The worst case acceleration power for the Rover 200 demands 65 kW for 10 s into the traction motor.

Suburban driving produces a power requirement profile with large, short duration peaks and almost equally large peaks of power output due to regenerative braking. A driving cycle used as a yardstick by Rover involves an urban/suburban journey to the centre of Coventry. The speed/power profiles obtained from driving the cycle with electric vehicles have been used to calculate the power requirements in urban driving. The power requirement profile for a Rover 200 series vehicle is shown in Figure 2. It will be observed that though the maximum power requirement is nearly 45 kW, it is required for a vanishingly small period of time. Some 90% of the cycle demands can be covered by a power of 20 kW.

If the fuel cell system train is sized to meet a continuous 65 kW, it will be wastefully large since, as explained in section 2, the size of the fuel cell stack and its peripheral systems depend on the peak power requirement. The situation then argues for a hybrid system using a buffer store as shown in Figure 3. This is a classical series hybrid which can be configured with a variety of energy converters and power storage devices.

The series hybrid arrangement has the following advantages-

i) The size weight and cost of the fuel cell, fuel processing and peripheral items are drastically reduced.

ii) The fuel cell and fuel processing stages can be run in steady states whilst the buffer store meets transient demands.

iii) The buffer store can provide power to cover activation and starting phases.

iv) The buffer store can accumulate power generated in regenerative braking.

The negative side of a series hybrid is that inefficiencies in succeeding stages accumulate and may give small overall efficiency. In addition, the extra complexity brings weight, weight volume and cost problems of its own.

Classically the buffer store considered has been a battery but in recent years, new candidates have appeared in the form of flywheels and supercapacitors.

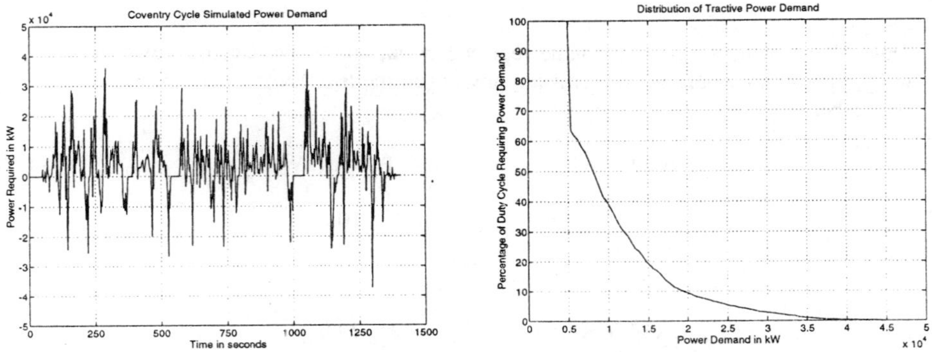

Average Power Over Cycle	3.40 kW
Maximum Tractive Demand	44.64 kW
Maximum Regenerative Power	37.80 kW
Energy Used	1.20 kWhrs

Figure 2 Power Requirements for a Rover 200 in the Simulated Urban Cycle

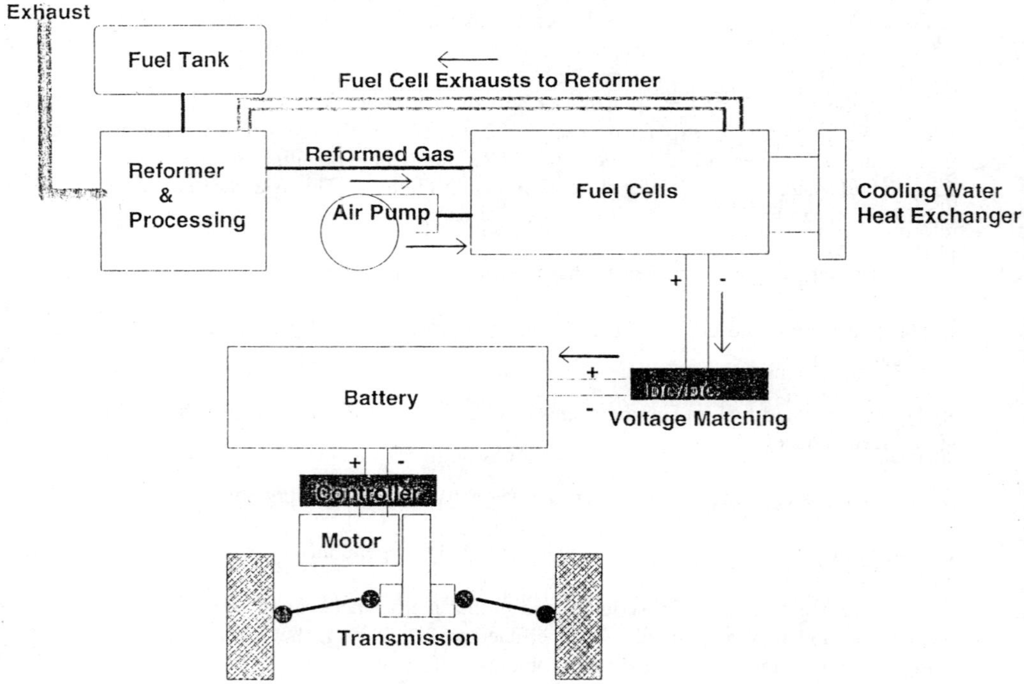

Figure 3 Fuel Cell Series hybrid System

4 HYBRID SYSTEM COMPONENTS

It is appropriate to consider the various components of the power system as required by the Rover 200 example and to examine the critical influences which will affect the practicality of this type of traction system.

4.1 The Fuel Processor

The purpose of the fuel processor is to take the available fuel and turn it into a hydrogen-bearing gas stream compatible with fuel cell operation. The simpler hydrocarbon and alcohol molecules are easily treated by the process of steam reforming. This involves reaction with water vapour on a copper/zinc oxide catalyst at temperatures around 250°C. Typically this is applied to methanol which reacts as follows.

$$CH_3OH + H_2O \rightarrow CO_2 + 3H_2$$

The resulting gas containing 75% hydrogen (by volume) may also contain some fractional percentage of carbon monoxide, a fuel cell electrode poison. It can be removed by further reaction with steam in the water gas shift reaction.

$$CO + H_2O \rightarrow CO_2 + H_2$$

It is also possible to add a controlled amount of air to the gas stream and to pass it over a selective oxidation catalyst. This can reduce CO to levels (~1 ppm) harmless to the fuel cells.

The reforming reaction is endothermic and heat input is required (this is, however, reflected in the higher thermal value of the output gas). The heat may be supplied by using the fuel gas and air exhausts from the fuel cell stack and burning the excess hydrogen in the reformer. A methanol catalytic burner is used for starting. Similarly the water for the reforming process is obtained by condensing the steam from the fuel cell exhaust gases. Intensys and CJB Developments have operated a methanol reforming system of this type in conjunction with a 10 kW fuel cell installation from Ballard (5).

Steam reformers are well understood as parts of large chemical plants but reducing them to the appropriate size for this application presents challenges. Reformers are also normally run in steady state conditions and are not really suited to load following applications. Fortunately the hybrid system uses its buffer store for load following and allows operation of the fuel processor at appropriate steady levels of output. In such favourable conditions, system efficiencies of over 70% may be attributed to the reformer and emissions from its exhaust contain 30 ppm CO and 0.5 ppm NO_x (6).

Figure 4 shows the elements of a typical reformer unit of tubular construction. Heat is applied in the central flue, around which the catalyst beds are disposed. Such units tend to respond relatively slowly to changes in loading and start-up times may be several minutes. Newer structures are being developed based on metal-supported catalyst matrices which have much faster response and shorter warm up times in addition to being physically smaller.

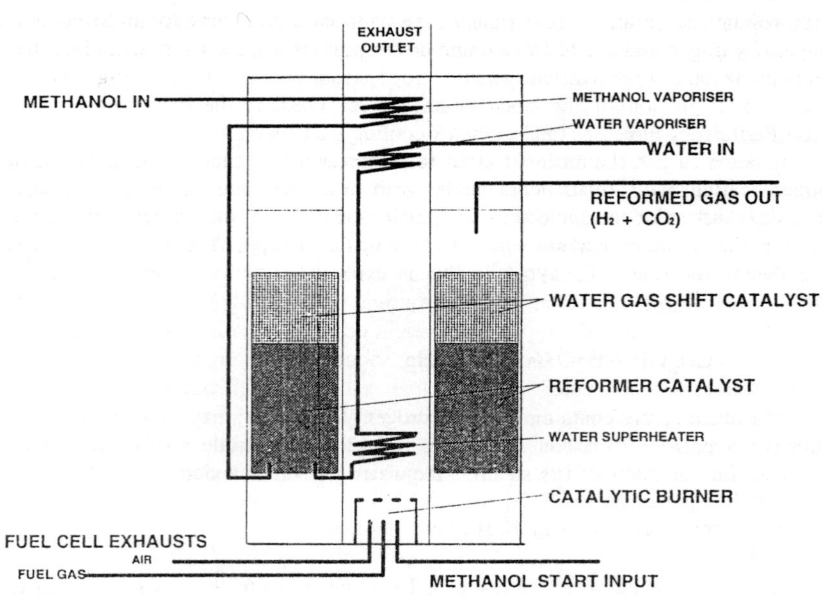

Figure 4 Tubular Steam Reformer Schematic

A methanol reformer for the Rover 200 would be required to support a net fuel cell output of 30 kW (39 kW total) and, the latest matrix version would occupy about 48 dm^3 and weigh about 72 kg.

A different fuel processing route involves partial oxidation of the vaporised fuel with carefully controlled amounts of air. In the case of methanol, the reaction is-

$$2CH_3OH + O_2 \rightarrow 2CO_2 + 4H_2$$

It can be seen that the hydrogen yield is less than with the steam reformer. However, no water input is needed and, since the reaction is exothermic, no external heat supply is required. Partial oxidation can be used with a variety of fuels including petrol. The operating temperatures are higher than for reformers and with lower hydrogen/carbon ratio fuels, carbon deposition in the reactor may be a problem.

The fuel and the fuel processor requirement is now a critical problem for transportation fuel cell systems. Discussions on appropriate fuels which could be available in the early 21st century involve a wide variety of political and economic as well as technical factors, but until the fuel situation becomes clearer, system designers will not be completely confident about their fuel processing arrangements.

4.2 The Fuel Cell Stack

Fuel cell stacks consist of the basic cells shown in figure 1 stacked up to contain the required voltage and power. Figure 5 shows the voltage/current curve for an SPFC cell running on hydrogen and air. The maximum power point is obtained at roughly half the open circuit voltage. Cell efficiency is also proportional to the cell operating voltage expressed as a fraction of the open circuit value. Thus as the cell load current decreases, efficiency increases (and the stack cooling load decreases).

As noted earlier, the airflow to the stack represents a significant parasitic load. According to Ballard (7), cells might be designed to allow water removal at the fuel electrode as well as at the air electrode, thus reducing air flow requirements. This results in reducing the compressor loss to 11% of the net output. Up to 60% of the loss can then be reclaimed via a turbine in the air exhaust.

The development of suitable electric/turbine driven compressors, operating at the 5 to 10 kW level, and of compact lightweight design is now a critical factor. It is important that noise is kept low and that the unit should have a long service life.

Further issues connected with the fuel cell stack are concerned with cost reduction. Until recently costs of over a $1000/kW were being quoted. Substantial efforts have been made to reduce the cost of the electrode catalysts (platinum metal based) and also the costs of the specialised polymer membrane electrolyte. A recent estimate (8) has put costs at $235/kW with about $12/kW being catalyst costs and $13/kW being apportioned to the electrolyte membrane. Target costs for the fuel stack are about $150/kW.

Figure 5 Polymer Fuel Cell Voltage/Current Characteristic

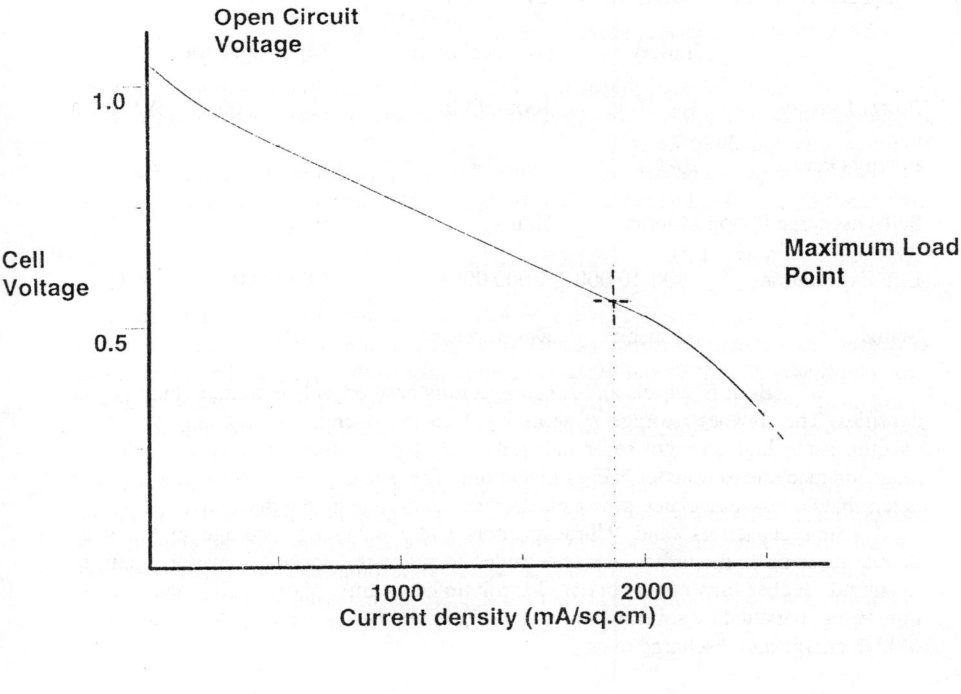

The fuel cell stack for the Rover 200 is expected to provide 30 kW (allowing 1 kW for auxiliary systems) at about 200-400V. In terms of the latest SPFC technology, this would occupy a space of about 60 dm^3.

4.3 Power Conditioning

The voltage variations noted in Figure 5 require matching to the input of a suitable power storage device. Compact multi-kilowatt DC/DC converters are available which may be 95% efficient at the rated load. It is important to achieve designs that will maintain high efficiencies at part load, since continuous full load operation would only be encountered in motorway cruising conditions.

The converter will need to cope with the fuel cell peak output (30 kW for the Rover 200) and to convert it from the 200-400V levels at the fuel cell to 250-300 V for the buffer store (battery) and the traction system.

4.4 Buffer Store

The buffer store in the case of the Rover 200 needs to provide 35 kW in maximum acceleration periods. In maximum regeneration it would require to absorb nearly 45 kW. If the buffer store has to support the maximum load on start-up, it may be required to provide 65 kW, unless a lower initial performance is accepted.

The actual energy required is small. Figure 2 shows that the actual energy required to drive a 20 minute suburban cycle is only 1,25 kWh. Thus the buffer store requires optimisation for maximum power output.

There are three candidate technologies available; batteries, flywheels and supercapacitors. The appropriate characteristics are summarised in Table 3.

TABLE 3-BUFFER STORE TECHNOLOGIES

	Battery	Flywheel Store	Supercapacitor	
Power Density	100-1000	1000-5000	1000-10 000	W/kg
Energy Density	20-150	5-100	2-5	Wh/kg
Self-Discharge Period	Months	Hours	Hours	
Life Expectation	1000-10 000	1 000 000+	1 000 000+	Cycles
Status	Available	Development	Trials	

Both flywheels and supercapacitors now provide extremely high power densities. The flywheel storage systems involved (9) operate at very high speeds in vacuum using high-strength rotor materials and incorporating some type of electro-magnetic machine to transfer energy in and out. The actual power transfer rate may be determined by the associated power electronics rather than by the machine itself.

Supercapacitors (and 'Ultracapacitors') rely on charge storage in the ionic double layer on high surface area electrodes to produce capacities several orders of magnitude higher than conventional electrolytic capacitors. Various trial samples are now being marketed by several companies (10). Properly designed versions are capable of high charge and discharge rates.

Both of the above devices are capable of higher power flows per unit weight and volume than batteries. However, the buffer store is expected to provide starting power perhaps after long periods of lay-off. Here the battery scores as a more reliable power store. Moreover the redesign of battery cells can produce constructions capable of very high power densities (kW/kg) with some sacrifice of energy density (11). The battery life in the expected duty is yet to be determined. It is known that battery life is exponentially dependent on the depth of discharge on regular cycling. Longer life on deep discharges is usually bought at the expense of added weight and volume.

In terms of available battery technologies, the 35 kW battery required for the Rover 200 would be as shown below.

Battery System	W/kg	Wh/kg	Weight(kg)	Volume(dm^3)	Capacity(kWh)
Lead-acid	250	40	140	70	5.6
Nickel-cadmium	600	45	58	40	2.6
Nickel-hydride	250	65	140	56	9.1

High temperature batteries such as sodium-sulphur and sodium-metal chloride are ruled out by long term thermal management problems. As noted earlier the buffer store may be required to activate and support the system after a long period of inactivity. The high conductivity of aqueous electrolytes and the wealth of experience with lead-acid and nickel-alkaline systems makes them more likely to be developed in a high power form specifically suited to this type of duty.

It can be seen from the figures above that the battery size is determined in this particular application by the power requirement. The resulting capacities are more than sufficient to cover two urban driving cycles. Optimisation of cell designs for highest power output would probably decrease the capacity available.

The charge input rate to batteries is more limited by electrochemical effects than by cell design and this may limit the effectiveness of charge return via regenerative braking.

4.5 Electric Drive Train

Motor and controller technology for electric vehicles is well advanced and is probably the least critical part of a fuel cell hybrid drive train. Developments in power electronics have enabled more efficient AC and brushless designs to take over from the traditional DC technology. The latest generation of electric vehicles (e.g. the Ford Ecostar and the GM Impact) use induction motors since that technology is familiar in other contexts. More advanced vehicles may use synchronous permanent magnet motors with neodymium-iron-boron magnet technology (12). The switched reluctance motor may well prove a more economic alternative if the cost of magnet materials remains high.

A 65 kW permanent magnet drive train for the Rover 200 vehicle considered is likely to weigh about 70 kg and to occupy 35 dm^3.

4.6 System Controls

A principal advantage of an electric power train is its potential simplicity at the driver interface. The type of hybrid system described requires individual control strategies for the fuel processing system, the fuel cell stack, the buffer store and the drive train (including braking). None of these pose difficult technical requirements but further operating experience is required before satisfactory controls can be fully developed. The individual systems would cover issues such as starting/shut-down, safety and efficiency. A main vehicle system controller would co-ordinate the individual system functions.

Some means of recognising the type of journey being travelled (e.g. urban, motorway cruise) might considerably improve the system efficiency.

5 THE ROVER 200 SYSTEM

As explained earlier, the Rover 200 has been used as a model vehicle for system development. Some elementary simulation of the components required has resulted in dispositions as shown in Figure 6. This is based on the use of methanol as fuel and uses a lead-acid battery as the buffer store. The overall weight of the power system is estimated at 350 kg and the volume at 290 dm^3.

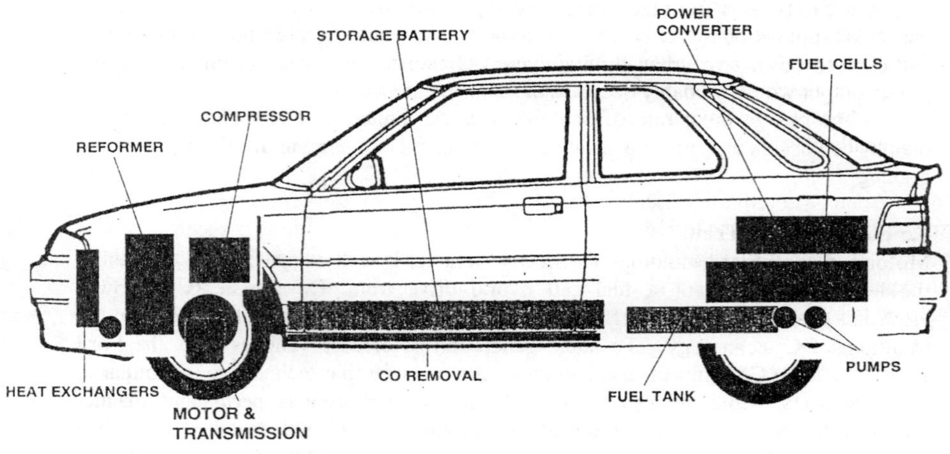

Figure 6 Disposition of Drive Train Components in the Rover 200

6 CONCLUSIONS-AREAS CRITICAL TO FURTHER DEVELOPMENT

Experience in configuring the drive train for the Rover 200 has highlighted the following critical areas.

Fuel and Fuel Processing
It is important to determine fuel availability and to design a hydrogen generator based on the fuels available. In the foreseeable future, these latter are likely to be those compatible with developed forms of spark ignition and diesel engines. The fuel processor should have the shortest warm-up time otherwise the buffer store may need increased power and energy to support this phase.

Fuel Cell Airflow Requirements
The large parasitic demand for air pumping needs to be reduced. This can be helped by making use of exhaust energy available but changes in the fuel cell design may be more important.

Buffer Store Technology
Batteries may be the best in terms of storing starting power over long periods but re-engineering to maximise power is needed to match the available systems to the expected duty.

Production to Automotive Cost Targets
Several of the major items involved have not yet been produced for automotive applications and the relevant production engineering has yet to be applied. The manufacture of technically functional items with an acceptable life-span and within the limits of car producers budgets is probably one of the largest challenges.

Apart from the fuel and processing issues, none of the above items are expected to contain major technical obstacles. Steady evolutionary developments could provide the necessary solutions. The potential of fuel cell power trains to provide convenient and efficient electric vehicles is such that an increasing effort is likely to be mounted to expedite their realisation.

REFERENCES

1. WALLACE, J R, 'Proposed Commercialisation Plan for the Electric Vehicle',Paper 1.05 Proc. 11th International Electric Vehicle Symposium, Florence, September 1992.

2. BUECHLER, L W, 'Fuel Cells for Vehicles', Proc. 17th Annual Power Sources Conf., New Jersey, USA, May 1963 (pub PSC Conf. Committee)

3. PRATER, K, 'SPFC Fuel Cells for Transport and Stationary Applications', 4th Grove Symposium on Fuel Cells, London 1995.

4. BENTLEY, J M, 'Hydrogen Storage for Transportation Fuel Cells',SAE Toptech on Fuel Cells for Transportation, Diamond Bar, California, March 1993.

5. SEYMOUR, C M, DAMS, R A, ADAMS, K and MOORE, S C, 'Development of a 10 kW Breadboard System Demonstrator and its Potential for Transportation', Fuel Cells for Transportation III, European Fuel Cells Group Workshop, Cambridge, 1994.

6. VAN DEN BROECK, H, 'Fuel Cells and Reformers', AVERE Contract P-001(European Community), EDS Study on Electric Drive Systems.

7. WILKINSON, D P, VOSS, H H AND PRATER, K, 'Water Management and Stack Design for Solid Polymer Fuel Cells', J. Power Sources, 1994, 49, 117-127.

8. BORRONI-BIRD, C, 'The Challenges Facing Fuel Commercialisation for Light Duty Vehicle Applications', 4th Grove Fuel Cell Symposium, London 1995.

9. HOWE, D, MELLOR, P H, 'High Speed Electro-Mechanical Energy Storage Systems for Electric Vehicles', Invited Paper SPT IS 03-2, Stockholm Power Technology Conference, Stockholm, June 1995.

10. DOWGALLIO, E J, 'Perspective on Ultracapacitors for Electric Vehicles, IEEE AES Systems Magazine, August 1995, 26-30.

11. GRIBBLE, D, 'The ARA Bipolar Battery', Batteries International, Issue 19, April 1994, 45.

12. WEST, J G W, 'DC, Induction, Reluctance and PM Motors for Electric Vehicles', Motors and Drives for Battery-Powered Propulsion, IEE Colloquium (P1/P6/C12 Groups), London, April 1993.

C498/29/079/95

Turbogenerators – a new way forward

M R ETEMAD PhD, CEng, MIMechE, **A FENOCCHI** BEng, **J HALL** BEng, and
K PULLEN PhD, CEng, MIMechE
Imperial College of Science and Technology, London, UK

SYNOPSIS

New impetus for compact, highly efficient, and environmentally friendly powertrains for automotive use has been prompted by public demand leading to Government legislation's as well as technological breakthroughs. The present contribution concentrates on new advances in material technologies as well as design paving the way for the realisation of such powertrains. The concepts of series and parallel hybrid are exposed and examples of existing prototype vehicles cited. The topology of a series hybrid powertrain is further detailed and the constituent elements of continuous power and energy storage discussed. For this configuration, an advantageous powertrain to provide continuous power is argued to be turbo-electric. This is a directly coupled high speed generator to a gas turbine engine. The contribution describes advances in magnetic materials and composites and argues these to be the prerequisite technologies for high speed permanent magnet generators. The two known designs of such devices are discussed and their basic and theoretical performance compared. A disc type generator, being presently developed at Imperial College, is described and its performance data presented. Finally, the high speed generator coupling to the gas turbine engine is discussed and possible solutions are described in the form of magnetic, gear and diaphragm couplings.

1. INTRODUCTION

Energy, efficiency, and the environment are the most important issues for the automotive industry today, promoted by technological advances as well as US legislation directives. The latter was formally triggered off by Californian 1992 legislation requiring three way catalysts to enable 0.39 g/mile Hydrocarbon emissions. This is to be reduced progressively to 0.07 g/mile by the year 2003 (an 80% reduction in twelve years). Similar reductions in Nitrogen oxide and carbon monoxide emissions are also being enforced. Improvements in vehicle emission standards are being pursued in tandem with outright legislation on sales. In California 2% of all vehicles sold by 1998 must be zero emission (ZEV). This is to increase to 5% and 10 % by 2001 and 2003 respectively.

The impact of such aggressive legislation has accelerated research and development in all areas of automotive technology, with industry and regularity authorities providing funding and direction. Areas being particularly targeted are technologies with direct ZEV potential, chemical and mechanical energy storage devices, as well as those with hybrid potential and benefiting from ultra low emission (ULEV) characteristics.

Variety of ZEV have been manufactured and tested with batteries. The pace of technological progress on battery research and development is however hindered by lack of realisation of suitable energy storage in chemical form, i.e. it is too big, too heavy, not cheap, with fast charging and road side infrastructure still a problem. Moreover, the motoring range is limited due to energy capacity by auxiliaries such as power steering and air conditioning, a necessity for most drivers.

There is another point of paramount importance. For urban use, motor vehicles only require about 5 kW to be driven at about 60 km/hr. Larger powers are however required for acceleration. It is therefore the short time power demands, synonymous with performance, in combination with long distance-high speed travel that is to this date unattainable targets of ZEVs.

The contender for replacing ZEV is ULEV. The term ultra low emission vehicle is sometimes taken to mean lower emissions than equivalent fossil fuel burning power stations. One way to realise this is through a hybrid configuration where the power from the powertrain in the vehicle is augmented by an energy storage unit. This configuration allows ZEV operation in urban areas as well as long distance-high speed motoring with ULEV.

The present contribution concentrates on the series hybrid configuration, allowing independent motoring through the energy storage and the powertrain as well as the two together. In this way on board recharging is realised allowing excess power from the powertrain to be diverted for energy storage recharge. The parallel hybrid misses this crucial element, demising the powertrain to the spare wheel syndrome, to be used as emergency back up only.

Most major European car makers are also pursuing the US objectives of lower emissions for motor vehicles at increased efficiency. Pilot studies on electric vehicles are too numerous to cite. One example, in full production, is the Fiat 500 Elettra (0 to 25 mph in 10 s, 50 mph max. speed, range 44 miles). Others such as Volvo [1,2] series hybrid concept car has been exhibited at various motor shows. Other advocates of series hybrid powertrain, [3,4,5,6], have concentrated on particular aspects of the technology such as the powertrain. The concept of using a gas turbine engine directly coupled to a high speed generator as the primary powertrain has been fully developed [7]. The European Commission funded research in this area was awarded to the project "TETLEI", turbo-electric taxi with low environmental impact. The three year programme, to be concluded in 1997, aims to develop a prototype vehicle powered by a dual gas turbine and battery electric propulsion system. This is to be demonstrated in the city of Athens where motoring pollution is a serious environmental hazard. The project team has been established to reflect a combination of component expertise, design analysis, vehicle Manufacturers and end users. The present contribution does not aim to provide a full synopsis of TETELI and its progress, but to describe one crucial element. That is the high speed generator which is directly coupled to a gas turbine engine to make up the primary power source for series hybrid motoring.

2. TURBOGENERATORS

The prime mover in a series hybrid configuration is required to provide electrical power directly to the electric motor drives. This necessitates a generator set. Traditionally, most sub-200 kW generator sets have been based around the diesel engine which not only gives relatively good fuel consumption figures, but also its speed/torque characteristic is ideally suited to conventional 3000 rpm a.c. generators. However, these are too large, heavy, and noisy for the automotive application. The concept of directly coupling a high speed generator to a gas turbine engine for hybrid automotive applications has become attractive due to its potentially low emission characteristics as well as its small size, [8].

Efficient gas turbine engines with low emissions is a target already achieved in the aeronautical field. Its application for mobile land applications have always suffered from conversion of high speed shaft outputs to suitable forms of electrical or mechanical energy. High speed gear boxes have been coupled in the past with conventional generators to take the power off the gas turbine shaft. Power losses, size, life and cost have made this solution un-attractive for the automotive application. A simple solution is to dispense with the high speed gear box by design and development of a generator capable of directly coupling to the gas turbine engine. This solution requires a compact generator to overcome stresses associated with high speed operation. It also needs to be highly efficient to reduce the problems of cooling at reduced size, [3].

It is generally accepted that a permanent magnet high speed generator would offer the best solution to this problem. The advent of new high power rare earth magnets such as Neodymium-Iron-Boron (five fold increase in energy products over the last 20 years) with remnant flux densities in excess of 1.2T and high coercive strengths (1000 kA/m) have helped to increase the viability of efficient air gap wound permanent magnet machines.

These machine types would, however, require a rotor capable of withstanding the high centrifugal stresses incurred for operating at matching speeds to the gas turbine engine, and this is most easily achieved by employing a high specific strength (strength/density) material such as carbon fibre composite as a reinforcement shell around the rotor. The retainment shell is to prevent the development of tensile hoop stresses in the magnets which have generally poor tensile properties.

The generator could be designed either as a conventional cylindrical, drum type machine with radial flux path, or as a disc machine with axial flux path. At these high operating speeds, when design constraints are considered, the drum machine quickly becomes a less viable proposition: If a design with a long, small-diameter rotor is to be avoided (undesirable dynamic characteristics), a thick containment shell is required around the rotor, which in the case of a drum machine, will need to lie in the magnetic air gap thereby lowering the flux density and efficiency. Drum type machines have been successfully used with gas turbines, but only at speeds of up to 60 000 rpm and efficiencies generally not higher than 80%.

The choice of a disc or drum type permanent magnet machine is therefore dictated by the operating speed, with the disc configuration having the following advantages for high speed use:

- Acute radial tolerances do not have to be adhered to.
- The retainment ring to hold the magnets in position does not lie in the air gap.
- Stator manufacture is easier.
- Magnet geometry is simpler with flat pole surfaces.
- All iron losses due to a.c. flux can be eliminated (rotating keeper).
- There is no constraint on magnet length.
- Cooling is facilitated by the larger surface areas.

There are additional issues such as vibration and balancing which are clearly easier for a machine with no radial growth restrictions. The modularity of the disc machine is also beneficial easing manufacturing problems. Figure 1 shows an schematic of a double rotor/ single stator geometry.

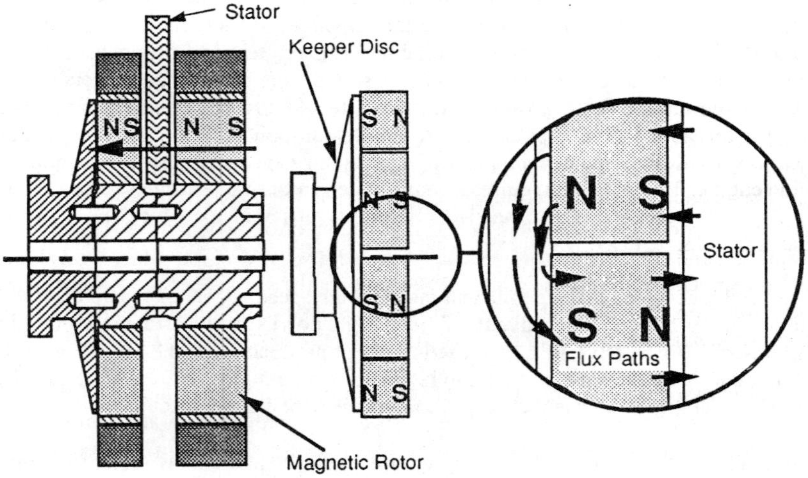

Figure 1 Generator Geometry

The basic design of the Imperial College generator is a disc-type, permanent-magnet machine with eight poles and producing alternating current in a three-phase coil-type winding. The stator winding may be tailored to a certain extent to produce the required voltage. Its construction and type of winding was specially developed to minimise eddy-current losses and to allow for good cooling. Together with the gas turbine engine and peripherals, a 50 kW generator set would not weigh more than 60 kg.

The design may be scaled in the radial direction to allow perfect matching to the speed of the gas turbine in order to obtain the maximum efficiency from the generator, and the number of stators/rotor modules can be increased or decreased according to the power requirements.

Single modules of the disc high speed generator is presently under test producing about 10 kW output.

3. GENERATOR PERFORMANCE AND COOLING

The compact generator must be highly efficient to reduce cooling requirements. A 50 kW high speed generator with a 90 % efficiency still means removing 5 kW of heating in a volume not much larger than an automotive alternator. Electrical losses, therefore, such as eddy current and hysteresis losses of stationary iron, must be minimised. The mechanical losses, to include windage and bearing losses, also need to be minimised. Windage losses are caused by the air around the rotating machine imparting a drag torque, resisting the rotation of the rotor. Earlier work were focused on experimental measurements of windage losses using large diameter samples tested at low speeds. Recent measurements have addressed the small disc - high speed problem, much more akin to high speed generators, [9]. Losses in both the disc and drum sections (axial and radial directions) require to be quantified. The imparting torque may be calculated using:

$$T_w = T_w(disc) + T_w(drum) = 1/2\, C_m\, \rho\, \omega^2 r^5 + \pi\, C_f\, \rho\, \omega^2 r^4 L$$

where C_m is the torque factor (= $0.062/[(s/r)^{0.25} R_e^{0.25}$ for the turbulent flow, merged boundary layer correlation, and $R_e = \rho\, \omega\, r^2/\mu$), μ is viscosity, ρ is the air density, ω is the speed, r is the disc radius, C_f is the coefficient of drum resistance, and L is the drum length. Using the above equation the effect of various parameters on the windage power loss ($P_w = T_w\, \omega$) can be investigated. For example in the axial direction, the exponent of stator gap to radius, s/r, is a quarter, and so, to halve the windage loss requires changing the value of this term by a factor of sixteen. Gas viscosity is found not to significantly affect windage losses. Density is a stronger function, but changes to density requires either the replacement of the air by another gas of lower density or a reduction of the air pressure within the air gap.

For electric machinery the most promising solution is to replace the air with hydrogen, since its density is 14 times lower than that of air at the same pressure and temperature, the viscosity is lower, and its volumetric heat capacity is similar to that of air. This solution has been used extensively for large 3000 rpm generators. The hydrogen gas is circulated in a closed system incorporating a heat exchanger. This is however not possible with rare earth magnets which strongly react with hydrogen.

The other two terms, rotor speed and diameter, are powerful in reducing the windage loss, having exponents of 2.75 and 4.5 respectively. However their reduction would also reduce the machine's output and/or affect its efficiency. This is because power output is also proportional to diameter and speed, e.g. for electrical disc machines power output is proportional to diameter square and rotational speed. It is noted here that drum losses for conditions under consideration were about 40% of the total. The total windage loss for a 50 kW disc high speed generator operating at 100,000 rpm were found to be nearly 8%. This number reduced to about 4% at 75,000 rpm.

Figure 2 shows the variation of windage loss for various diameters at a constant length of 10 mm. This is for constant air gaps to the stationary members of 10 and 1 mm in the radial and axial directions respectively, and at the operational speed of 60,000 rpm. The governing equation with a coefficient of correlation of 0.998 (unity signifies a perfect fit) is as follows:

$$P_w = 704140\, r^2 - 43715\, r + 730.9$$

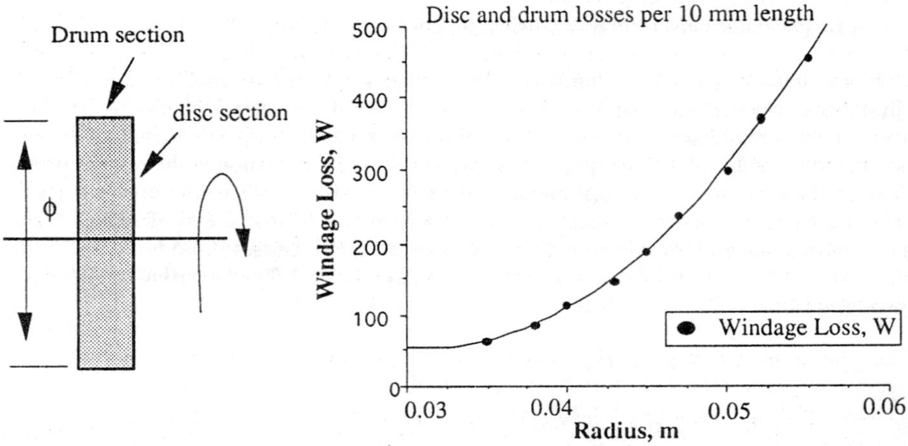

Figure 2 Windage losses (disc and drum) per rotor.

4. GENERATOR COUPLING

The design strategy to couple the gas turbine engine with the high speed generator has been two pronged. The first, more easily identified with the Garratt and Volvo drum machines, advocate a single shaft arrangement with shared bearings, [10]. In this arrangement the generator is usually placed in between the compressor and turbine and suffers from heating problems from the turbine end. The balance and dynamics of the components, as well as the total system, on the single shaft also tend to be problematic.

The other approach advocated here is independent development of the gas turbine engine and the high speed generator. The primary reason for this has been the diverging requirements of the two technologies. The prime example for this is that contrary to the gas turbine engine, the generator continuous safe operation is synonymous with low operating temperatures. Moreover, the axial and radial clearances effect gas turbine engine efficiency much more profoundly. It is therefore not surprising to find different requirements for the two devices in terms of balancing and shaft dynamics.

This route has the added advantage of allowing separate development programmes. For the gas turbine engine, an off the shelf unit for the automotive application is not available, and the procurement of an efficient, ultra low emission gas turbine engine at low cost is presently being pursued. For the high speed generator attention is focused on novel designs as well as best use of material advances and numerical design tools.

The penalty for developing the two required technologies independently is the requirement for a coupling with stringent requirements such as high speed operation, torque capacity, as well as catering for axial and radial misalignments. Three types of coupling have been investigated to this date, namely magnetic, gear and diaphragm.

The non contacting nature of the magnetic coupler, using high strength magnets, was thought to be a useful feature, with non transfer of modes and frequencies and potential for good alignment. Theoretical calculations gave designs for an axial magnetic couplers with disc and sector type magnets (sector type: inner radius of 26 mm, outer radius of 40 mm, disc thickness of 10 mm, dead space angle between magnets of 11.6°, and residual induction 1.1 Tesla). These were manufactured and tested in specially developed testing apparatus. Couplers were operated at speeds in excess of 60,000 rpm, producing torques of 6 Nm with an air gap of 5 mm and at half sector angle, i.e. an angular displacement of 20°. The axial load, however, was found to be around 260 N and a burden for the thrust bearings.

The gear coupler is essentially a thin hollow drum gear floating between two geared flanges at the gas turbine engine and the high speed generator ends. It out performed the magnetic coupler, but requires lubrication. As the high speed generator is coupled to the gas turbine engine at the cool end, compressor end, its use requires oil seals to keep the compressor inlet clean.

The latest coupler, in the form of thin discs sandwiched by studding between an 'H" section and two end flanges, is lubrication free and performed equally well. Schematics of all three designs is shown in Figure 3.

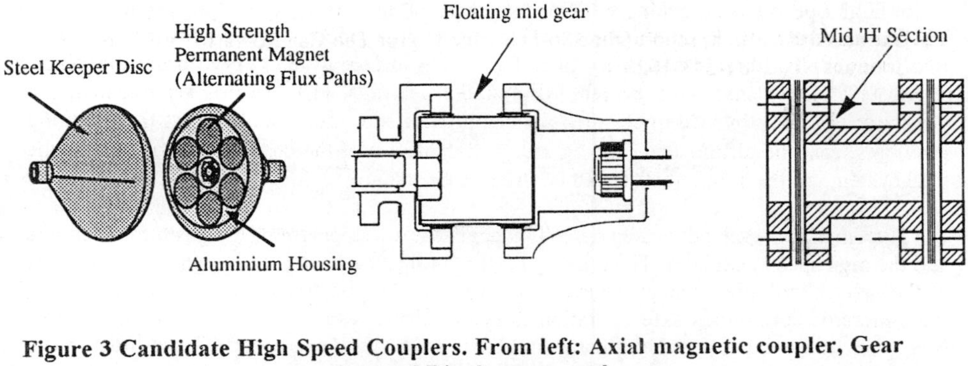

Figure 3 Candidate High Speed Couplers. From left: Axial magnetic coupler, Gear coupler, and Diaphragm coupler.

5. CONCLUSIONS

Design and development of high speed generators, small gas turbine engines, and associated technologies have progressed to an advanced stage. Its realisation for automotive use in a series hybrid configuration is within grasp. Whilst further technical improvements in all areas are still required, attention is now focussed on performance measurements and their optimisation of the fully integrated system.

REFERENCES

1. Wallman, S. "A Volvo Concept Car- For Future More Severe Emission Standards". The Sir Henry Royce Memorial Lecture, The Institution of mechanical Engineers, London, February 1993.
2. ABB Report 9/93, A Hybrid Drive for the Car of the Future, 1993.
3. "Gas Turbine Goes Under the Bonnet". The Engineer, 15/22 August 1991.
4. Fenocchi, A., Etemad, M. R. S., Besant, C. B., Ristic, M., and Baines, N. C., "Turbo-electric transmission: A new concept for vehicles". In the proceedings of the 24th "ISATA" Symposium in Florence, Italy, 20th to 24th May 1991, pp. 177-184.
5. Baines, N. C., Panting, J. R., Etemad, M. R. and Besant, C. B., " A Gas Turbine-Electric Vehicle Concept". In the Proceedings of FISITA (Automotive Technology Serving Society) 92 organised by the Institution of Mechanical Engineers, London, June 7-11, 1992.
6. Besant, C. B., Etemad, M. R., Pullen, K., Ristic, M., and Fenocchi, A. "Hybrid Turbogenerator vehicle (HTV)". Paper 93 EL021 pp. 343-350 in the proceedings of the 26 ISATA Symposium, Aachen, Germany, 13-17 September 1993.
7. Etemad, M.R., Fenocchi, A., Hall, J., and Pullen, K.R., "An Evaluation of High Speed Generators". Paper 94EL063, pp. 629-637 in the proceedings of the 27 ISATA Symposium . 1-4 November 1994, Aachen, Germany.
8. Pullen, K.R., Martinez, R., and Leontopoulos, C. "Factors Influencing the Choices of Ga Turbine Engines for Series Hybrid Electric Vehicles". Paper 94EL041, pp. 515-523 in the proceedings of the 27 ISATA Symposium . 1-4 November 1994, Aachen, Germany.
9. Etemad, M. R., Pullen, K., Besant, C. B., and Baines, N. "Evaluation of Windage Losses for High Speed Disc Machinery". Proc Instn Mech Engrs, 1992, Vol 26, pp149-157.
10. Garratt AirResearch, report (June 86-February 87) for The Gas Research Institute, Contract No. 5084-243-1028.

C498/29/110/95

The gas turbine series hybrid vehicle – low emissions mobility for the future?

A G W DAVIS PhD and **P FAITHFULL** BEng
University of Warwick, UK
K LILLIE FIEE
Rover Group Limited, UK

SYNOPSIS

A gas turbine coupled to a high speed generator together with a battery block are the energy sources on a prototype hybrid electric vehicle. The gas turbine is operated at fixed power output, and the battery block is required to either source or sink electric power according to the traction power requirements. The electric traction motors sink power during driving and source power through regenerative braking. A power management system controls the flow of power between these elements.

INTRODUCTION

The object of this paper is to introduce TETLEI (Turbine Electric Taxi for Low Environmental Impact), the collaborative program of work which, with EC support under the Brite Euram scheme, is in the process of designing, building and evaluating a gas turbine/battery hybrid vehicle. The prototype vehicle will be based on the latest Range Rover model with the responsibility of vehicle integration ultimately resting with Rover Group. Rover is also responsible for the project co-ordination as a whole. The profile of the project consortium in terms of companies is as follows, two vehicle manufacturers (Rover and Renault), two gas turbine manufacturers (Turbo-Energie and OPRA), two Universities (Imperial College of Science, Technology and Medicine and The National Technical University of Athens) a battery manufacturer (Hawker Energy Products) and a battery charger and monitor manufacturer (Oldham France). An end user organisation, the Municipality of Athens, is also involved. The plan is to integrate the developed technologies into a fully operational vehicle and to evaluate it in the environment of the city of Athens, a city which suffers from high levels of air pollution at ground level and therefore has a particularly strong interest in the benefits of a low emission vehicle application.

The quantified objectives are to:

- improve on ULEV (Ultra-Low Emissions Vehicle) equivalent emission limits for the standard european drive cycle.

-achieve these limits substantially independent of cold start-up and machinery wear out.

-meet ZEV (Zero Emission Vehicle) emission limits by providing a battery powered electric vehicle with a 50 km range for use at city/suburban speeds.

-demonstrate that vehicle acceleration, speed and drivability are not compromised by using a gas turbine/ battery hybrid vehicle.

BACKGROUND

Hybrid vehicles containing a heat engine and an electric motor have been suggested as a way in which the environmental impact of a passenger car can be reduced in the future.

In the series hybrid vehicle, a heat engine is coupled to an electrical generator and the electrical power from this generator is used to propel the vehicle (see figure 1). In this configuration the engine is decoupled from the road wheels and is therefore free to run at the most appropriate speed. Hence the range of possible engine types is greater than for conventional vehicles or parallel hybrids.

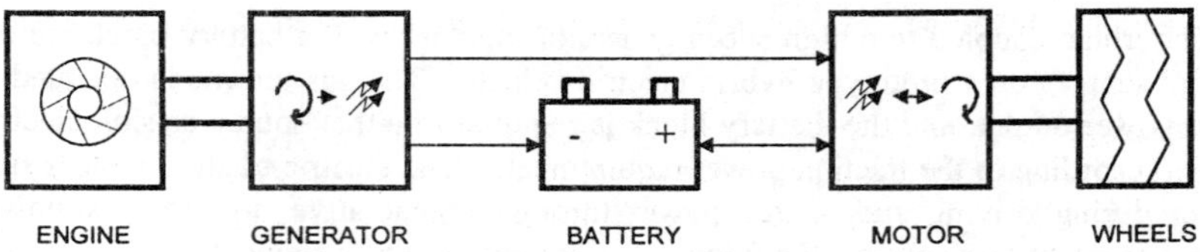

Figure 1. Series Hybrid Vehicle Schematic

Gas turbines have previously been investigated as a source of power for vehicles but were rejected on the grounds that they suffered from high fuel consumption at light loads, had poor dynamic response and the high rotational speeds of the engine made power extraction difficult. The series hybrid powertrain proposed allows us to overcome these problems.

Firstly significant strides forward have been made in improving the efficiency of the small gas turbine by the use of heat recovery devices, making their variable load operation efficiency comparable to that of other heat engines. Notwithstanding this, if operation of the turbine is confined to a fixed load point, the power flow into and out of the battery block can be managed according to the tractive power requirements. This also overcomes the problems of poor dynamic response, the electric traction motors propelling the vehicle give excellent dynamic response and the batteries are specified to meet the dynamic power requirements. The power extraction mechanism envisaged for this project directly connects the turbine shaft to a high speed generator rather than to the road wheels via a gearbox.

The primary motivation for the project is to investigate the potential low emissions capability of the gas turbine due to its continuous combustion nature.

OVERVIEW OF TECHNICAL PROGRAMME

The main research thrusts will be directed at development of the key components themselves, their interrelationships and their integration in the vehicle. The project is structured in a number of tasks and these are now briefly described.

Task 1: Specification and simulation

Main partner involved: Rover

The duty cycle of a vehicle in the Athens environment together with general user requirements for the vehicle will be established in detail. Consideration will be given to the power electronics (DC/DC converter) required to control the power flow within the vehicle. The specifications of the main vehicle components will be determined.

Task 2: Gas turbine

Main partners involved: Turbo-Energie, OPRA, National Tech. Univ. of Athens (NTUA).

The radial gas turbine will be designed and tested. Turbo-Energie company will design, instrument and manufacture the small gas turbine for use in the test vehicle. Optimisations through computational modelling will be provided by NTUA. Since Turbo-Energie has previously concentrated on developing methods to improve the efficiency of the machine, this will not be optimised for low emissions. For this reason the OPRA combustor will be fitted to the Turbo-Energie machine which will then be optimised for low emissions. Careful consideration will be given to the interface between the OPRA combustor and the Turbo-Energie machine and this will be achieved by detailed technical collaboration.

Task 3: High speed generator

Main partner involved: Imperial College

The design required to meet the needs of the vehicle will be implemented and the techniques required for manufacture developed. The coupling to the turbine, in collaboration with Turbo-Energie, will be developed and tested.

Task 4: Battery block, charger and monitoring system.

Main partners involved: Hawker Energy Products, Oldham France.

The development of the battery to be used in the block, the charger and the battery monitoring system will be designed to meet the requirements of the vehicle. The battery itself will be developed by Hawker Energy Products and the battery monitoring system and battery charger will be developed by Oldham France. Tests in the laboratory which integrate these components will precede the testing in the vehicle.

Task 5: Electric motors

Main partner involved: Renault

Specification and procurement of the electric motors and their controllers.

Task 6: Vehicle design and integration

Main partner involved: Rover

Design work necessary to incorporate the components into the vehicle dynamic environment will be undertaken and implemented. The vehicle management system will be developed and integrated with motor controllers, battery monitoring system, turbine/alternator control system and DC/DC converter for power flow control.

Task 7: Evaluation and viability analysis

Main partner involved: Rover

The vehicle will be evaluated in an operational environment and compared to predictions. User feedback will be obtained and the viability of the vehicle assessed. The results of the project will be disseminated at this stage.

Task 8: Project management

Main partner involved: Rover

Responsible for the day to day running of the project.

GENERAL REMARKS

There are a number of areas of technical challenge in the project which determine the eventual success.

1. The potential for low emissions is the motivation behind the project, thus the combustor (to be developed by OPRA) in the gas turbine (to be developed by Turbo-Energie) is a crucial element since this determines the amount of pollution created. Preliminary laboratory experiments (by OPRA) have indicated that a turbine is capable of better than ultra low emissions (ULEV), of the order of 7 ppm NOx. This performance needs to be achieved on the very small turbine to be supplied by Turbo-Energie.

2. The high speed capability of the gas turbine and alternator are important. The gas turbine must run at high speed (over 60,000 rpm) in order to minimise size and weight and the alternator must also be able to operate efficiently at high speeds since it is coupled directly to the output shaft of the turbine. Thus it is important to ensure the high speed integrity of the turbine and alternator components.

3. In order for the vehicle to operate efficiently a number of independent control systems are co-ordinated collectively to achieve the desired overall effect. For instance, the turbine may be run in an on/off mode so that power is produced when it is operating most efficiently, at rated conditions. This places high performance demands on the integrating control system.

VEHICLE DESIGN OUTLINE

A schematic of one of the designs being considered for the vehicle system is illustrated in figure 2. The initial task for Rover is to specify the system components and this can be assisted through system simulation. A model for the simulation has been produced which uses the vehicle properties and circuit analysis techniques providing the information required to specify the power, voltage and current requirements of the system components as well as to provide a basis for the powertrain controller algorithms. Figure 3 illustrates this model and the sign conventions used.

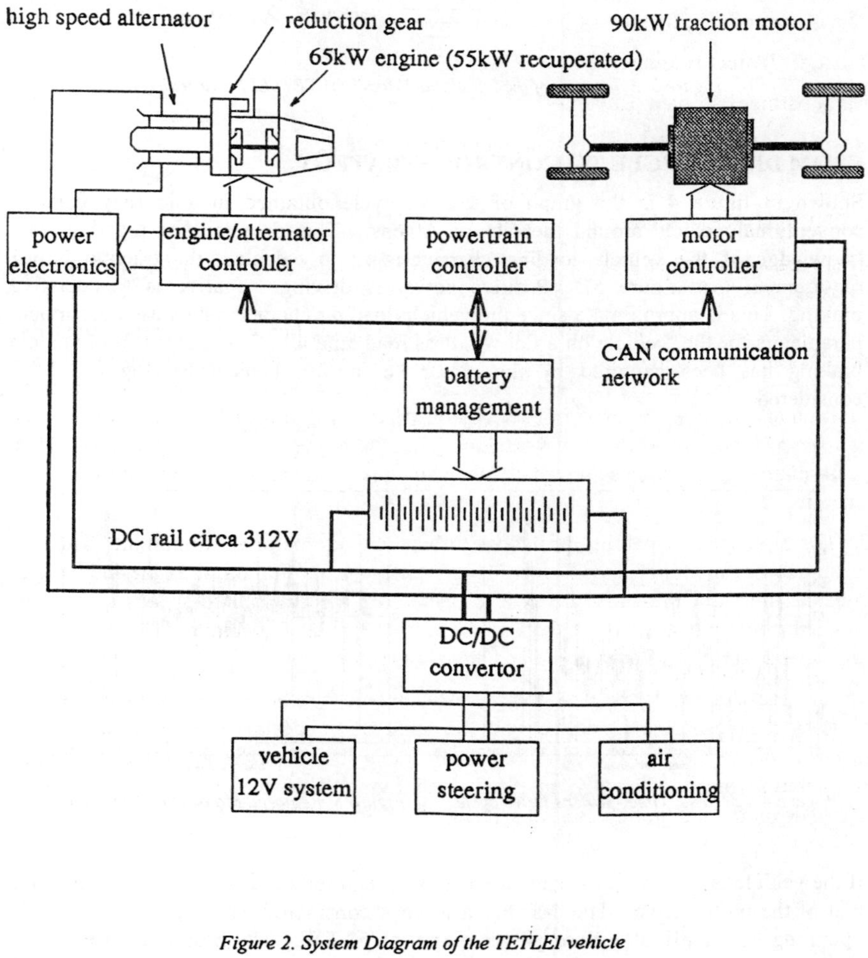

Figure 2. System Diagram of the TETLEI vehicle

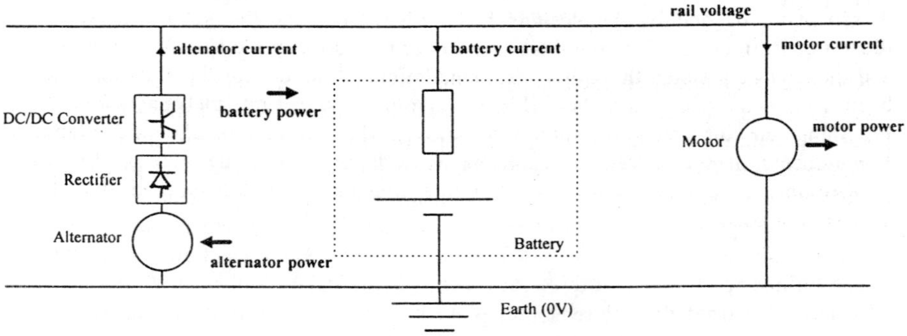

Figure 3. Schematic of the Simulation Model for TETLEI Powertrain

FROM DRIVE CYCLE TO CONTROL STRATEGY

Shown in figure 4 is the graph of a drive cycle obtained in a journey using a conventional vehicle around the city of Athens. This data is used together with knowledge of the vehicle loading characteristics to calculate the tractive power requirement (see figure 5). Positive power is driving, negative is regenerative braking. This is approximate since the vehicle loading characteristics used are model parameters for the vehicle on a flat windless road, and all of the energy from vehicle braking has been captured by the motor i.e. no mechanical braking has been considered.

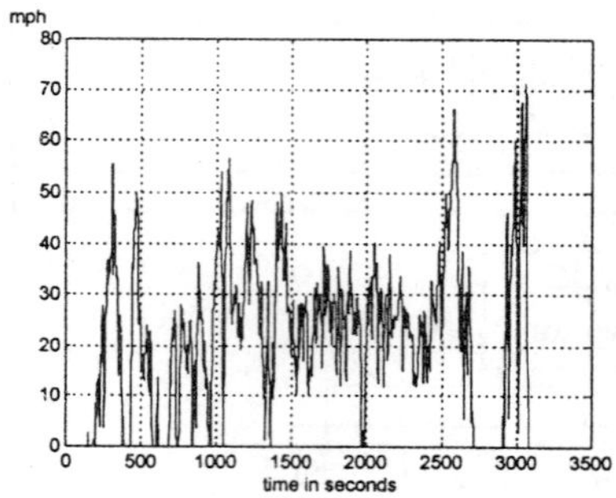

Figure 4. Speed vs. Time - Athens Drive cycle

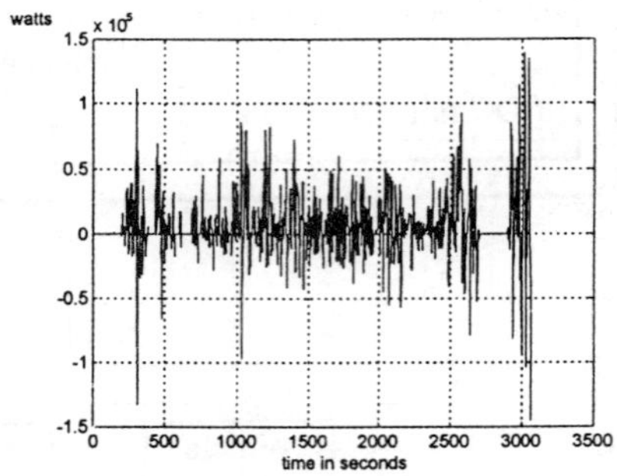

Figure 5. Required Motor Power vs. Time

If the vehicle is in electric only mode the power requirement from the battery matches that of the motor drive. The case has also been considered where the gas turbine is operating constantly at a *fixed* power output of 50 kW. The power required to be sourced or sunk by the battery is as shown in figure 5 and this is calculated by realising that the alternator power of 50 kW must be constantly used. That is, where the tractive power requirement is, say 30 kW, then the power sunk by the battery must be 20 kW.

Knowing the internal characteristics of the battery such as internal resistance, open circuit voltage and state of charge, it is possible to calculate the battery voltage and current required to meet the battery block power flow requirements. This is shown in figures 6 and 7 respectively. As the positive power rail is common to the power electronics on both the motor and the alternator, results from these simulations can be used in the specification of both of these devices. The variation of state of charge over the journey time is shown in figure 8 for electric only and figure 9 for electric and turbine operation. It is this data which is of critical importance in developing the powertrain control strategies. It allows the prediction of a necessary recharge of the batteries. An accurate model for the state of charge allows the use of a control strategy which prolongs the life of the batteries.

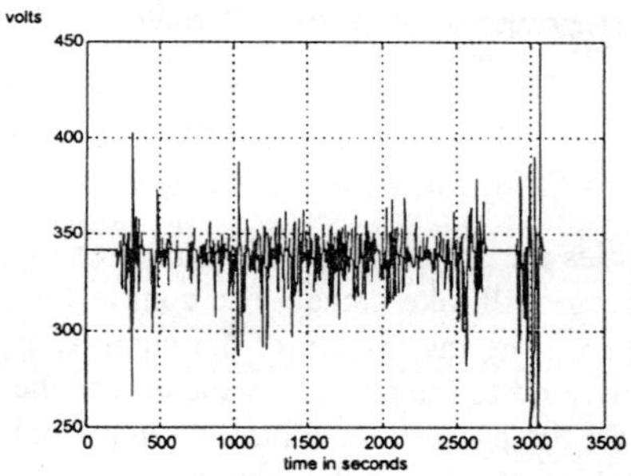

Figure 6. P ower Rail Voltage vs. Time

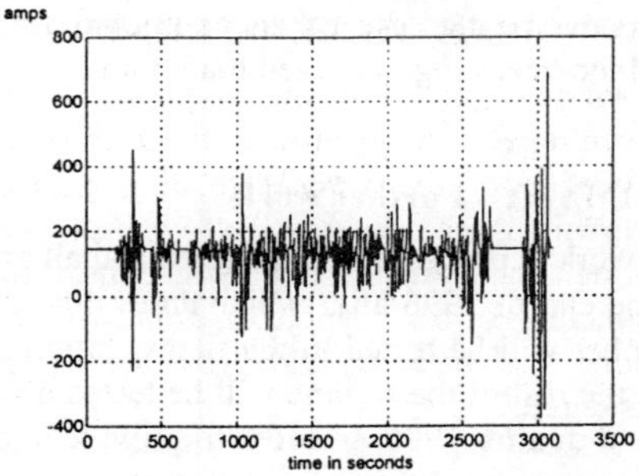

Figure 7. Battery Current vs. Time

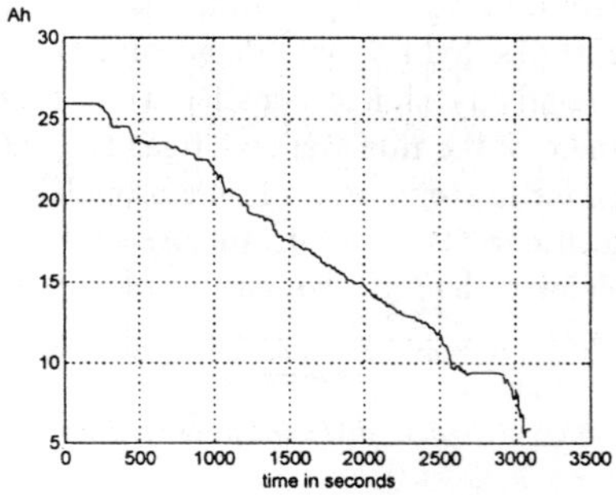

Figure 8. Battery State of Charge vs. Time

(Turbine off)

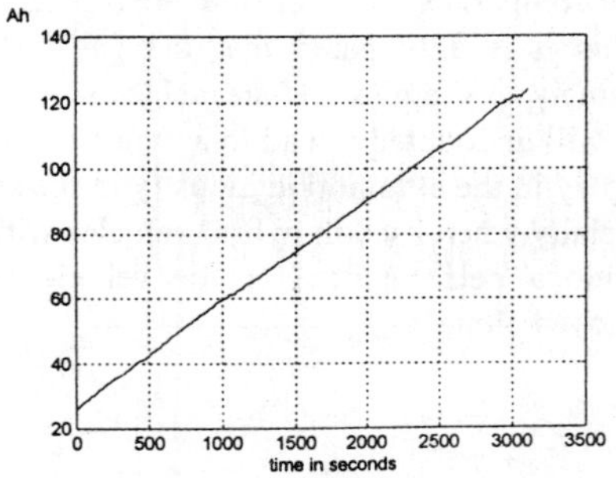

Figure 9. Battery State of Charge vs. Time

(Turbine on)

DISCUSSION

The development of a valid simulation requires that a number of factors are fully considered. As all of the data stems from an initial calculation of the battery current, it is important that this value is accurate. Small variations in the internal resistance of the battery can cause large variations in the rail voltage (square of variation). It is important to have an accurate model of the battery which considers both the variation of internal resistance and open circuit voltage of the battery at different states of charge. Modelling the drive cycle at the two extremes of battery operation (80% and 20% discharge) gives a good indication of the range of values over which the voltage and currents in the system may vary for specification purposes. The next step is to introduce a more realistic battery model and practical limits on the power sourced from regenerative braking.

During electrical regenerative braking, there is a practical upper limit to the voltage permitted across a battery in order to avoid 'gassing'. This restriction may be overcome be the use of an additional form of power sink if the alternator is to continue operating at a fixed load point.

SUMMARY - Further work

The work is proceeding as planned and all activities are on target to deliver the vehicle by the end of 1996 after which an evaluation phase will take place. The combustion chamber will be tested in the first quarter of 1996. The integration of the combustor with the rest of the turbine will be tested and delivered to the project by the end of the second quarter 1996. In order to provide a test for the powertrain control strategy and the integration of the main power electronic equipment prior to the installation of the gas turbine the Range Rover will first be built as a diesel/battery series hybrid by the second quarter 1996. It is hoped that tests performed on the series hybrid vehicle will allow comparison with an equivalent parallel hybrid vehicle. Apart from the benefits of this it is anticipated that the gas turbine series hybrid will demonstrate that extremely low emissions are achievable from relatively simple, small gas turbines. This will undoubtedly lead to a major reassessment of the role that small gas turbines can play in the automotive industry in the search for low emission vehicle propulsion, especially when used in hybrid vehicles with electric traction motors, the use of which provide a performance for the vehicle far greater than if the gas turbine were employed alone.

C498/29/204/95

Hybrid drive concepts for emission free city travel – a summary of BMW developments

G FISCHER
BMW AG, Germany

SUMMARY

Correctly designed hybrid drive concepts combine the benefits of conventional drive principles, namely the utilisation of existing infrastructures and the availability of large operating ranges, with the advantage of electric drive concepts, which allow localised emission-free travel in an all-purpose vehicle. However, in order to achieve the desired fuel consumption and emission advantages, further improvements to the efficiency, weight and operating behaviour of the individual components are still necessary.

1 INTRODUCTION

Emission-free travel, at least in city centres, is an increasingly important challenge to motor vehicles. Electric vehicles entirely satisfy this requirement but in practice their operating radius is limited to urban areas. Electric vehicles are therefore typically urban concepts which are tailored to this type of use (1-2). Their use by private owners will with few exceptions be limited to that of a second car. Hybrid vehicles, on the other hand, can be used as the main vehicle which covers all requirements. They combine in principal the benefit of electric vehicles - localised emission-free travel in urban areas - with the advantages of conventional drive systems, namely the fact that operating range is limited only by the capacity of the fuel tank, short refuelling times, comprehensive infrastructure and the necessary engine power for inter-city travel in a fully-fledged all-purpose vehicle.

1.1 Classification of hybrid drives

Drive concepts in which at least two different forms of drive energy are available, either permanently or for limited periods, are referred to as hybrid drives. In addition to systems with an internal combustion engine and hydraulic system, which have not yet achieved widespread acceptance, systems involving a combination of internal combustion engine and electric components are currently being investigated and constitute the main topic of discussion below. The choice of electric motor as the second power source is based on electricity's ease of storage and its reversible action, notwithstanding the fact that its specific storability is much poorer than that of the liquid hydrocarbons used. By way of comparison, to store the energy

content of 1 kg of liquid fuel in the form of electrical energy, approximately 100 kg of battery is needed. Moreover, the energy store occupies considerably more space.
Virtually any form of hybrid system can be envisaged and described. They are always based on two fundamental concepts (3):

- Parallel hybrid: here, the internal combustion engine and electric motor act in unison, i.e. in parallel, on the driveline.
- Series hybrid: here, the internal combustion engine with alternator and the electric travel motor are located one behind the other in the driveline, i.e. in series.

BMW is conducting intensive research into one version of each of these basic concepts, both on the test rig and in test vehicles.

2 BMW RESEARCH VEHICLE WITH PARALLEL HYBRID CONCEPT

A 5 Series BMW is the basis for this research work. The driveline consists of a standard 4-cylinder spark-ignition engine and a manual gearbox. An asynchronous motor is fitted in parallel with the internal combustion engine, acting on the driveline between the clutch and the gearbox input. In this instance, the asynchronous motor acts both as a power source and an alternator. For emission-free travel using only electrical energy from the battery, an additional clutch controlled by the engine management system disconnects the internal combustion engine from the driveline once it has been shut down. This layout permits the two drive systems to interact in numerous ways. After preliminary simulation, the following operating strategy was devised:
In the combined mode, the spark-ignition engine runs continuously, and the driver encounters a familiar engine-speed characteristic. When accelerating rapidly, for instance when overtaking, the electric motor cuts in and temporarily boosts the peak output required. With this superimposed torque, the BMW with parallel hybrid drive has reserves of power which are equivalent to a high-displacement internal combustion engine, developing 295 Nm at 1700 rpm. This in turn permits a much higher overall gear ratio, which reduces engine speed to a level equivalent to a standard 525tds.

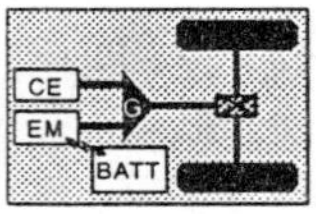

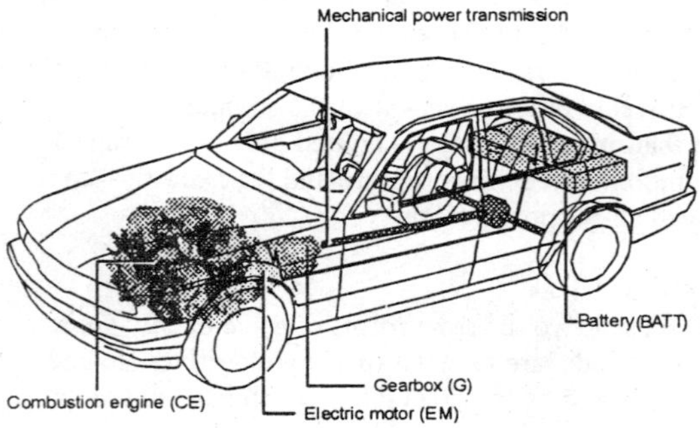

Fig. 1 Parallel hybrid

In addition, the vehicle battery can be recharged by the electric motor in the alternator mode when the car is on the move. Since this imposes a greater load on the internal combustion

engine, its efficiency is increased. A further advantage is that the deceleration energy normally lost in the form of braking heat can be recovered at least in part as electrical energy and channelled back into the battery.
In the electric mode, the driver can use the downstream gearbox to improve the vehicle's dynamic behaviour. At urban speeds, performance is comparable to driving on the internal combustion engine.

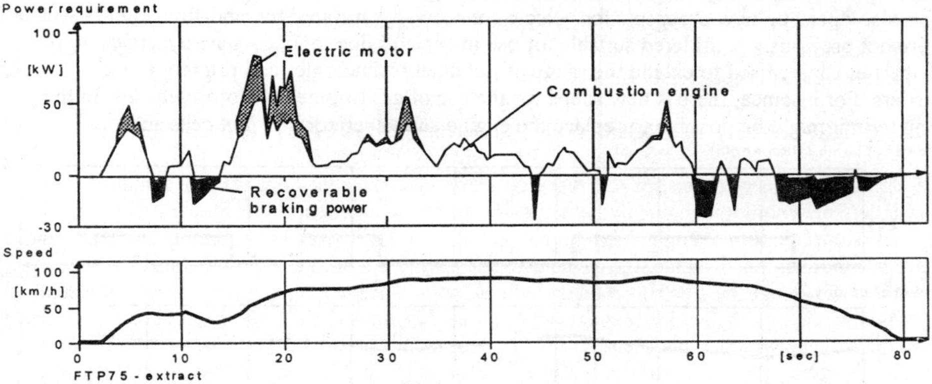

Fig. 2 Operating strategy with parallel hybrid

3 BMW RESEARCH VEHICLE WITH SERIES HYBRID CONCEPT

An electric BMW 3 Series car of the type currently being tested on Rügen island serves as the basis for research. In addition to the purely electric drive system, there is an internal combustion engine linked directly to a generator. The special aspect of this drive version is that the energy generation and utilisation stages are entirely independent of each other, offering entirely new prospects:

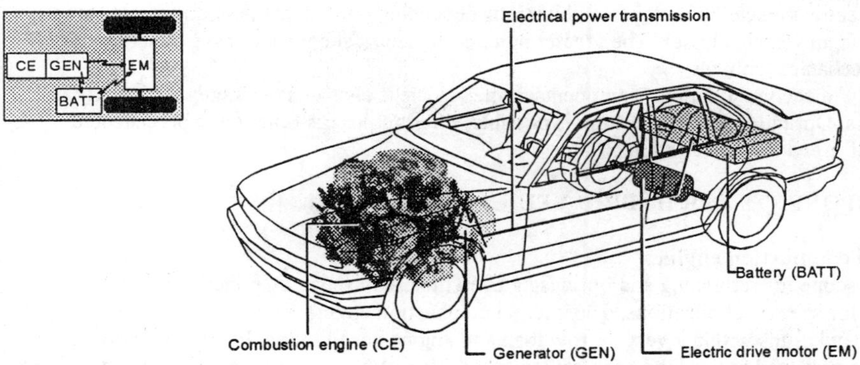

Fig. 3 Series hybrid

The internal combustion engine can be run within a very narrow speed range or even at a single constant speed. Uneconomical operating points such as idling or the lower part-load range are thus avoided, and the engine can be optimised for the range that affords maximum efficiency. On the one hand this keeps the internal combustion engine simple, and on the other very low emissions can be achieved with little extra effort.
The internal combustion engine only cuts in automatically if the battery is no longer capable of meeting the actual energy requirements or if its charge falls below a certain level.
This also opens up new prospects for vehicle concepts: for instance, combustion engines which were not previously considered suitable for use in vehicles due to inadequate dynamic properties can be used to extend the range of potential technologies and primary energy carriers. For instance, there is new scope for the use of gas turbines in motor vehicles. In the long term it might be possible to replace the engine and generatorwith fuel cells and

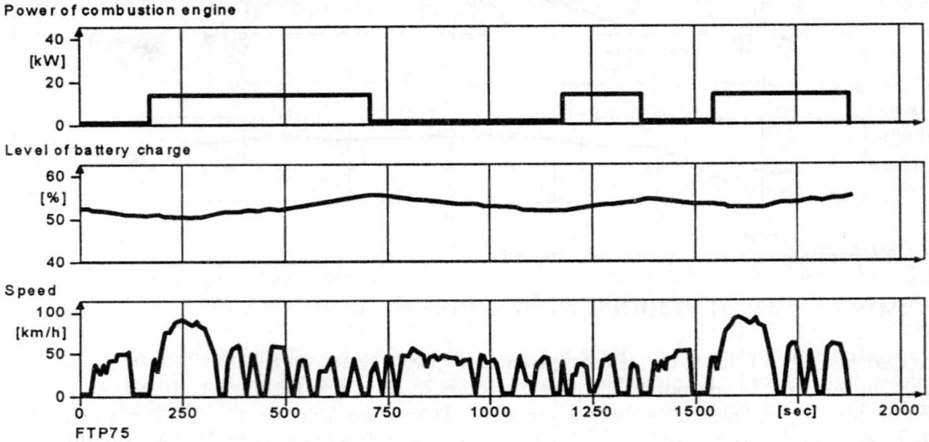

Fig. 4 Operating strategy with series hybrid

therefore to convert chemical energy directly into electrical energy.
Since the individual components of the drive system are only connected electrically, as in a purely electric vehicle, numerous combinations depending only on actual requirements are possible in all vehicle classes. The greater number of electrical connections is offset by the lower mechanical content.
Flexibility in accommodating components in the vehicle is also an advantage, particularly on small cars. Optimum space utilisation is possible, offering the customer an unprecedented degree of freedom for variation.

4 COMPONENTS FOR HYBRID VEHICLES

Internal combustion engines

There is scope for rethinking and optimising even the established components of hybrid vehicles for specific applications. The internal combustion engine on the series hybrid, for instance, only runs within a very narrow load and engine-speed range. Idling, operation across extensive part-load ranges and peak engine speeds do not occur at all, and there could even be only a single operating point. The areas for optimisation of vehicle engines are therefore completely new. They include all engine components, such as the pistons and their rings, or the choice and design of the charge cycle control and exhaust treatment systems. The auxiliaries, too, should be redesigned; the generator performs the task of the starter motor in the internal combustion mode, the coolant circuit and oil supply can be adapted to the new situation and the engine consequently simplified; at the same time, emissions and fuel consumption improve.

4.1 Electrical machines

Electrical machines in the hybrid vehicle are basically subject to the same requirements as on purely electric vehicles, but the following requirements are particularly important:

- Maximum power density with compact dimensions
- Minimum electro-magnetic radiation as an integral part of the concept
- High efficiency across a wide range of running speeds and loads
- Reduced costs.

In conjunction with partner companies, BMW is currently developing a new asynchronous motor with integral power electronics that is intended to satisfy these requirements (4). If one considers the position of operating points in practical driving cycles such as the FTP cycle, it will be noted that the quality of the electrical machines depends not so much on their efficiency at the optimum point as on the spread of efficiency in the lower speed and load range. Attention is being focused on this range rather than on spectacular but not particularly representative individual points. However, the development of synchronous machines is also being pursued further (5).

4.2 Batteries

Sufficient capacity, high effective power output, low weight and dimensions and scope for recharging while the vehicle is in motion are the stringent requirements that electrical energy stores on the vehicle have to satisfy. All systems commonly used on electric vehicles are being investigated, with the exception of the non-rechargeable zinc-air battery. Unlike electric vehicles, the priority in assessing batteries for hybrid vehicles is power output, not capacity. In contrast to internal combustion engines, capacity and output are closely linked, which complicates the choice of ratings. When in motion, high peak power is required during the acceleration phase in addition to an adequate continuous power output. Scope for recharging also requires particular attention. The battery should be capable of receiving high outputs such as those generated when braking. In the medium term, sodium-nickel chloride and nickel-metal hydrid systems are considered to offer the best prospects for success (6). These systems can be used in both electric and hybrid vehicles. However, in addition to price, weight and required installation space, all battery systems suffer from a serious disadvantage: during the acceleration and braking phases, high short-term outputs occur when in motion, but the battery packages designed for these peak values are unacceptablelarge, heavy and costly. Moreover, such high peak values impair the batteries' operating life.

4.3 High-performance store

For this reason, BMW is investigating additional storage systems with properties which complement those of the batteries: high performance from a limited energy content. These systems can cover peak demand when accelerating, and store braking energy when decelerating. In addition to improved accelerating behaviour, this type of store protects the battery and consequently extends its operating life. BMW is investigating two different systems, the supercapacitor and the flywheel.

Supercapacitors are capacitors with very high capacitance. However, they are not yet available with the storage capacity needed for a vehicle (6). They also necessitate very complex converter circuits which must likewise still be developed. Their use in the vehicle as a booster therefore does not appear to be feasible in the foreseeable future.

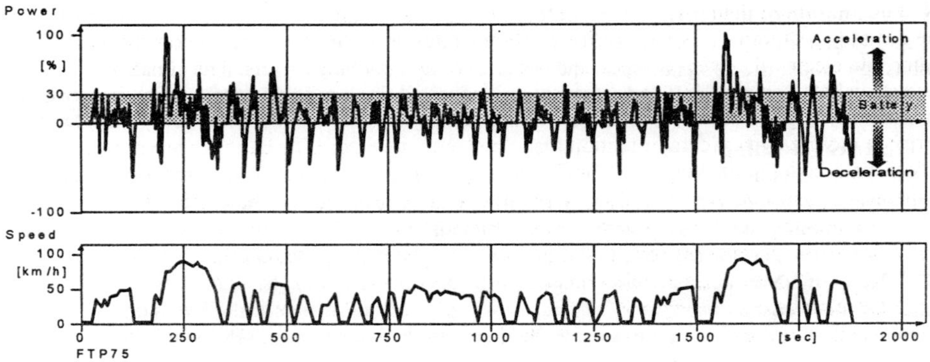

Fig. 5: Power demand in FTP 75 cyclus

Flywheels on the other hand, are further ahead in terms of development (7). This principle involves converting electrical energy into rotational energy by means of an electrical machine, and storing it (8, 9). Progress in the field of power electronics offers scope for developing suitable drive systems. Composite fibre materials which permit high rotating speeds are now available, though the radial forces in the flywheel ring in particular are problematic. However, there are prospects of solving these problems by arranging the fibres appropriately. The bearings are a further focal point of development work. The aim is to produce a system with electro-magnetic bearings, these having the advantage of being absolutely contact-free and non-wearing and also permitting their rigidity and damping properties to be varied during operation, according to load and speed. The effects of mechanical phenomena, such as critical engine speeds, can consequently be influenced with the aid of software in such a way that they no longer cause any noticeable interference in practical operation. In a joint research project conducted together with its development partner UTC, BMW is investigating the safety aspects of such a system on the vehicle (10). A comprehensive series of tests is being prepared to clarify the failure mechanisms, with the aim of influencing energy conversion in such a way that in the event of damage, energy transfer to the surroundings takes place in a controlled manner. These practical experiments with the flywheel system will put BMW in a position to assess the scope for flywheel systems in the vehicle by means of substantiated measurements. They will also supply answers to the question of whether the "mechanical battery" that has been the subject of intense discussion for some time could solve the problem of how to store electrical energy in the vehicle.

Table 1
Flywheel Technical data

Effective capacity	600 Wh
Maximum speed	35,000 rpm
Output	25 kW
Weight	70 kg

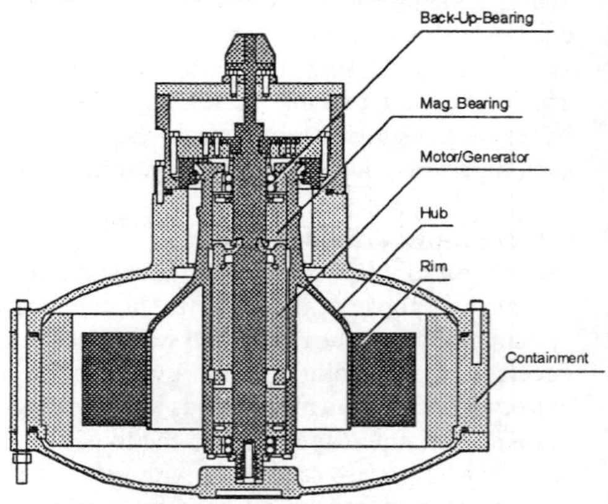

Fig. 6 Flywheel Surge Power Unit

4.4 Thermal management

The differing requirements presented by the temperature levels of the internal combustion engine and electrical system components necessitate two cooling circuits. The constant fluctuation between charging and discharging modes and the longer operating times resulting from overcoming the battery's limited range impose higher loads on electrical components and therefore require even greater attention than on electric vehicles. The batteries are especially critical, with high thermal capacities and thermal inertia levels and the restricted permissible temperature ranges rendering control more difficult. Whereas high-temperature batteries need to be kept at their operating temperature and therefore have a high energy consumption, the relatively narrow permissible temperature range is a serious problem on other systems. Nickel-metal hybrid batteries, for example, can at present only operate unrestrictedly within a temperature range of about 0 to 30° Celsius; by way of comparison, standard versions of series-production vehicles are designed for unrestricted operation from -28 to +35° Celsius.

4.5 Auxiliary assemblies

Hybrid vehicles can also operate exclusively in the electric mode, so the auxiliary assemblies need to be equipped with suitable drives. In the travel mode at low speeds, the auxiliary assemblies have a power demand similar to that for propulsion. Here again, the question of energy demand and efficiency is highly relevant, as these factors have a considerable influence on range in the electric mode.

5 OPERATING EXPERIENCE WITH HYBRID VEHICLES

BMW is using both of the vehicles described to conduct fundamental research. The interplay of all components in practical operation is of particular significance. Subjectively "test-driving" the various operating strategies is a further key task. With the aid of a suitable control system, all driving statuses and operating conditions can be monitored conveniently on the vehicle without the driver having to intervene in any way. The fuel consumption values measured with these experimental models are equivalent to those of the series-production models. The significant reduction in fuel consumption that has been forecast has not yet materialised. The reason is that the assumed high efficiency levels of the electrical components in particular are only achievable at isolated points, whereas the values for the remaining points on the map fall short by a significant amount.

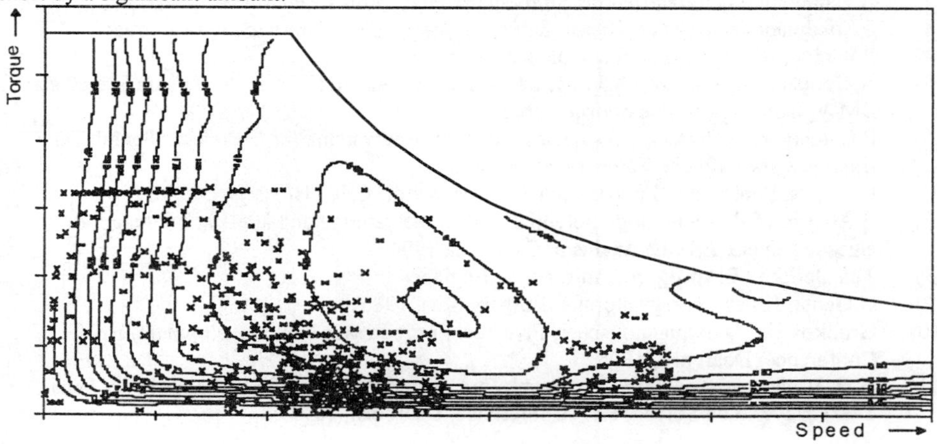

Fig. 7 Electric Motor Efficiency Map with operating points in FTP 75 cyclus

If for example the point of optimum efficiency for an electric motor is 93 %, in test operation an average efficiency of only 80 % will be achieved. Significant improvements can only be expected if the efficiency levels of the electrical components, and the electrical machines in particular, can be improved significantly in the part-load range. Restricted dynamic behaviour, few operating points within the map range and flexibility when warming up permit a significant reduction in emissions, particularly with the series layout.

6 OUTLOOK

Hybrid vehicles and electric vehicles alike offer much scope for making individual transport more environmentally acceptable. Considerable development work is still needed on all driveline components. Efficiency improvements to the operating range of all components are a key area. Operation and thermal management and the optimisation of auxiliary units are further areas of development work. However, the potential of conventional drive systems running on gasoline (petrol) or diesel fuel should not be ignored. They will remain significant, including or especially in combination with new fuels, such as natural gas, which is already available, or hydrogen, which will become available in the future. These "clean" fuels will gradually reduce the carbon content to zero, provided they can be produced and distributed without generating emissions of hydrocarbons. It will thus become possible not only to eliminate virtually all hydrocarbons which are responsible for causing smog, but also to cut CO_2 emissions drastically. Hydrogen and electricity are ideal secondary energy carriers for powering zero emission vehicles. In the long term they are capable of satisfying the demand for mobile energy from regenerative primary sources (11).

REFERENCES

(1) K.Scheurer et al : Derzeitiger Stand der Entwicklung bei Elektro-PKW am Beispiel des BMW E1, Elektrizitätswirtschaft 1992

(2) H.H. Braess, Fahrzeugtechnik der Elektro-PKW - Problembereiche, Lösungsmöglichkeiten, VDI Report 985, 1992

(3) A.Kalberlah, Hybridantrieb für Straßenfahrzeuge, Volkswagen AG

(4) S.Friedmann, G. Fischer, Hybrid Drive - Taking a Dual Approach with a Great Potential, BMW Engineering Conference 1994

(5) A.Goubeau, K.Scheurer, S.Juraschek, Advancements in Electric Propulsion Systems at BMW, Convergence,Dearborne, 1992

(6) P.Catchpole, A.Prassek,U.Wagner., Neue Batteriesysteme für Elektrostraßenfahrzeuge, Energiewirtschaftliche Tagesfragen, 8/1993

(6) P.Zegers, Fuel Cells, Batteries and Supercapacitor, CEC, Brussels, 6/1992

(7) B.Anerdi, et al, Technology potential of flywheel storage and application impact on electric vehicle, ESV12 Anaheim, California 1994

(8) The Oerlikon Electrogyro, Automobile Engineer

(9) G.Genta, Kinetic energy storage, Butherwoths, 1985

(10) Grudkowski, Olbermann,Fischer, Flywheel Surge Power Unit Development, EnV Conference, Dearborn, 1995

(11) H.H. Braess, S.Friedmann, W. Strobl, Energy and Road Traffic of Tomorrow - What Tasks Have to be Solved, EnV Conference, Dearborn, 1995

C498/5/200/95

Prediction of urban bus cycle emissions

C J T van der WEIJER MSc and **R S BAERT** PhD, MSc
TNO Road Vehicles Research Institute, The Netherlands

SYNOPSIS

TNO has carried out extensive research to analyse the differences and find a relation between the emission results of the European 13-mode certification test and the emissions of an urban bus in practice.

This is achieved by testing two modern buses on a transient chassis dynamometer using the urban bus driving cycle and by measuring mapped emission results for the same engines on a steady-state engine test bed.

Based on the results of these measurements, a prediction method has been developed for deriving fuel consumption and emission levels in units of [g/km] from measured 13-mode values and some vehicle parameters.

1 INTRODUCTION

When acquiring new urban transport buses, fuel consumption and the level of exhaust emissions play an important role in the decision-making process.

In Europe, the emission of harmful exhaust gases is certified by the ECE R49 European 13-mode test (see Figure 1). Here, engine operating parameters and exhaust emissions are measured covering a schedule of 13 steady-state engine running points and summed to form weighted emission and fuel consumption values. These values are expressed in units of [g/kWh].

13-Mode Test Diesel Engines
EEC R49, EEC/88/77 EEC/91/542

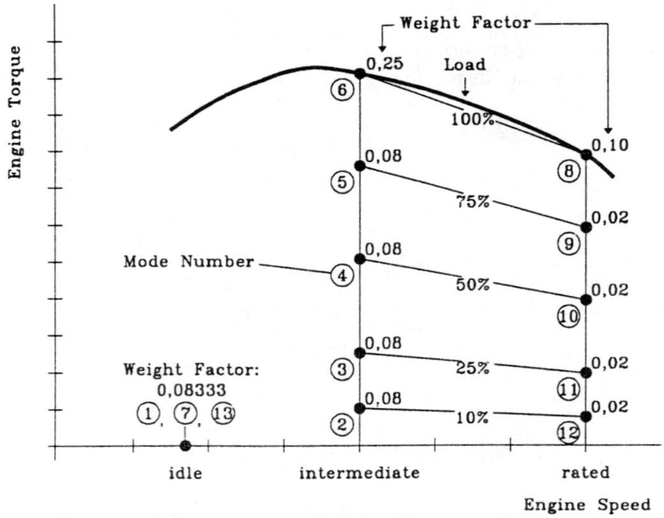

Fig 1 ECE R49 13-mode test

However, the relation between the 13-mode values and the emissions actually produced on the road in units of [g/km] is largely unknown. Differences are caused primarily by deviations in the engine speed-torque distribution patterns and effects due to transient operation. In order to analyse these differences and, more importantly, to attempt to find a relation between these results, TNO has carried out extensive research.

The actual on-the-road performance is characterized by the Urban bus driving cycle, as shown in Figure 2 [1]. This cycle has been developed on the basis of measurements in several cities in the Netherlands. Differences in driving style, routes followed, time of day, day of the week, et cetera, are all being taken into account. The resulting data are combined to form a single reference cycle covering 1700 running hours. Special statistical methods have been developed to produce a 15-20 minute cycle from these measurements which is representative of average urban bus use.
Meanwhile, by comparing several European cycles, it could be established that the cycle can also be seen as a good European average.

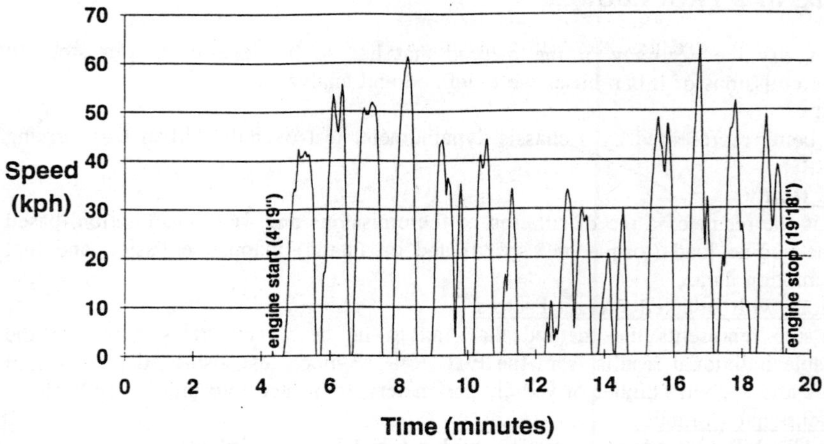

Fig 2 The Urban Bus Driving Cycle

Performing a transient chassis dynamometer test on every type of heavy duty vehicle, using this urban bus driving cycle, would yield the optimum basis for legislative purposes. Similar legislation tests are conducted on passenger cars, where every model has to be tested on an ECE driving cycle and emissions have to meet certain limits in grams per distance.

This would, however, not be feasible for heavy duty vehicles. First, the testing equipment required to perform such a test is very expensive. Second, for both buses and trucks there exist an almost infinite number of different models. This is the main reason why for heavy duty vehicles legislation is based on the engine instead of on the vehicle itself.

So, on the one hand, there is the 13-mode legislation test which is relatively easy to perform, but is inaccurate. On the other hand, there is the accurate but very expensive solution of performing a dynamometer cycle test on each new bus model.

In this paper, a method is presented that predicts emissions in practice, without having to perform a chassis dynamometer test.

This alternative method is based on individual 13-mode results that are normally available from the engine manufacturer. In this method, the weight factors of the 13-mode are recalculated so as to be representative of the urban bus driving cycle. In addition, the energy use at the flywheel [kWh/km] is calculated in order to convert the [g/kWh] data into [g/km] data. It is expected that this method will enable emissions and fuel consumption to be calculated at low extra cost and at only a small sacrifice of accuracy. For these calculations, a small number of vehicle parameters will be used.

The project has been commissioned by the ministry of Transport of the Netherlands as part of the programme "silent, clean, and energy efficient transport in urban areas".

2 METHODS AND PROCEDURES

In order to verify the accuracy of the method that had to be developed, four steps to determine the emissions of urban buses were defined and analyzed:

1. Reality
   Here being represented by a chassis dynamometer test over the Urban Bus Driving Cycle.
2. Static reality
   This is the result of a the calculation of the emissions and fuel consumption, based on the engine load (torque and speed) and on (static) mapped emission and fuel consumption data.
3. 13-mode with Adapted Weight Factors
   This step represents the method that had to be developed. It is based on the available individual results from the ECE R49 13-mode test, using adapted weight factors and a small number of vehicle parameters, translated into units of [g/km].
4. Conventional 13-mode
   The ECE R49 13-mode test (possibly with a translation into [g/km])

Figure 3 shows a graphic representation of these four steps of emission determination.

The four steps show the following main differences:

- The difference between step 1 and step 2 is caused by the effects of the dynamic load in step 1.
- The difference between step 2 and step 3 occurs because of a limitation of the number of analyzed engine load points from infinite (in the maps) to 13.
- The difference between step 3 and step 4 is due to the fact that in step 3 the weight factors of the 13-mode test are adapted to the bus load in practice.

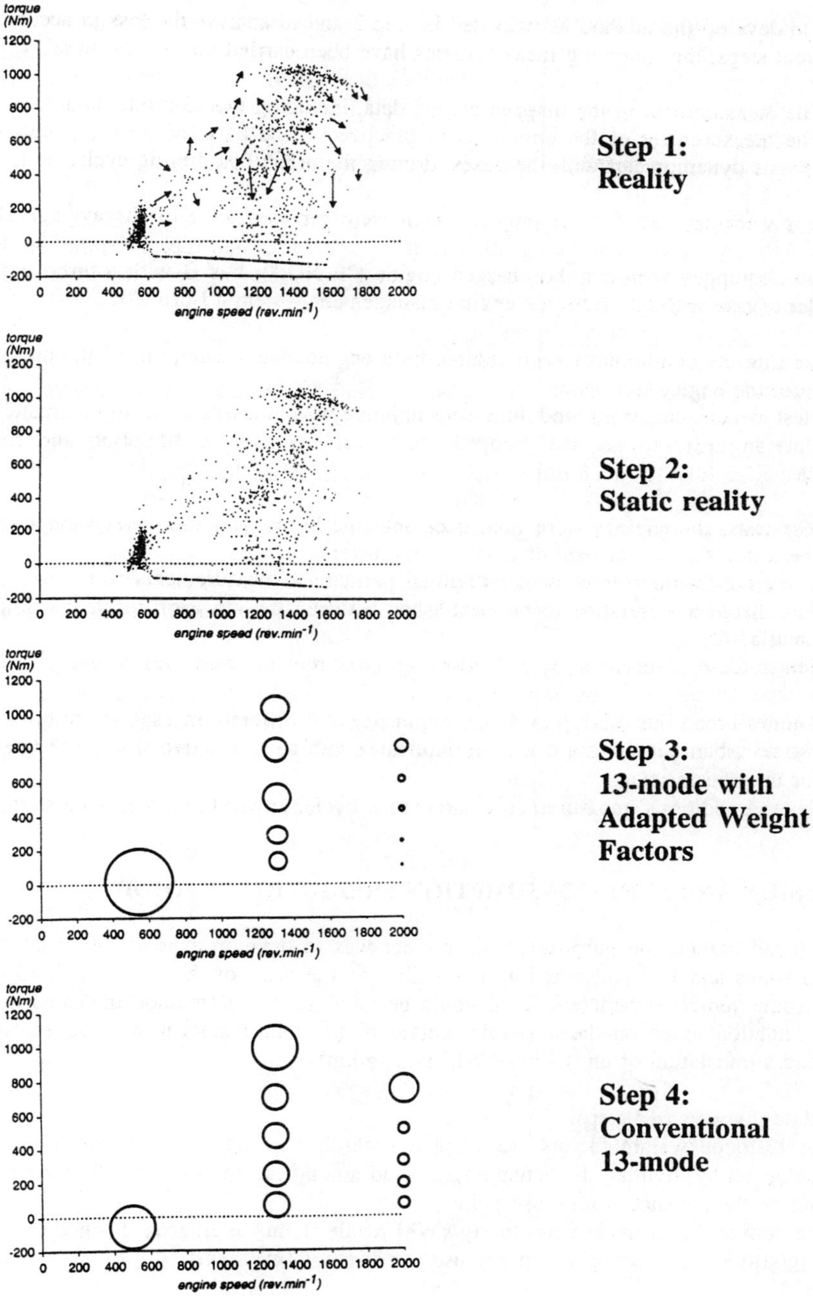

Fig 3 The four different steps for the determination of bus emissions.

In order to develop the method as indicated in step 3 and to analyse the loss in accuracy in the different steps, the following measurements have been carried out on two buses.

- The measurement of the mapped engine data, including the 13-mode points.
- The measurement of the emissions in practice on the basis of measurements on a chassis dynamometer with the buses, driving the urban bus driving cycle.

Two buses were selected for the project which were representative of the average Dutch park.
Bus A was equipped with a turbo charged engine ("9 NO_x"), bus B with a turbo charged intercooler engine with an electronic engine management system ("Euro 2").

After checking the condition of each engine, both engines were taken out of the buses and prepared for the engine test bench.
On the test bench, emissions and fuel consumption were measured at approximately 50 load points in order to get the mapped charts for the fuel consumption and for all emissions.

After these tests, the engines were built back into the buses, and the buses were taken to Graz to be tested on the transient chassis dynamometer.
In Graz, several 13-mode tests with individual particulate measurements were carried out in order to enable a correlation to be established between the emission results as obtained by Graz and TNO.
In addition to these 13-mode tests, the following measurements were carried out:

- 4 times urban bus driving cycle (in duplicate) at 4 different masses of the bus.
- 1 time urban bus driving cycle (in duplicate) with an alternative shifting programme for the transmission.
- Several additional measurements (alternative cycle, power limitation, cold start,...)

3 EMISSION AND FUEL CONSUMPTION PREDICTION METHOD

As mentioned earlier, the purpose of the project was to develop a prediction method for cycle emissions and fuel consumption, if possible on the basis of the results of a 13-mode test and some vehicle parameters. This would be obtained by the method indicated as step 3, the calculation based on the available individual 13-mode results using adapted weight factors and a translation of units of [g/kWh] into [g/km].

The method involves two steps.
First, the 13-mode weight factors are adapted according to the load patterns of the bus. This is achieved by dividing the actual engine load among the surrounding 13-mode points, depending on the distance from those points.
Second, a conversion is made from the [g/kWh] result of this alternative 13-mode test into units of [g/km] by calculating the energy use per distance [kWh/km].

The measurements showed that both calculation steps could be carried out as a function of the following parameters:

* Mass of the bus (=mass of the empty bus + average load).
* The maximum power of the bus.
* Data about the application of an alternative energy-saving shifting programme for the automatic transmission.

Additionally, the individual 13-mode data for fuel consumption and emissions in units of [g/kWh] are necessary.

The method does not apply to buses with an unconventional drive line concept (e.g. Continuously Variable Transmissions or hybrid drive lines).

Table 1 shows the formulas concerning the adaptation of the weight factors.

13-mode point	Weight factors		
	Conventional	Adapted	Adapted (average urban bus)
Idle	.08333	.196 - .00034*m/Pmax	.1710
Intermediate 10%	.08	.040 - .00017*m/Pmax	.0275
Intermediate 25%	.08	.123 - .00085*m/Pmax	.0605
Intermediate 50%	.08	.164 - .00068*m/Pmax + 0.01*ESP	.1140
Intermediate 75%	.08	.095	.0950
Intermediate 100%	.25	-.020 + .00170*m/Pmax + 0.03*ESP	.1050
idle	.08333	.196 - .00034*m/Pmax	.1710
Rated 100%	.10	-.066 + .00153*m/Pmax - 0.03*ESP	.0465
Rated 75%	.02	.018 - 0.01*ESP	.0180
Rated 50%	.02	.037 - .00034*m/Pmax	.0120
Rated 25%	.02	.018 - .00017*m/Pmax	.0055
Rated 10%	.02	.003	.0030
idle	.08333	.196 - .00034*m/Pmax	.1710

"m" stands for mass of the bus [kg], Pmax for the maximum engine power [kW].
"ESP" stands for Energy-saving Shifting Programme of the automatic transmission and is 1 if applied or 0 if not applied. ESP is the programme that shifts as early as possible in such a way that the bus can still be driven properly.

The last column shows the adapted weight factors for a standard bus with an average load (12 500 kg, 170 kW), without ESP. The most distinct difference occurs in the intermediate 100% point (maximum torque at low engine speed) and especially at idle. In the cycle, the engine is running idle during 28.4% of total running time. In view of the fact that negative torque points are not included in the re-division of the weight factors, and because some of the load points between idle and the 13-mode maximum torque region are partly added to the idle weight factor, the new alternative weight factor for the three idle points combined exceeds 50 % (from 25% in the conventional 13-mode).

After the 13-mode weight factors have been adapted, a new weighted 13-mode result becomes available according to the conventional 13-mode formula:

$$E\,[g/kWh] = \frac{\sum_{i=1}^{13} E_i\,[g/h] \cdot WF_i}{\sum_{i=1}^{13} P_{engine,i}\,[kW] \cdot WF_i}$$

Where: E = Weighted emission [g/kWh]
E_i = Emission from the 13-mode point concerned [g/h]
WF_i = (Adapted) weight factor of the 13-mode point concerned [-]
$P_{engine.i}$ = Power from the 13-mode point concerned [kW]

This (newly) weighted emission is still expressed in units of [g/kWh]. In order to translate this into units of [g/km], the following formula for the energy use per kilometre has been developed:

$$[kWh/km] = (c_o + c_1 * m + c_2 * m^2 + c_3 * m^3 * (\frac{Pmax-170}{2*Pmax+170} - ESP*0.043))$$

Where: [kWh/km] = energy use per kilometre [kWh/km]
m = Bus mass [kg]
c_0 = -0.06476
c_1 = 1.733 E-04
c_2 = -2.27 E-09
c_3 = 5.12 E-13
Pmax = Maximal engine power [kW]

When the individual 13-mode values are not available, it is also possible to use the weighted 13-mode results and translate them to a [g/km] value with the formula above. This would however return a lower accuracy (see chapter 4).

4 ACCURACY OF THE METHOD DEVELOPED

The method as described in chapter 3 has two marked differences with respect to reality. First, the engine load is dynamic in reality. Second, the engine load in reality is not limited to only 13 load points.

For each of the approximately 15 measurements that were carried out on each bus in Graz, the emissions as measured at the Technical University of Graz were compared with the emissions that were predicted by using the methods all in the 3 other steps, mentioned in Figure 3.

Figure 4 shows the differences measured between:
Step 1: Reality, and
Step 2: Static reality

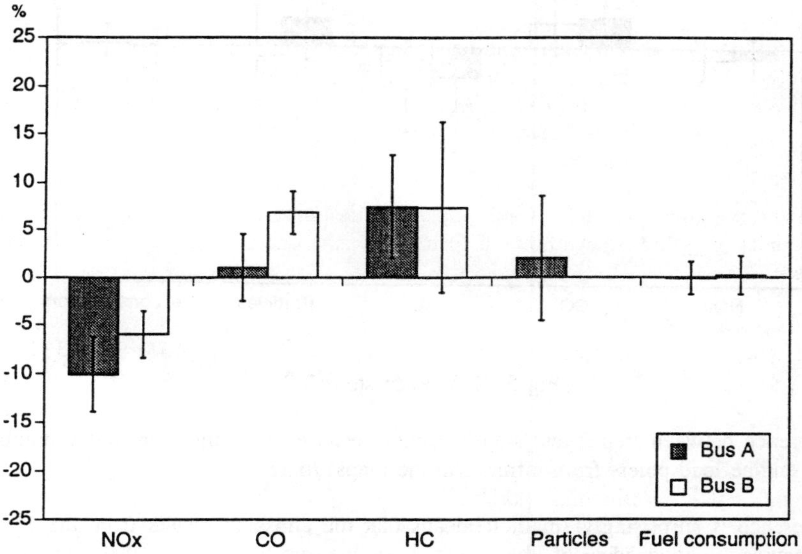

Fig 4 Difference steps 1-2 (static-dynamic)

The difference between these two steps is caused by the effects of the dynamic load in step 1.

It appears that the deviation for both these buses is comparable, in spite of the fact that the engines are of a different generation. This difference very much depends on the type of engine. It can be expected that with the introduction of systems like EGR, air injection during acceleration, et cetera, this figure may turn out to be quite different. It will therefore be very difficult to attach a fixed transient factor to the method.

Figure 5 shows the differences measured between:

Step 2: Static reality, and
Step 3: 13-mode with Adapted Weight Factors

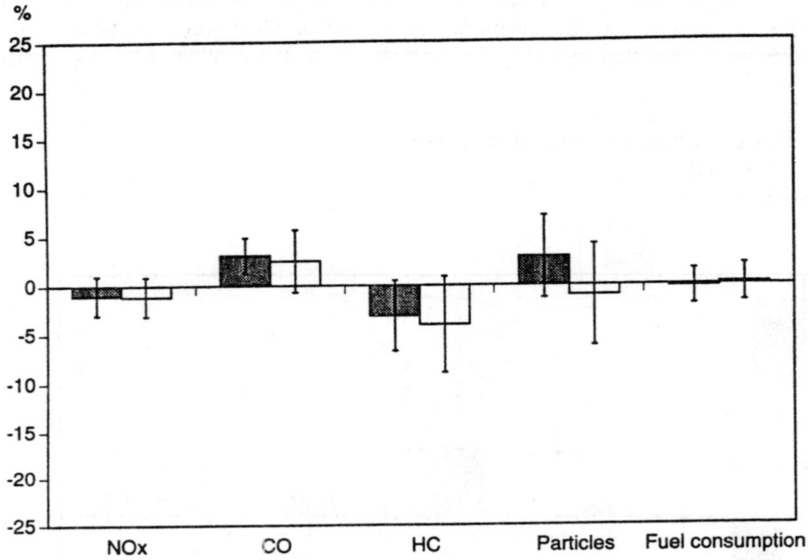

Fig 5 Difference steps 2-3

The difference between step 2 and step 3 occurs because of a limitation of the number of analyzed engine load points from infinite (in the maps) to 13.

This difference is surprisingly small. It means that the emission results from the 13-mode points provide a good idea of the mapped engine emissions, and that therefore the limitation to 13 data points is not a particularly great problem.
A problem would occur, however, when so-called "cycle-beating" is applied. This means that the engine behaviour greatly differs outside the 13-mode points.
This did not appear to be the case with the buses tested in this project. However, in the future, this could prove to be more often due to the more stringent emission limits and the increasing possibilities of electronic engine management systems.

When the individual 13-mode values are not available, and weighted 13-mode values are used to calculate the static reality, bigger differences occur.
This is indicated by the difference between:

Step 2: Static reality
Step 4: Conventional 13-mode test

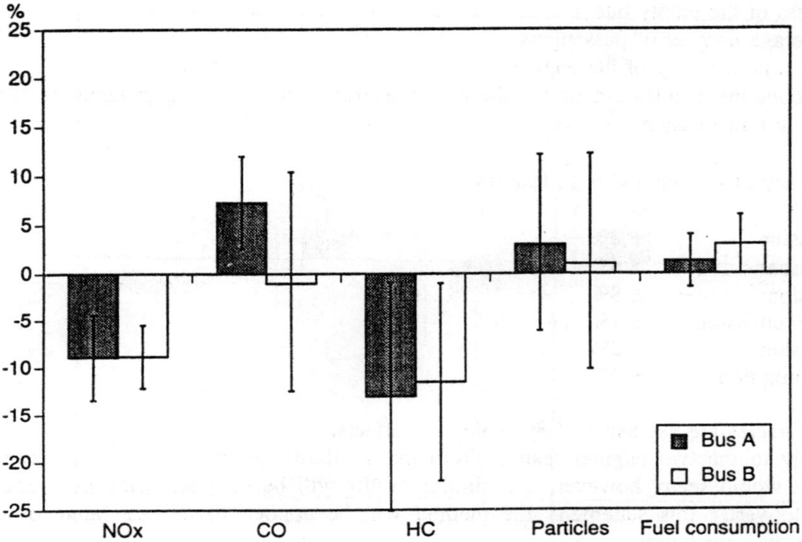

Fig 6 Difference steps 2-4

It appears from the difference in accuracy between the Figures 5 and 6 that the adaptation of the weight factors results in a substantial benefit in terms of accuracy. This is achieved by a relatively small effort.

It can be concluded that the method can provide a very accurate picture of the static reality. This accuracy is so defined independently of the influence of the dynamic behaviour.
However the influence of the dynamic load is comparable and not very high for the engines that were tested, they cán differ greatly for different engine concepts. It therefore does not seem reasonable as yet to include the dynamic factors in the model. This will require additional research.

The accuracies reproduced above apply to the two engines tested. From the similarity of the results from these two engines, it would seem, however, that similar results will be obtained with other engines. In order to verify this statement, the method will be applied to a wide range of other engines in the near future.

5 CONCLUSIONS

The method as developed in this project calculates the emissions and fuel consumption for the on-the-road reality in units of [g/km]. This is achieved on the basis of the following data:

- Individual 13-mode data for emissions and fuel consumption
- The mass of the empty bus
- The average number of passengers
- The maximum power of the engine
- Data about the application of an alternative energy-saving shifting programme for the automatic transmission

The accuracy of the method is as follows:

NO_x emission:	± 3%
CO emission:	± 6%
HC emission:	± 8%
Particulate emission:	± 7%
CO_2 emission:	± 2%
Fuel consumption:	± 2%

These accuracies are exclusive of dynamic load effects.
They apply to the two engines tested. From the similarity of the results from these two engines, it would seem, however, that similar results will be obtained with other engines. In order to verify this statement, the method will be applied to a wide range of other engines in the near future.

It appears that the 13-mode results provide a good basis for the calculation of fuel consumption and emissions in reality, provided that the weight factors are adapted for the purpose, and no cycle-beating is applied to the engine (optimization at the 13 testing points).

REFERENCES

[1] VAN DE WEIJER, C.J.T., VAN DER GRAAF, R., HENDRIKSEN, P. and VERBEEK, R.P. Urban bus driving cycle, 4th international EAEC conference, 16-18 June, 1993, Strasbourg.

[2] PISCHINGER, R., ANKOWITSCH, M., OSPELT, W., Instationäre Erfassung der Emissionen von LKW-Motoren im US-Transient test und ihr Vergleich mit stationäre messungen.

[3] CORNETTI, G.M., KLEIN, K., FRÄNKLE, G.J., STEIN, H.J., US Transient Cycle versus ECE R49 13-mode Cycle. 1988. SAE paper 880715, Detroit Michigan.

Diesel Engine Developments

C498/2/001/95

Design and development of variable geometry diesel engines

R ZIARATI BSc, PhD, CEng, FIMechE, FIMfgE
University of Central England, Birmingham, UK
A COLE PhD, BSc, MIMechE
MIRA, UK

SYNOPSIS

There are a number of major problems in matching fuel injection equipment and a turbocharger to a diesel engine particularly when the engine is expected to run at different operating conditions. The authors have been involved in the development of the Holset variable geometry turbocharger and the Lucas high pressure fuel injection equipment projects. The intention of the former project was primarily to improve the transient response of a highly rated diesel engine and the purpose of the latter to reduce specific fuel consumption as well as to reduce exhaust emissions.

In this paper a simple diesel engine model as well as existing experimental results were used to develop a unique variable geometry diesel engine. A series of specific tests were carried out to verify and validate the result of initial theoretical studies.

The paper examines the basis for a clean diesel which comprises of a tested variable geometry turbocharger system and a proven variable geometry fuel injection equipment. Based on the recent experimental and theoretical studies, a number of recommendations are made to improve engine performance and/or to reduce exhaust emissions in future diesel engines.

NOTATION

A	Area
BMEP	Brake Mean Effective Pressure
C_θ	Tangential Component of Flow Velocity
FIE	Fuel Injection Equipment
h	Width
M	Mach Number
MFF	Mass Flow Function
P	Pressure (suffix 0 relates to stagnation state)
r	Radius
R	Gas Constant
SFC	Specific Fuel Consumption
T	Temperature (suffix 0 relates to stagnation state)
α	Absolute Flow Angle
γ	Specific Heat Ratio

1 INTRODUCTION

The matching of main components such as turbochargers and fuel injection equipment to an engine is a costly and a time consuming operation. Experimental testings and mathematical modelling offers useful solutions in shortening the design and development time of an engine. Theoretical investigations will also allow the effect of various parameters to be basically and systematically studied, offering interesting possibilities in improving existing and/or developing new engine systems. There are many diesel engine computer simulation and mathematical models (**1**, **2**, **3**). In this study a mathematical engine model developed by (**4**) was used to carry out a series of preliminary investigations. The experimental results obtained by (**5**) were used to validate the result of preliminary theoretical studies and support the development of the mathematical model. Available test results on turbochargers (**6**, **7**) fuel injection equipment (**8**, **9**), combustion (**10**, **11**, **12**) and overall diesel performance characteristics (**1**, **2**, **3**) have been used to further develop the engine model to ensure that each part of the model and mathematical calculations are based on physical realities. The model has been crucial in identifying two major parameters for developing a variable geometry turbocharger and a variable geometry fuel injection equipment. In addition, a one-dimensional turbine flow program and a one-dimensional compressor program were developed (**7**) to study the effect of selected variable geometry devices on the performance of turbocharger turbines, compressors and the combined units respectively.

The theoretical studies (**7**) clearly indicated that with a given operating pressure ratio and temperature, total efficiency for both turbocharger turbine and compressor is fixed for any turbine or compressor by the designers choice of blade pattern and rotor tip speed. Since power from a turbine and a compressor depends on the flow rate and the isentropic head, which in turn depends only on temperature and pressure ratios, therefore, with fixed temperature and pressure ratios, the power of a given turbine (and a compressor) can be varied only by changing the flow. In an extensive investigation (**13**) the performance of a Perkins engine (T6.354) was measured using different turbine casings (**Figure 1**). The tests were repeated with a Leyland 520 engine. It was evident that, as the effective flow area of the turbine volute was reduced, brake mean effective pressure of the engine increased at reduced engine speed. Since both theoretical and experimental results were indicating that control of the effective throat area of compressor and turbine have substantial effect on engine performance, a systematic investigation was instigated to develop a variable geometry turbocharger.

At a later date based on a number of investigations (**5**), it was clear that engine performance could substantially improve if the engine has a high average injection pressure as well as a variable geometry fuel injection equipment, rather than the conventional types used on most engines. The theoretical studies clearly indicated the advantage of raising fuel injection pressure and having a facility to reduce and vary injection timing. To this end, a series of specific tests were carried out on a Ricardo and a Mirlees engine. A number of limited tests were also carried out on Perkins and Lister single cylinder engines involving various FIE at different engine loads and speeds.

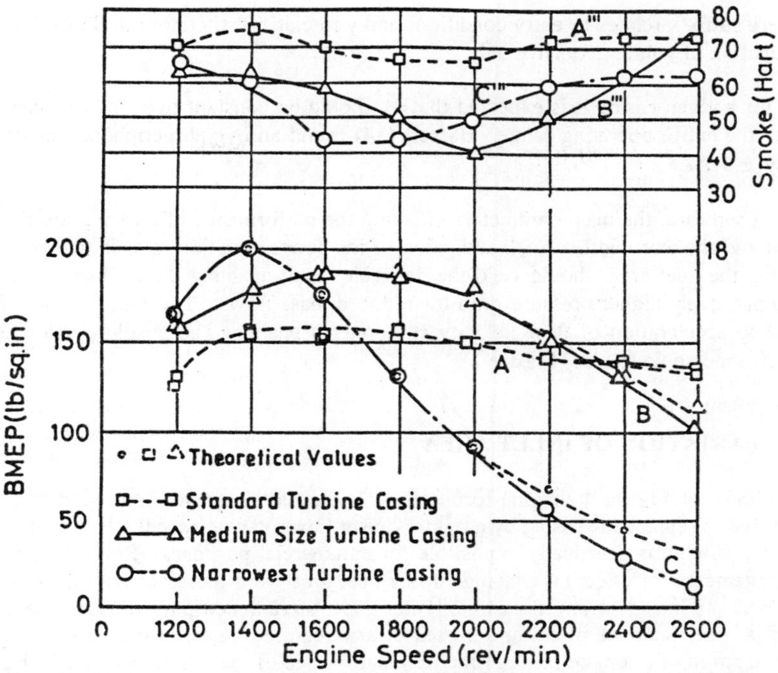

Fig 1 Performance of Perkins T.6354 Using Three Different Turbine Casings

2 TURBOCHARGER STUDIES

The angle at which the fluid reaches the rotor of a turbine or a compressor greatly influences the overall performance of a turbocharger. This angle in a nozzleless turbine and compressor is determined by the geometry of the volute of the turbine and/or compressor volute. In larger turbocharger units, use of a swivelling nozzle unit facilitates the angle at which flow should reach the rotor. In automotive turbocharger units, use of swivelling nozzles is prohibited due to a lack of space. This paper will not discuss the application of variable geometry turbochargers with swivelling nozzles.

Theoretically, considering incompressible flow, and conservation of angular momentum in the case of a turbine, it can be shown that the angle of flow at exist from the volute is uniform around the periphery and is constant for the whole operating range of the turbine. By combining the momentum and continuity equations, it can be shown that:

$$\therefore \alpha_y = \mathrm{Tan}^{-1}\left(\frac{h_y}{F_x} 2\pi r_x\right)$$

where suffix x relates to entry conditions and y to exit. It, therefore, follows that y is a function of volute geometry only.

So for a given casing it is expected that α_y should be constant over the periphery and over the entire operating range. Bassett **(14)** found an overall peripheral variation of only ± 3½%.

Therefore, the main parameters affecting the performance of a casing are the inlet area and the exit width. An examination of the above equation reveals that reduction of Fx, the inlet area, should yield the desirable result enabling the turbine to build up comparatively higher pressure ratio for reduced mass flows. The reduced mass flows lead to acceleration of flow velocity components enabling the absolute velocities to reach supersonic speeds (**Figure 2**).

3 VARIATION OF INLET AREA

As shown in **Figure 1**, engine torque back up would be greatly improved if it were possible to replace the casing by a tighter one at lower engine speeds. It is evident that such a scheme is practically impossible for commercial purposes. Several alternative arrangements for effecting a change in the inlet area were therefore considered by (7) such as the introduction of a carefully designed, streamlined plug through the casing wall at inlet, capable of being adjusted at any required depth in the passage, or the introduction of a tongue through the terminator edge of the volute casing which could be slid into the volute casing in order to reduce the inlet flow area F_x.

4 VARIATION OF EXIT WIDTH

The most appropriate method for the reduction of the exit width was considered to be the introduction of a 'flat plug' from one side of the casing, in the parallel sided wall section, which could be moved to close the exit width of the volute to the extent required.

The 'flat plug' tests proved to be unsatisfactory due to the effect on the exit flow angle α_y. The technique was extended by the application of a sliding nozzled plug (and later by double sliding and inter-meshing nozzle plugs). This second approach consisted of a sliding plug with nozzle blades cut at its end . With the sliding plug fully withdrawn, the nozzle blades cover the entire exit passage width thereby providing a fully nozzled volute. As the sliding plug is moved into the flow passage the nozzle blades enter slots machined into the opposite side of the casing and the plug provides a restriction to the flow.

The nozzled section of the plug ensures that the flow is fully guided by blades and the exit angle is dependent on the casing's geometry, therefore, the major drawback of the 'flat plug' is removed. It should be noted that both devices in practical terms have few problems such as leakage. **Figure 3** shows the effect of 'nozzle plug' restrictions on engine performance.

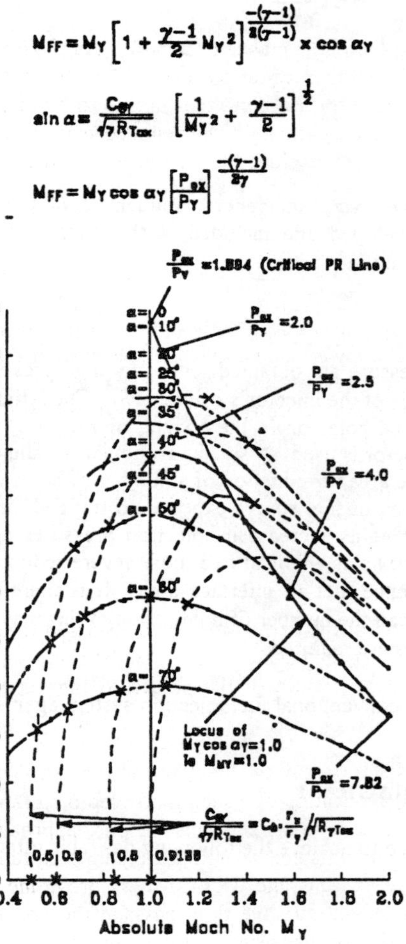

Figure 2 Stationary duct flow characteristics

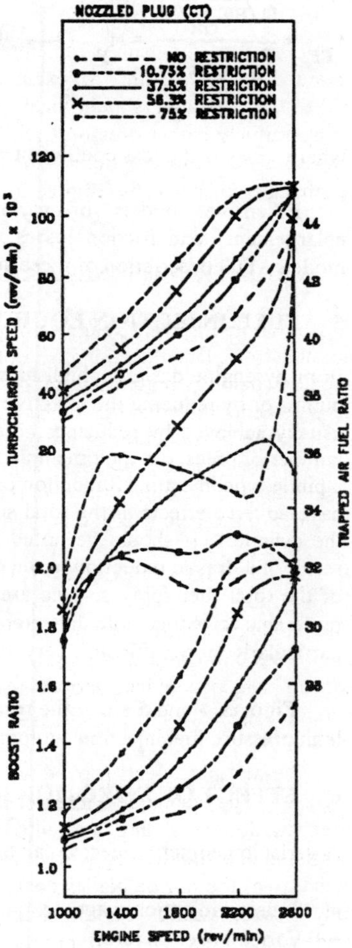

Figure 3 Nozzled Plug - Constant Torque Curves

In order to assess the effect of the sliding plug, including increased losses due to the sudden contraction and enlargement, a number of theoretical models were developed. The models were based on a unified approach method developed by (7). Each passage of the turbine and compressor were represented by:

$$M_{FF} = \frac{\dot{m}\sqrt{RT_{ox}/\gamma}}{F_y \, P_{ox}} = M_y \left(1 + \frac{\gamma-1}{2} M_y^2\right)^{-\frac{1}{2}\frac{\gamma+1}{\gamma-1}} \left(1 - \frac{\gamma-1}{2} \zeta_{xy} M_y^2\right)^{\gamma/\gamma-1}$$

where ζ_{xy} is the loss coefficient.

Extensive models of various losses viz, incidence, sudden contraction, enlargement, skin friction losses were developed and included in the mathematical models. Full description of these are given in (7).

5 FUEL INJECTION EQUIPMENT

In many engine designs, high injection pressure are obtained either by high pressure pumps or by reducing the effective flow area of the injector's orifice **(15)**. The latter is usually achieved by reducing the area of the holes on in the nozzle, or reducing the number of holes, if a multiple hole type injector is used. A similar argument is valid for a pintle type injector. Reduction of the hole area or reduction of a number of holes can have adverse effects in the total surface area of fuel sprays exposed to trapped air in the cylinder. It should be noted that the higher increase in injection pressures as a result of a drastic reduction in the number of holes will lead to a more severe reduction of the total fuel spray surface area and hence fuel-air interface area. It is therefore preferable to reduce hole diameter rather than the number of holes, though this may be particularly more difficult if very small holes are required.:

Figures 4 and **5** show the results of a conventional fuel injection system against a high pressure fuel injection equipment.

6 EFFECT OF INJECTION RATE DIAGRAM

A variable geometry injector can be designed to achieve the following desirable effects

- Squared injection diagram.
- Varied injection rates

Different shape rate diagrams have been fully examined by (**16**). While the latter reference investigated the rate of injection diagrams in order to achieve the best possible efficiencies, others mainly (**17**) considered the effect of the shape of these diagrams on other limiting factors such as smoke. Further, (**18**) has advocated that with a given mechanical limitation, the ratio of mean injection pressure to maximum fuel pressure should be maximised for the best performance. To this end, a brief investigation was conducted. **Figure 6a** shows the rate of injection diagram for the high pressure system for full load and at 1500 rev/min. Fuel line pressure diagram is also shown.

It was then assumed that the rate diagram is that represented by dotted lines, as demonstrated and a theoretical study was conducted. Surprisingly this study indicated

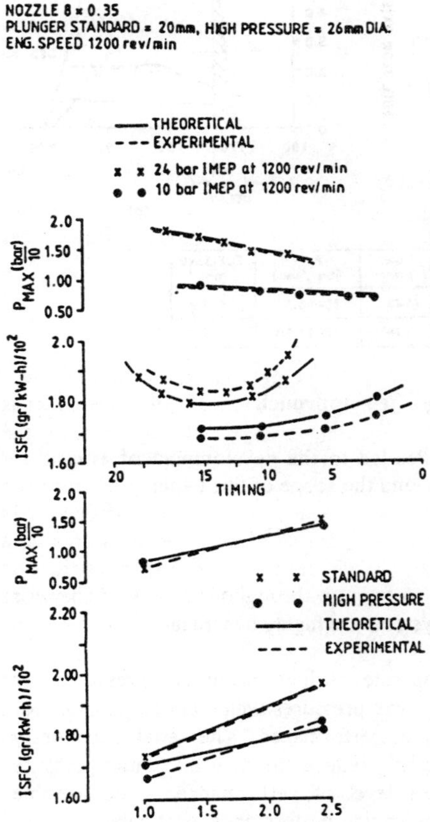

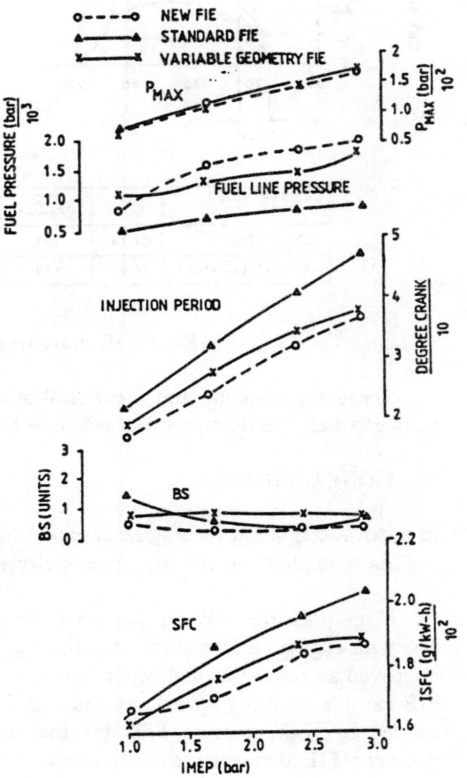

Fig 4 Standard FIE vs High Pressure FIE

Fig 5 Performance of Variable Geometry FIE

no significant degradation of engine performance. A series of experimental tests were commissioned to verify the theoretical findings. The new diagram **Figure 6b** was achieved using a variable geometry device. The engine performance parameters as shown in the table below **(Figure 6)** show no sign of performancedegradation despite substantial reduction of peak injection pressure. As shown the peak line pressure ie 1095 bar is well below the 1730 bar produced by the actual system.

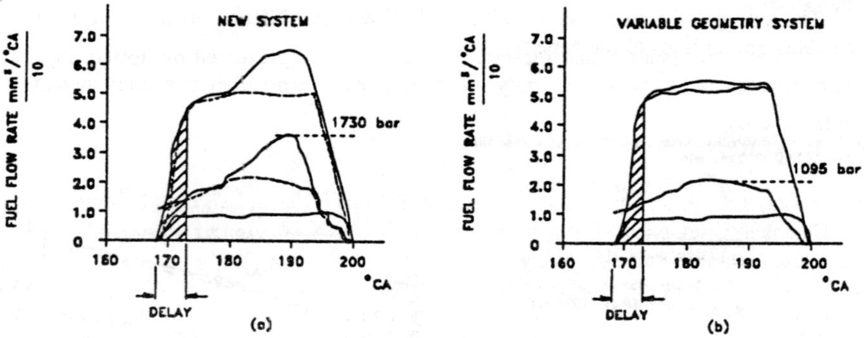

DIAGRAM SHAPE	IMEP bar	ISFC g/kW-h	P max bar	P/θ (bar/sec)	FUELLING mm³
NEW SYSTEM	24.14	184	145	25403.51	1316.4
VARIABLE GEOMETRY	24.00	186	146	27017.20	1320.4

Fig 6 FIE matching - new approach

These experimental and theoretical studies led to the development of a variable geometry FIE, the description of which, is beyond the scope of this paper.

7 CONCLUSIONS

In applications when the engine is expected to operate throughout a range of speeds and loads, application of variable geometry systems are highly beneficial.

High pressure FIE do not need to operate at high maximum pressures for optimum engine performance. It is the high mean pressures which are responsible for improved air utilisation and hence overall engine performance. The variable geometry FIE can maintain high pressures and significantly reduce the high maximum pressure needed by high pressure FIE, for the same level of performance. The variable geometry FIE also was proved to improve the engine performance, particularly at part load.

The ability of the overall model to predict engine performance parameters to within acceptable error 'band' for well over fifty cases at different test conditions with and without variable geometry devices must be considered very encouraging.

8 ACKNOWLEDGEMENTS

The authors with to thank Dr A Veshagh, Warwick Manufacturing Group for his valuable contributions to the paper.

8 ACKNOWLEDGEMENTS

The authors with to thank Dr A Veshagh, Warwick Manufacturing Group for his valuable contributions to the paper.

9 REFERENCES

1. Janota, J.S., et al, The Prediction of Diesel Engine Performance and Combustion Chamber Component Temperatures using Digital Computers. Proc., I.Mech.E, 1968, Vol. 182, Part 3.L. No. 9.
2. Wallace, F. J., and Smyth, R., Comparative Performance Assessment by Digital Computers of Various Compression-Ignition Engine Configurations in Combination with Compressors and Turbines, Proc. I.Mech.E, 1966-67, Vol. 181.
3. Wallace, F.J., and Ziarati, R., Variable Geometry in Turbocharging for Transport Engines, I.Mech.E. Conference, 1982-3.
4. Ziarati, R., Veshagh, V., Matching of Fuel Injection Equipment to Diesel Engines, Proc IUTAM Symposium, Taipei/Taiwan, 1991, Book Title: Aero-Thermodynamics in Combustions, Springer-Verlag Part IV, Page 265, ISBN 0-387-55404-1.
5. Ziarati, R., An Investigation into Fuel Injection Equipment and Combustion Performance of Medium Size Diesel Engines, Proc IMechE/MEP Seminar, Practical Limits of Efficiency of Engines, 1986, ISBN 0852986181.
6. Watson, N., Turbochargers and Turbocharged Engines, Proc SAE Special Publication SP 442, 1979.
7. Ziarati, R., Mathematical Modelling and Experimental Testing of Variable Geometry Inward Radial from Turbines, PhD Thesis, 1979, Bath University.
8. Glikin, P.E., Fuel Injection in Diesel Engines, Proc. I.Mech.E., 1985, Vol. 199, No. 3.
9. Little, C.M.D. and Scott, P.E., Fuel Injection Equipments for Optimum Engine Performance, Proc CIMAC, 1981, Helsinki.
10. Shahed, S.M., Chiu, W.S., and Lyn, W.T., A Mathematical Model of Diesel Combustion in Combustion Engines, 1975, Proc I.Mech.E., p119.
11. Gilchrist, J.M., Chart for the Investigation of Thermodynamic Cycles in Internal Combustion Engines and Turbines, Proc. I.Mech.E., June, 1947, P.335.
12. Wakuri, Y., et al., An Experimental Study on the Fuel Spray Distribution and Combustion Characteristics in a Diesel Engine with Side Injection System, Kyushu University Paper, March 1985, Vol. 45, No. 1.
13. Ghadiri-Zareh, M.S., Variable Geometry Single-Stage and Two-Stage Turbocharging of High Output Transport Diesel Engines, PhD. Thesis, University of Bath, 1976.
14. Basset, R.W., Pressure Loss Tests on a Model of a Turbine Volute, NRC, Canada, Test Report MET 328 A/969.
15. Ziarati, R., Veshagh, A., Hawksley, G., Injection Investigation and Combustion Modelling, ED Non-Nuclear Research Programme, Lloyd's Register Project/Reports Nos A7 and C7, 1989.

C498/2/014/95

The performance and emission characteristics of an indirect injection, naturally aspirated, diesel engine in the presence of corona discharges

S H NASSER BSc, MSc, PhD, CEng, MInstE, MRAeS, **S MORRIS** BSc, MSc, and **M RENTSCHLER** BEng
University of Hertfordshire, UK

SYNOPSIS

Results from an ongoing investigation into a novel technique to improve the performance and emission characteristics of vehicle engines are presented. It involves generating intensive corona discharges in the combustion air stream of a four cylinder naturally aspirated indirect injection diesel engine. The coronas were created using an array of eight circumferentially equi-spaced sharp electrodes positioned in a plane perpendicular to the combustion air stream. The presence of such discharges in the combustion air increased ionisation and excitation levels of atoms and molecules, creating new radicals and highly oxidising species such as atomic oxygen (O) and ozone (O_3). These played a key role in modifying some of the chemical reactions involved in the combustion of the liquid hydrocarbon fuel. Measurements of engine torque, power output, various temperatures, fuel consumption and exhaust emissions were obtained at various engine loads and throttle positions under both normal running conditions and in the presence of coronas. Preliminary results show promising improvements in engine emissions and performance, demonstrating clearly the need for further research in this field.

1. INTRODUCTION

With the need to conserve the world hydrocarbon resources and reduce their ecological and environmental impact, it is imperative that the internal combustion engine operate under maximum efficiency with minimal noxious emissions. In recent years both public and legislative interest has focused towards road transport vehicles, with particulates and gases such as carbon monoxide (CO), carbon dioxide (CO_2), oxides of nitrogen (NO_x), unburned hydrocarbons (UHC) and sulphur dioxide (SO_2) emitted in the exhaust gases being partially blamed for causing acid rain and the greenhouse effects (1-2). Considerable efforts have contributed towards reducing these emissions and improving fuel economy. Diesel engines in particular have benefited from refinement to their overall design along with the use of electronics to control operations such as fuel injection and exhaust gas recirculation. However further research and development to meet future obligatory exhaust gas legislation is necessary. Consequently both the motor industry and academic institutions have been forced to investigate ways to further refine current engine technologies while exploring new and innovative ideas.

The work reported here is an experimental investigation in to a novel technique to improve the performance and emission characteristics of diesel engines. The technique involves the generation of corona discharges in the combustion air stream which in turn produced ionised species and radicals such as O and O_3. The presence of these radicals in the combustion chamber improved the overall fuel combustion process, ultimately leading to improved engine performance.

2. THEORETICAL OVERVIEW

A corona is one form of gas discharge or electrical breakdown generated around electrodes at high voltage and low current. It is characterised by the ionisation and drift regions found between the electrodes, and by faint luminous manifestations at the highly stressed electrodes (3,4).

The most common corona discharge electrode configuration is the point-plane gap as used throughout the current investigations. The asymmetrical nature of this configuration results in the drift velocity of electrons being much greater than that of ions due to their low mass and relatively elastic impacts with ions. The presence of high energy electrons around the electrodes ultimately produces atoms and molecules of higher vibrational and ionisation energy levels. In air this leads to the creation of various species including O and O_3 according to the following reactions;

$$e^- + O_2 \rightarrow 2O + e^- \quad (1)$$

$$O + O_2 \rightarrow O_3 \quad (2)$$

Both O and O_3 are highly reactive oxidising species and of particular importance in the combustion of hydrocarbons. The presence of O in the combustion chamber at concentrations higher than that expected from equilibrium considerations enhances the chemical reactions in the flame front (5). In a hydrocarbon-air mixture, it is generally accepted that the following elementary reactions are responsible for the primary chemi-ionisation (6-7);

$$CH + O \rightarrow CHO^{+} + e^{-} \quad (3)$$

$$CHO^{+} + H_2O \rightarrow H_3O^{+} + CO \quad (4)$$

$$H_3O^{+} + e^{-} \rightarrow H + H_2O \quad (5)$$

The presence of electrons and chemi-ions such as CHO^{+} and H_3O^{+} at elevated temperatures enhance the overall combustion process in the combustion chambers (8-9).

It must be noted that atomic nitrogen (N) also forms as a result of initiating coronas in air according to the following reaction;

$$e^{-} + N_2 \rightarrow 2N + e^{-} \quad (6)$$

This, however, effects the concentration of NO_x in the exhaust gases as discussed later.

The generation of atomic hydrogen (H) through the primary reactions (reactions 3,4,5) and the creation of O by the corona discharge can have important effects upon subsequent combustion. Both O and H are key species as they produce the all important hydroxyl radical (OH) which is crucial in hydrocarbon oxidation. It is generally accepted that the chemical reactions involved in the oxidation of liquid hydrocarbon fuel are extremely complex (10). However attention is drawn here to some of the relevant reactions involved in the oxidation of a simple hydrocarbon (CH_4);

$$O + H_2 \rightarrow OH + H \quad (7)$$

$$H + O_2 \rightarrow OH + O \quad (8)$$

$$CH_4 + OH \rightarrow CH_3 + H_2O \quad (9)$$

$$CO + OH \rightarrow CO_2 + H \quad (10)$$

It is clear from the above reactions that any increase in the concentration of OH radicals increases the rate of reaction of both CH_4 and CO.

Bradley (11) pointed to other mechanisms associated with corona discharges which can lead to improved chemical reaction rates such as an increase in the translational temperature of the air surrounding the electrodes and an increase in the concentration of active species. Again, these are to be discussed later.

3. EXPERIMENTAL FACILITIES AND PROCEDURES

Investigations were performed using a naturally aspirated Ford 1.6 litre four cylinder diesel engine. This was mounted on a comprehensively instrumented Froude hydraulic dynamometer with facilities to measure engine torque, power output, fuel consumption and various engine temperatures and pressures. An on-line ADC infra-red combustion analyser system was used to determine the concentrations of CO and CO_2 in the exhaust gases. This experimental arrangement is shown schematically in figure 1.

Cathodic corona discharges were generated in the combustion air stream prior to entry into the engines intake plenum. Eight stainless steel electrodes were embodied in adjustable nylon screws to facilitate adjustment of electrode spacing. These were mounted circumferentially and equi-spaced in a plane perpendicular to the combustion air flow in an insulating nylon case. A single central earth electrode was positioned along an axis parallel with the combustion air flow as depicted in figure 2. The gap between the axial electrodes and the central electrode was adjusted prior to installation onto the engine so that a visible corona was observed on all eight electrodes when connected to the high voltage generator. This was a simple aircraft ignition unit type WIPAC/C8198 providing a potential of 10kV DC. It was necessary to employ multiple electrodes in order to generate uniform corona discharges throughout the bulk of flowing air.

A series of tests were performed using four throttle positions; 25%, 50%, 75% and 100% at engine speeds ranging from 1000 to 3500 r.p.m. under various loading conditions. To ensure consistency of results, each test was performed following an initial warm-up period at the pre-determined throttle and loading condition in order to allow a steady quiescent cooling water temperature to be reached. The engine was subsequently run firstly under normal running conditions and then in the presence of coronas. This allowed a direct comparison between the engines performance with and without corona discharges. This procedure minimised the effect of varying atmospheric conditions and any experimental or instrumental error. However only results corresponding to 75% throttle position are reported hereafter. Measurements of both NO_X and particulate concentrations in the exhaust gases were not measured.

4. EXPERIMENTAL RESULTS AND DISCUSSION

Figure 3 shows the concentration of CO_2 versus engine speed and clearly demonstrates an increase in CO_2 emission levels in the presence of corona discharges. This increase is more significant (approximately 7.5%) at moderate engine speeds (2000 r.p.m.). It is evident that the generation of O and O_3 by the corona discharges increased the oxidation rate of CO and UHC which were subsequently converted to CO_2. However it was not possible to measure accurately the change in the concentration of CO as in general diesel engines are characterised by low CO emission levels.

Figure 4 displays the change in specific fuel consumption (SFC) at 75% throttle position. It is clear that at moderate speeds improvements ranged between 1% and 5.5% in the presence of coronas. This can only be attributed to the improved combustion process within the engines cylinders which is complemented by the increase in emitted CO_2 and higher exhaust gas temperatures, as can be seen in figure 5. A change is particularly evident at higher engine speeds with an increase of approximately 30°C at 2000 r.p.m. Again, this can be attributed to improvements in the combustion process and to the increased conversion of the fuel latent to thermal energy. It must also be emphasised that an increase in the temperature of the engine cooling water was also recorded in the presence of corona discharges although this is not reported here.

The calculated engine break thermal efficiency at 75% throttle position and various engine speeds is depicted in figure 6. It can be seen that engine efficiency is higher in the presence of coronas with an improvement of approximately 5% at 1500 r.p.m. This can be explained by the reduced fuel consumption and improved combustion process.

Bradley (11) has pointed to several reasons for the increase in the reaction rate of some key reactions involved in the combustion hydrocarbon fuel in the presence of corona discharges. These may be summarised as;

(i) an increase in the concentration of electrons and ionised species
(ii) an increase in the translational temperature of the gas in the vicinity of the electrodes
(iii) excitation of atoms and molecules to higher internal energy levels

The effect of an increase in the electron population in the vicinity of the negative electrodes is a crucial factor in producing O and O_3 as discussed previously, the consequence of which is the production of the all important OH radical. This ultimately resulted in an improved combustion process, manifesting as a lower consumption of fuel and higher CO_2 emission levels.

An increase in the translational temperature of air surrounding the electrodes due to ionisation, excitation and electron heating can ultimately lead to an increase in the temperature of air-fuel mixture in the cylinders. However an increase of only 2-4°C in the flowing air temperature was detected due to the corona discharges (12-14). This increase in the inlet air temperature is thought highly unlikely to significantly influence the combustion process.

It must be stressed that the increase in the population of excited and ionised nitrogen atoms due to corona discharges may ultimately lead to an increase in the concentration of NO_x in the exhaust gases. In addition, due a combination of an increased concentration of O and O_3 in the combustion air and a reduction in fuel consumption, it is strongly believed that the particulate concentration in the exhaust gases should decrease in the presence of coronas. However, due to the preliminary investigative nature of this work, levels of both NO_x and particulates in the exhaust gases were not measured.

5. CONCLUSIONS

1. Preliminary experimental results have shown that improvements in engine performance were possible through the use of corona discharges generated in the combustion air stream. These improvements resulted in lower SFC, marginally higher overall engine efficiency, higher CO_2 emission levels and higher exhaust gas temperatures.

2. These improvements were attributed mainly to the generation of highly oxidising species O and O_3 by the corona discharges.

3. Although the concentration of NO_X and particulates in the exhaust gases were not measured, it is suspected that the concentration of NO_X might increase and that of the particulates decrease.

4. Although the work presented here is a primary qualitative investigation into a new technique, results necessitate further comprehensive testing on all types of vehicles engines.

6. REFERENCES

(1) The way to cut global warming, The Chemical Engineer, 25th June 1992.

(2) Howard, D. Energy and transport: the problem and solutions, Energy World, March 1991, No.186, pp 6-10.

(3) Loeb, L.B., Electrical Coronas, 1965, University of California Press.

(4) Kunhardt, Erich E. and Luessen, Lawrence H., Electrical Breakdown and Discharges in Gases, Part B, 1981, NATO Advanced Science Institutes Series, Plenum Press.

(5) Bradley, D. and Nasser, S.H., Electrical coronas effects in combustion shear layers, Eight International Conference on Gas Discharges and Their Applications, 1984, St.Catherines College, Oxford University, pp 526-529.

(6) Green, J.A. and Sugden, T.A., Proc. 9th International Symposium on Combustion, 1963, Academic Press, New York.

(7) Calcote, H.F., 9th Symposium (International) on Combustion, 1963, Academic Press, New York, pp 622-637.

(8) Bradley, D. and Ibrahim, S.M.A., Electron-gas energy exchange in a.c. fields and their relevance in lasers, Combustion and Flame, 1976, vol.27, pp 353-362.

(9) Ali, M., Bradley, D. and Gupta, M.L., A theoretical study of energy transfer at a negative electrostatic probe, Journal of Physics D; Applied Physics, 1978, The Institute of Physics, vol.11, pp 1639-1652.

(10) Hucknall, D.J., Chemistry of Hydrocarbon Combustion, 1985, Chapman and Hall Ltd, London.

(11) Bradley, D., The effects of electrical fields on combustion processes, Advanced Combustion Methods, edited by Weinberg, F.J., 1986, Academic Press, London.

(12) Nasser, S.H. and Rentschler, M., Improving the diesel engine performance and emission characteristics by corona discharges, SITEV 94, Turin, Italy, 15-18 Nov.

(13) Morris, S., Rentschler, M. and Nasser, S.H., A new technique to improve the performance and emission characteristics of diesel engines, Proc. 9th World Congress on the Theory of Machines and Mechanisms, Milano, Italy, 29th August 1995.

(14) Nasser, S.H., Morris, S. and Tee, K.L., The performance and emission characteristics of a multi-valve direct injection gasoline engine in the presence of corona discharges, 2nd International Seminar on High Performance Spark Ignition Engines for Passenger Cars, November 23-24, 1995, Milan, Italy.

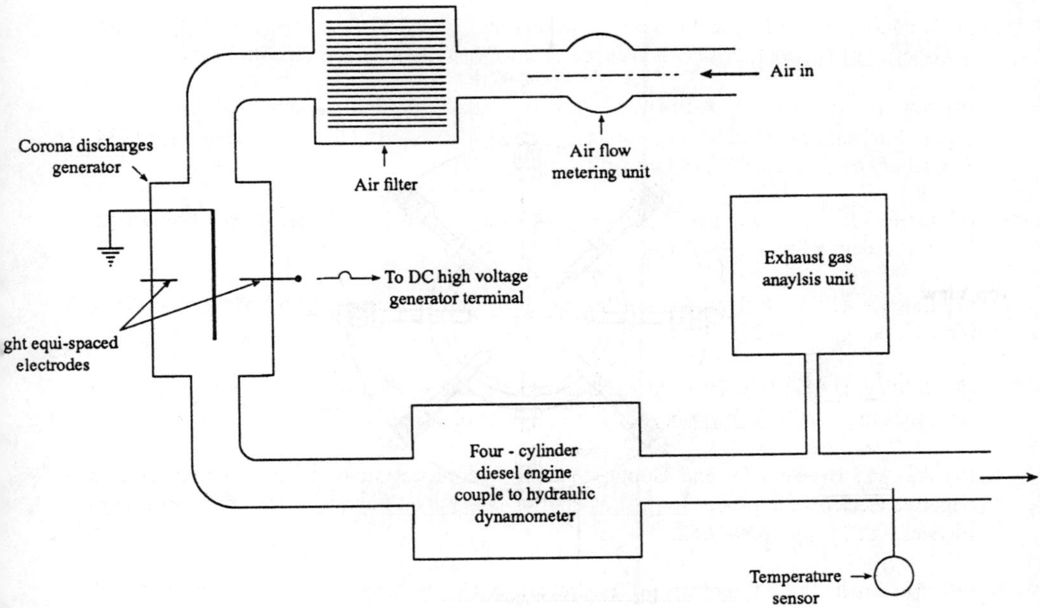

Fig.1 Schematic diagram of the testing facility

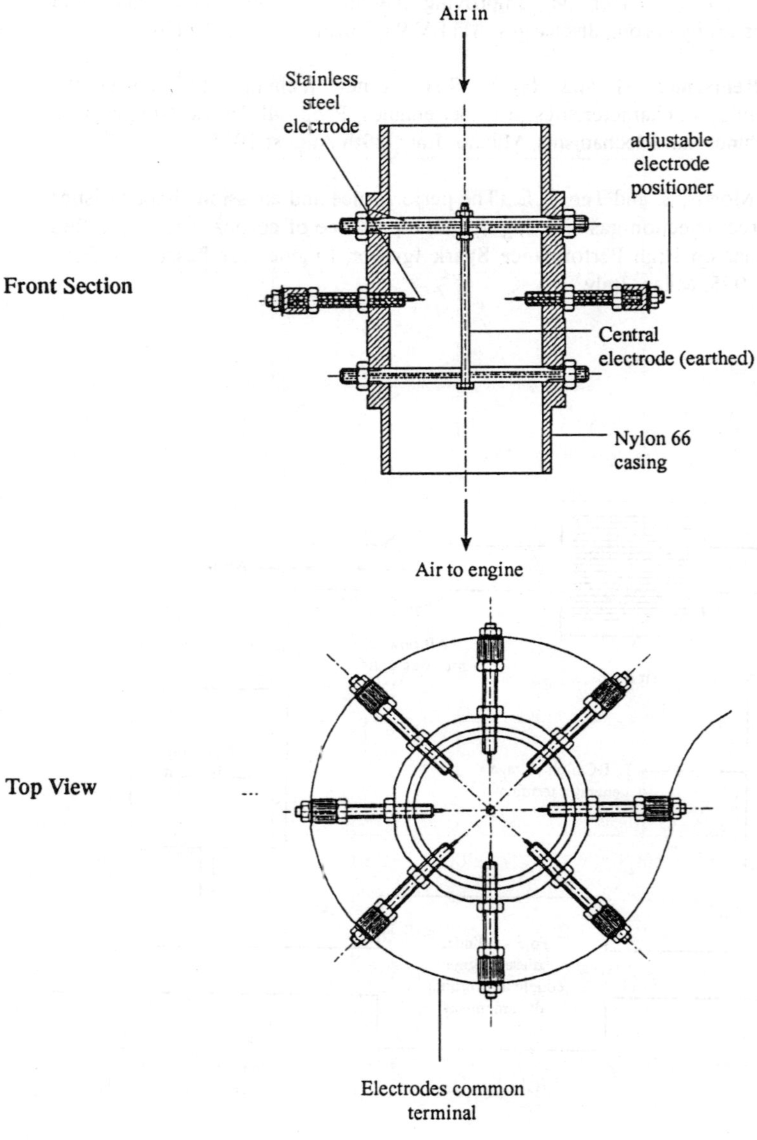

Fig. 2 General view of corona discharge generator

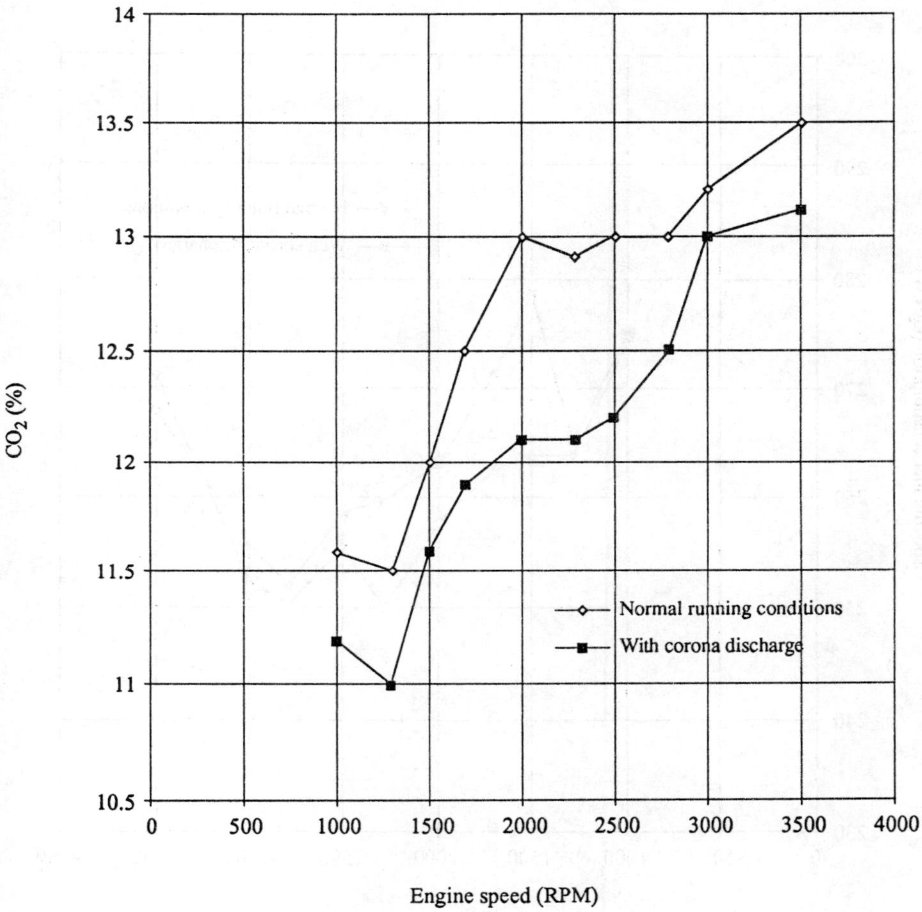

Fig. 3 Carbon dioxide (CO_2) concentration vs. engine speed at 75% throttle

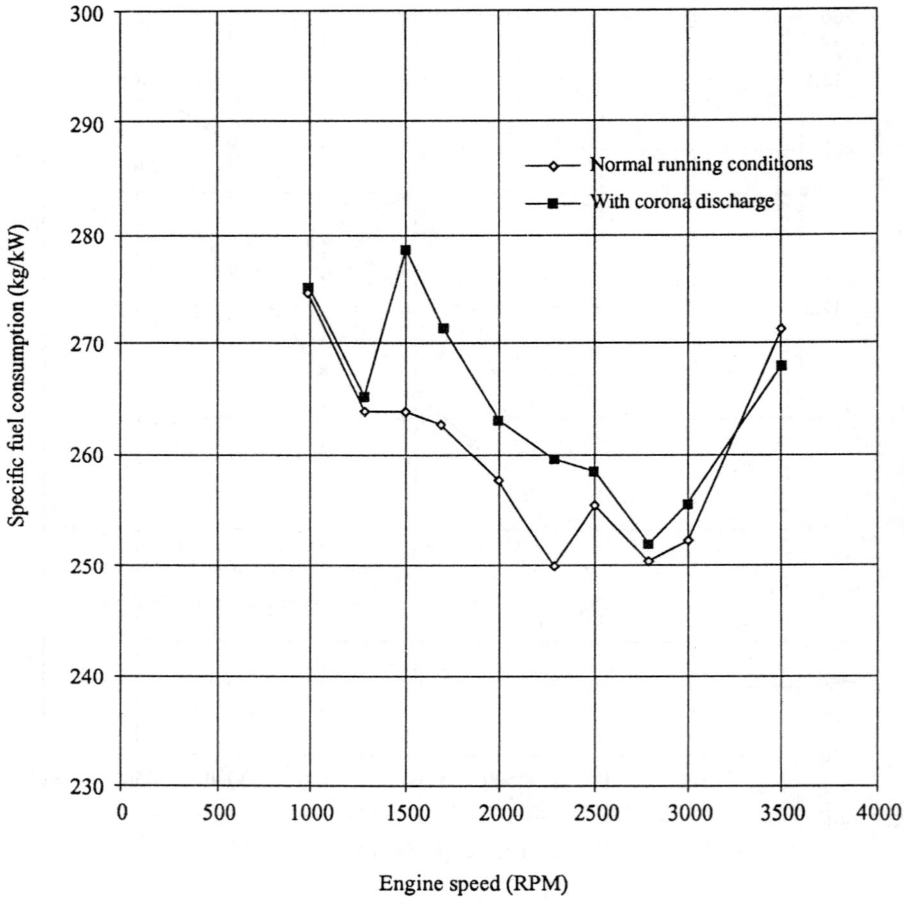

Fig. 4 Specific fuel consumption vs. engine speed at 75% throttle

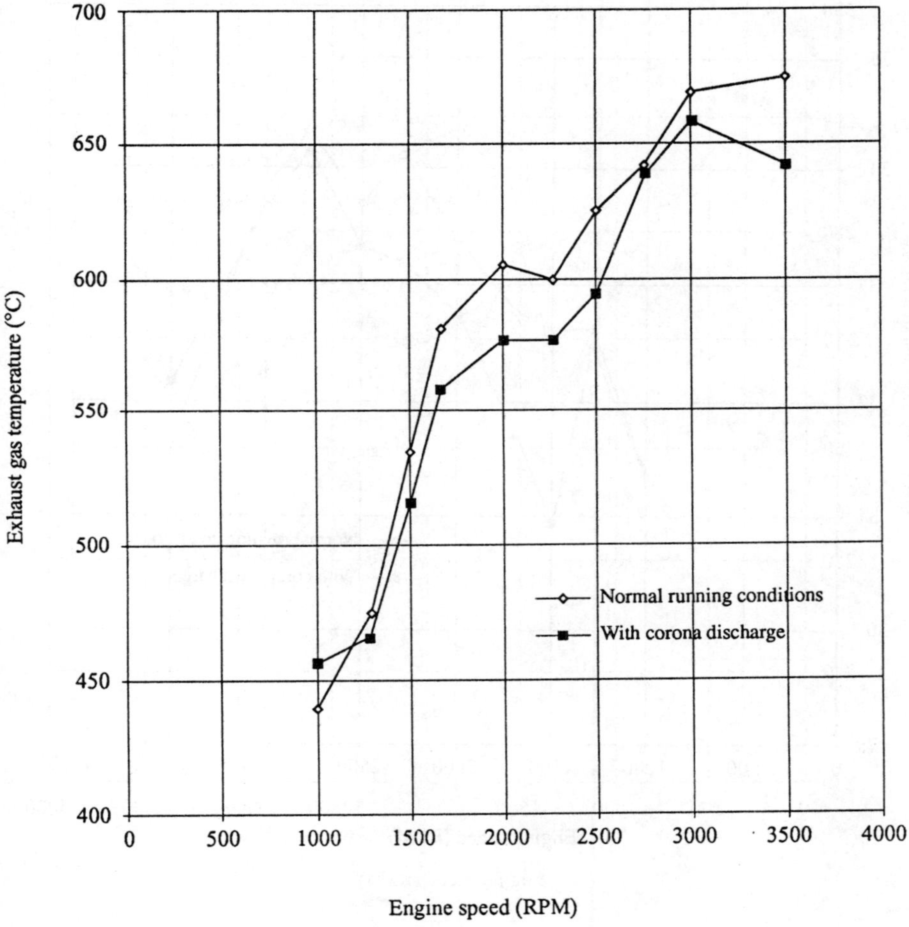

Fig. 5 Exhaust temperature vs. engine speed at 75% throttle

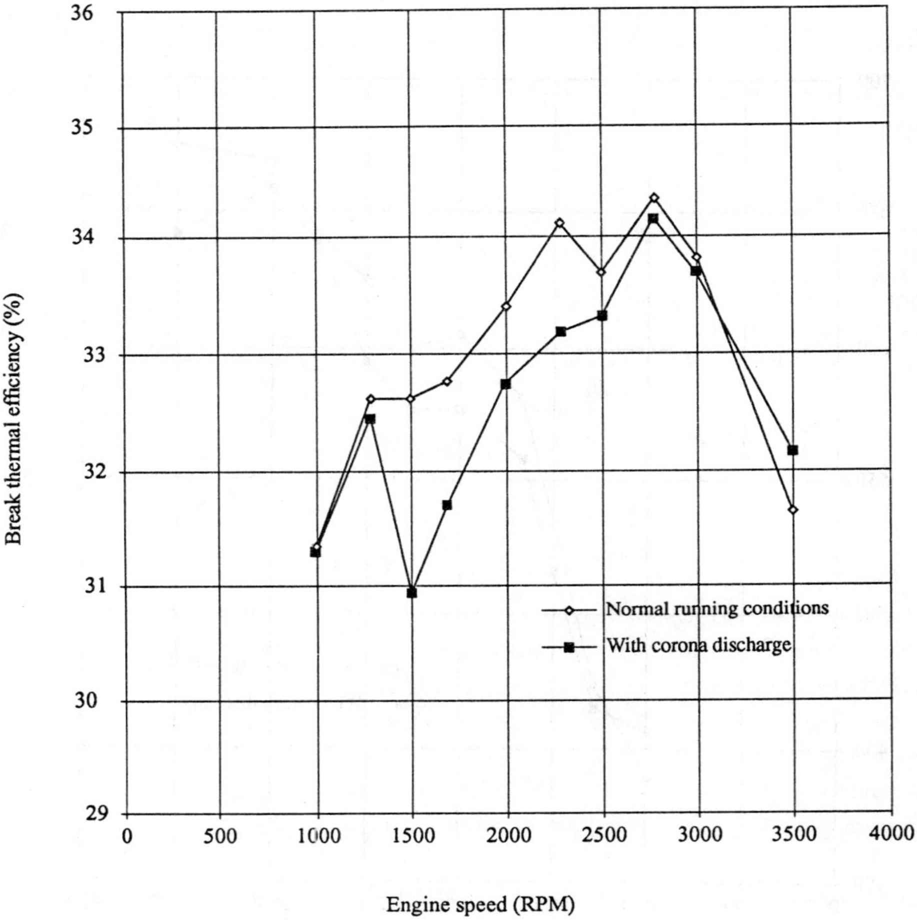

Fig. 6 Break thermal efficiency vs. engine speed at 75% throttle

C498/2/038/95

The determination of the hydrocarbon species present in the particulate SOF, from a turbocharged diesel engine

P W YATES BEng, **D DOUCE** BSc, and **P W FOSS** BSc, MSc, PhD, CEng, MRAeS
Sheffield Hallam University, UK
D H TIDMARSH BSc, PhD, CEng, FIMechE, MBIM, MIMgt
University of Central England, UK
M WILLCOCK BEng, PhD
AE Goetze Limited, Bradford, UK

Abstract

A direct injection diesel engine using reference fuel and oil was subjected to steady state emission testing. The methods and equipment used are discussed. The particulate matter was collected using a dilution tunnel and the Soluble Organic Fraction(SOF) extracted and separated into functional groups by liquid chromatography. Gas chromatography with mass spectrometry was used to identify the hydrocarbon species present in each group. The results show that as the air fuel mixture increases, the distribution and quantity of hydrocarbon species present in the SOF diminish for all functional groups analysed.

Test Equipment

The engine used to generate the particulate matter was a in line 4 cylinder, 4 litre, non inter cooled direct injection engine with a turbo charger. The fuel pump was a Lucas DPA with the fuel injection timing set to 12° BTDC. Throughout all tests the fuel and oil used was of reference quality and from the same batch to retain the same hydrocarbon species distribution, the oil being Silkolene RL 139 and the fuel CCRF03. The chromatogram for the fuel is shown in figure 1. Gas phase exhaust gases were sampled from the exhaust system at a point 1m from the turbo charger exit. These were transported by a heated sample line at 180°C to a multi gas analyser to determine the CO, CO_2 and NOx. The uHC levels were obtained using a flame ionisation detector. To allow repeatability of test data the engines cooling water, exhaust gas, inlet manifold, oil temperatures and the cylinder pressures were measured and recorded. The cylinder pressure data was recorded by piezo electric pressure transducers placed adjacent to the injector nozzle.

Origins of Hydrocarbon Emissions form a Diesel Engine.

The hydrocarbon emissions from diesel engines(C.I.) are typically lower than those from a spark ignition (S.I.) engine, but with exhaust after treatment the exhaust gas from a S.I. engine that enters the atmosphere contains comparable levels. Oxidation catalysts (3 way) that are fitted to S.I. engines will not function on C.I. engines because of the lower exhaust gas temperatures, the possibility of blockage through soot and the quantity of oxygen present in the exhaust gas. Oxidation catalysts have been developed for diesel engines, but any sulphur dioxide(bp -10°C) present in the exhaust gas is oxidised further to sulphur trioxide(bp 44.7°C), thus any sulphur compounds collected by a dilution tunnel would be 25% heavier and in greater quantities because the boiling point of the compound is nearer the maximum allowable sampling temperature (52°C).

Combustion in a C.I. engine is in three general stages:

1. Delay (the time between the start of injection and the start of combustion).
2. Rapid combustion (the burning of the fuel injected during the delay period) .
3. Controlled combustion (fuel burning as it is injected).

The unburnt hydrocarbon(UHC) formation mechanisms for each of these stages are different. The speculated routes have been shown by Yu(1), figure 2 does not show the effects of the system boundaries, i.e. wall quench and adsorption desorption of the fuel into the lubricating oil. Daniel and Whitehouse(2, 3) used in-cylinder sampling on a S.I. engine to determine the distribution of hydrocarbons. Their tests showed that as the sample point neared the cylinder wall the hydrocarbon concentrations increased, indicating wall quench. Multi spray C.I. diesel engines delivering high fuel quantities can allow the fuel spray to penetrate too far into the combustion chamber so that it impinges onto the cylinder wall or piston crown. Conditions in this area are relatively cool with poor or no air swirl and the fuel will not ignite. Lyn and Valdmanis(4) varied injector hole size, nozzle type and its geometry, with none of these affecting the ignition delay. When the hole size was doubled and the injection pressure remained constant, the fuel flow rate increased by a factor of four and the droplet size increased by approximately 30 percent.

A source of UHC also associated with the injector is that of the fuel remaining in the injector nozzle after injection. This is most significant at low fuel injection volumes where there is little overmixing. Greeves(5) found that a 1mm^3 nozzle sac volume gave 350 ppm C_1, and he also stated that 1mm^3 of fuel gave 1660 ppm C_1. He postulated that the volume may not be filled with fuel and the higher boiling point fractions of the fuel remained in the nozzle. The relationship between nozzle sac volume and exhaust HC is linear, with its intercept for zero HC at -1.8mm^3 for a speed range of 1700 - 2800 rpm. The rate of fuel injection in the majority of production diesel engines is assumed constant after the injector needle has attained its fully open condition. Vollenweider(6) completed a series of tests using a two spring injector. This style of injector allowed a small amount of fuel to be injected before the main quantity and this reduced the heat release within the premixed combustion stage. The net result in emissions was a reduction of HC and NOx but an increase in particulates. In a study of "in use" motor vehicles Williams(7) reported that injectors that had a lower opening pressure than the standard specification produced higher levels of emitted particulate matter. This is because the lower kinetic energy of the particle allows it to slow in the air stream, removing its supply of oxygen.

The highest concentrations of soot within the combustion chamber of a diesel engine are in the core region of each of the fuel sprays. This is where the local equivalence ratios are very rich. The formation of the soot particle initiates with a fuel molecule containing 12 to 22 carbon atoms and a H/C ratio of approximately 2 and progresses through the two stages proposed by Haynes(8),

i) Particle formation - the first condensed phase of the material formed by oxidation and pyrolysis. The typical components of these are unsaturated hydrocarbons i.e. acetylene and multiples of $C_{2n} H_2$, and polycyclic aromatic hydrocarbons (PAH). These particles are very small(d < 2nm).

ii) Particle growth - by surface growth (gas phase material condensing and being incorporated) and agglomeration of smaller particles.

Studies(9) with combustion bombs have shown three different paths to soot formation depending on combustion temperature. At the lowest temperatures (>1700 °C) only aromatics or highly unsaturated aliphatic compounds of high molecular weight are very successful in forming solid carbon through pyrolysis. At intermediate temperatures typical of diffusion flames (<1800 °C), all normally used hydrocarbon fuels produce soot if burned at a significantly rich mixture. At very high temperatures (above the range found in diesel engines) a nucleation process utilising carbon vapour is the primary formation mechanism.

Particulate Sampling

The exhaust gas immediately leaving the tailpipe of a diesel engine is composed primarily of elemental carbon and organic compounds, and at normal in cylinder temperatures these exist separately. The purpose of a dilution tunnel is to simulate the mixing of the exhaust gas with the external air cooling the exhaust and allowing the organic compounds to condense onto the elemental carbon. The maximum permissible sampling temperature for a dilution tunnel is 52°C. This is the mid point between the boiling points of pentane(Bp 36.1°C) and hexane(Bp 68.7°C). To reduce the sample gas temperature further to the boiling point of pentane, the dilution ratio would have to be increased to a region above that of the optimum ratio for the collection of the extractable fraction. The dilution ratio for the collection of the maximum amount of the extractable organic fraction is approximately 5:1. The reasons for this are that, as the dilution ratio increases from unity, the effect of the decreasing temperature on the number of active sites dominates the condensation process. When the dilution ratio becomes higher the decreasing partial pressure causes the extractable mass to fall again. Condensation will occur whenever the vapour pressure of the gaseous phase hydrocarbons exceeds its saturated vapour pressure. Increasing the dilution ratio decreases the hydrocarbon concentrations and hence the vapour pressure. The total exhaust flow from the engine at high speeds and loads was in excess of the quantity required by the Weslake dilution tunnel for collection of the maximum quantity of the SOF, therefore a diverter system was used to remove the unwanted exhaust gas.

Solvent extraction.

Because of the small amount of material collected for the majority of the test points, the optimum method to determine the amount of material removed by the extraction process is to weigh the filter before and after the process. Because the filter can contribute to the weight difference, blank filters will also have to be subjected to the extraction process.

The overall calculation for the soluble organic fraction by weight loss will be :-

SOF(filter weight loss) = (Wbe - Wae - Avb)
Wbe = Weight before extraction
Wae = Weight after extraction
Avb = Average weight loss of blank filters

The choice of solvent used for extraction influences the group or groups of hydrocarbons removed from the sample, dichloromethane was used for the results presented in this paper, a more aggressive solvent favoured by some researchers is a 4:1 mixture of distilled benzene and methanol[10]. To remove the solvent and to stop any mutation or oxygenation of the remaining matter prior to weighing, the papers were dried in a nitrogen atmosphere. Before any weighing of the filter papers can be undertaken they must be subjected to the same ambient conditions. It is recommended that the papers are stored in temperature and humidity controlled conditions for four hours to equilibrate.

Isocratic Eluant	Group
Hexane	Alkanes
Dichloromethane	Polycyclic Aromatic Hydrocarbons (PAH)
Acetronitrile	Polar PAH & Nitro PAH
Acetronitrile & Methanol (3:1)	Very polar PAH

Table 1

SOF analysis

Clean up procedure[11].

A high performance liquid chromatography(HPLC) system using a alumina column-activated, basic, Brockmann 1 (150 mesh, 250 x 5 mm). This was used to separate the hydrocarbons of interest, i.e. C_8 to C_{30} from any other emitted by the engine, i.e. aromatic and substituted aromatic compounds known to be released by incomplete combustion. Fractionation was achieved on a Varion isocratic HPLC system using a UV detector set at 254 nm. To separate the sample into these groups, low pressure liquid chromatography utilising different elutants was used. Table 1 shows the eluant with the corresponding group

Capillary GC/MSD Analysis

The residual weights of the test samples varied with the speed load conditions i.e. depending upon the contribution to the total collected weight of the soluble fraction, these samples were then diluted with Dichloromethane to give similar concentrations. A 1µl. sample of this was then used for analysis, the instrument used was a Hewlett Packard 5890 (Series II) coupled with a HP 5971A Mass Selective Detector. Electron impact mass spectra were obtained at 70eV and processed using a HP ChemStation data system. The GC conditions used are shown in table 2.

Gas Capillary Chromatographic Conditions	
Carrier gas	Helium
Analytical column	18m x 0.25mm (id), 5% biphenyl fused silica
Film thickness	0.25µm
Oven temperature program	50°C (2 mins) ramped at 10°C/min to 280°C (5 mins)
Volume injected	1µl. splitless

Table 2

The identification of the abundance peaks on the chromatograms produced was initially by the comparison to the retention times of a standard. The hydrocarbon species present in the exhaust and not the calibration sample were identified using a library search on the data system.

Results

The chromatograms shown were obtained by separating the sample into four groups by liquid chromatography. For ease of analysis the group that has undergone the closest scrutiny is the alkanes. This is the first group extracted from the organic fraction and thus has the least chance to degrade or mutate. The compounds that are recognised as the most harmful are in the second group, these being benzo[*a*]pyrene and possibly 7,12-dimethylbenz[*a*]anthracene. These compounds were recognised as being carcinogens in 1933 and are both polycyclic aromatic hydrocarbons. These compounds are not cancer forming until they form a diol-epoxide.

The chromatogram plots show that as the engine load increases the quantity of carbon atoms present in the most abundant peak also increases. When the load is increased beyond half rated power the two lightest species diminish below a level that can be accurately quantified. The alkane species that are presented are the lightest of those shown on all plots. There is a greater distribution of species present at lower speeds, i.e. there are both heavier and lighter species detected at idle than those detected at 75% of rated power. As the combustion temperatures and pressures increase so does the degree of fuel pyroliysis, therefore, the heavier fuel molecules break down into lighter components; these components have lower vapour pressures and auto ignition temperatures and are oxidised easier. The size of the most abundant species also increases with the temperature and pressure. A simplified illustration of the chromatograms for the abundance of n-alkanes is shown in figure 3. The concentrations of gas phase uHC's present just prior to the exhaust turbine inlet and a 1 meter from the outlet are shown in figure 4 for comparison. The survivability's of the PAH do not follow this trend (fig 5), the abundance of PAH increases with load and the most probable cause is the increase in oil consumption giving a higher initial quantity of PAH present in the exhaust. The speciation of the lubricating oil by gas chromatography does not produce clear and identifiable peaks, the oil with usage produces clearer peaks this is probably due to fuel contamination[10].

Conclusions

The identification and quantification of the species of hydrocarbon present in the particulate SOF will assist in determining the methods that exhaust pipe uHC has avoided oxidation or pyrolysis. The higher temperatures with increasing load reduce the quantity and carbon number distribution of the species present in the SOF but there is a greater quantity and larger spectrum of the PAH present. The speciation of the uHC present in the exhaust gas stream will be a fundamental method of achieving the ever tightening emission laws.

Acknowledgements

The authors wish to thank AEGoetze for the funding, technical assistance and provision of resources received during this research period, and the School of Science at Sheffield Hallam University for the use of their equipment to analyse the particulate matter.

References

1) Yu R.C., Wong V.W. and Shahed S.M.: " Sources of hydrocarbon emissions from direct injection engines. S.A.E Paper 800048 1980.

2) Daniel, W.A., Wentworth, J.T.: " Exhaust gas hydrocarbons Genesis and Exodus". S.A.E. 486B ,1962.

3) Whitehouse, N.D., Clough, E., and Uhunmwangho, S.O.: " The development of some gaseous products during diesel engine combustion". SAE Paper 800028, 1980.

4) Lyn, W.T., and Valdmanis, E.: "Effects of physical factors on ignition delay". S.A.E. 680102, 1968.

5) Greeves, G., Khan, I.M. , Wang, C.H.T. and Fenne, I.: " Origins of hydrocarbon emissions from diesel engines". SAE Paper 770259.

6) Vollenweider, J.:" Emission-optimised fuel injection for large diesel engines-expectations, limitations and compromises ". Proc IMechE C448/003, 1990.

7) Williams, D.J., Milne, J.W, Quigley, S.M.and Roberts, D.B.: "Particulate emissions from "in use" motor vehicles-II Diesel Vehicles". Atmospheric Environment Vol 23, No12, pp 2647-2661 1989.

8) Haynes, B.S. and Wagner, H.G.: "Soot formation" Prog Energy Combustion Science, Vol 7, pp 229 - 273, 1981.

9) Amann, C.A., and Siegla, D.C.: "Diesel particulates-What they are and why." Aerosol Science Technology, vol , pp 73-101, 1982.

10) Abbass, M.K., Andrews, G.E., Ishaq, R.B. and Williams, P.T.: "A comparison of the particulate composition between turbocharged and naturally aspirated D.I. diesel engines". SAE Paper 910733, 1991.

11) Ciccioli, P et al. Journal of Chromatography, 351 (1986) pp 451 - 464

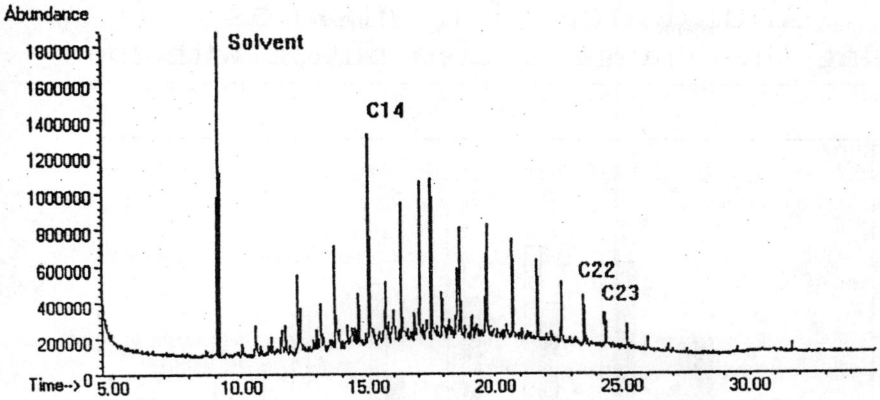

Chromatogram showing all hydrocarbon species

Figure 1

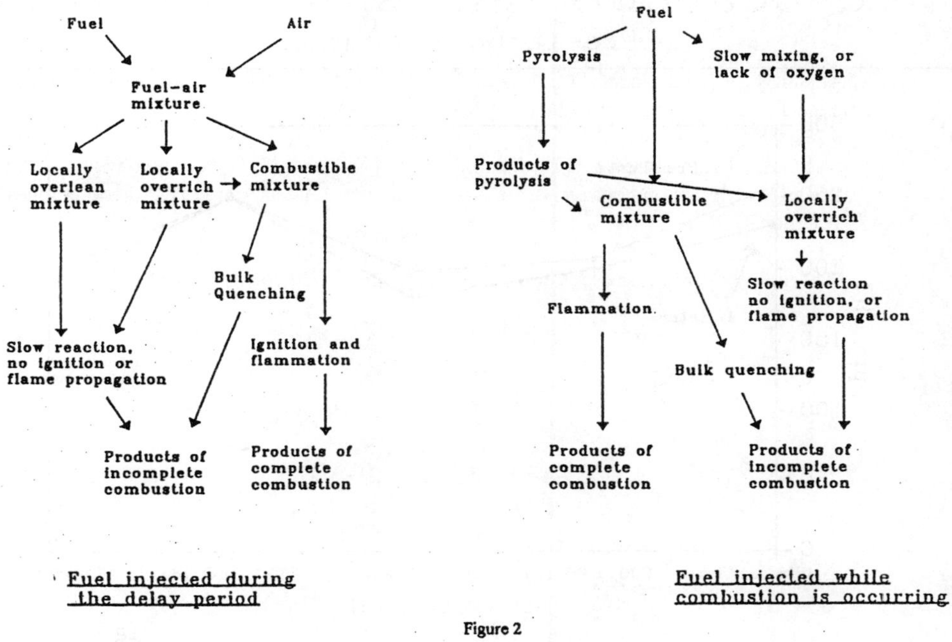

Figure 2

Abundance of n-Alkanes

Showing the change in distribution with load

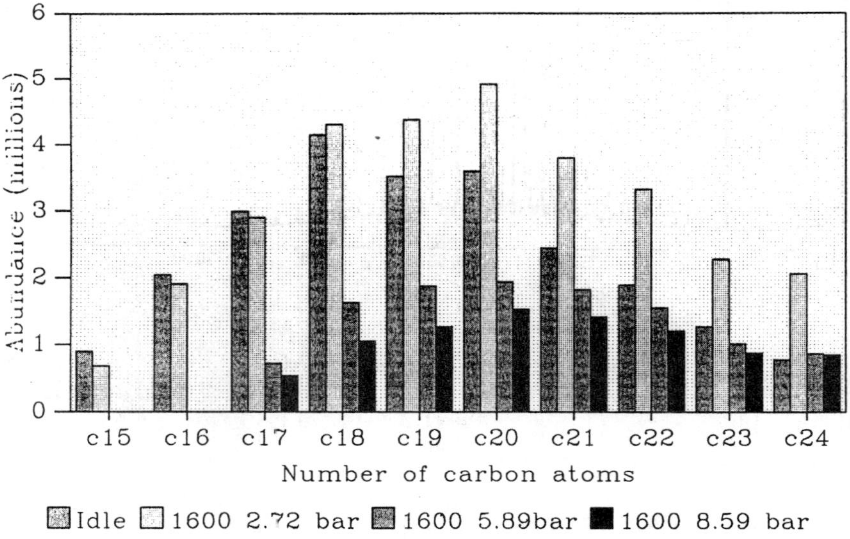

Figure 3

Hydrocarbon Emission with Load

Pre-turbo + 1 meter

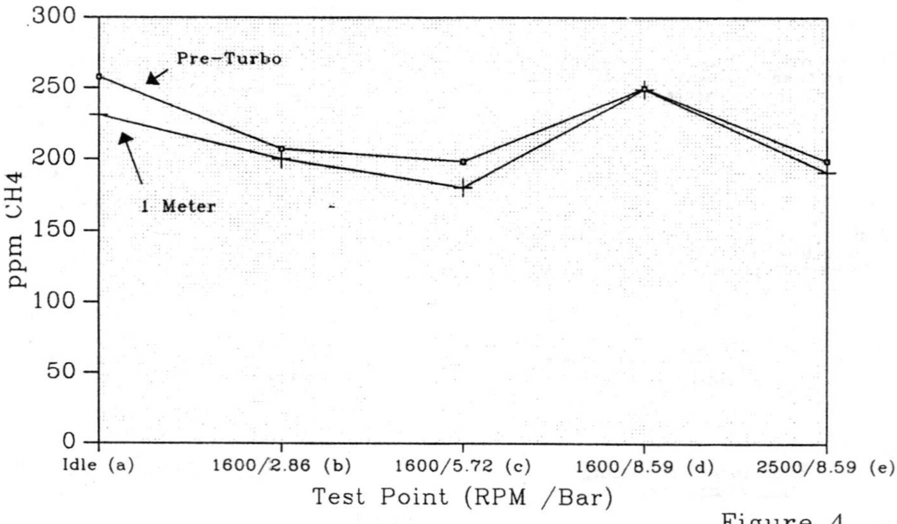

Figure 4

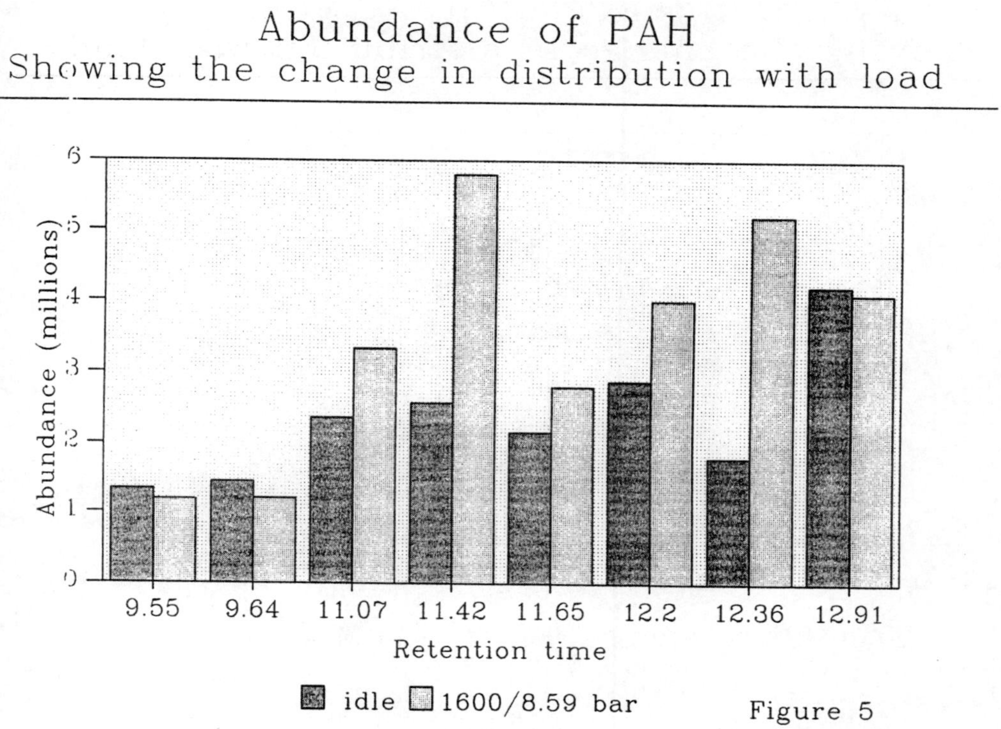

Abundance of PAH
Showing the change in distribution with load
Abundance (millions)
0
1
2
3
4
5
6
9.55
9.64
11.07
11.42
11.65
12.2
12.36
12.91
Retention time
idle
1600/8.59 bar

Figure 5

C498/2/039/95

Initial development of an experimental uniflow two-stroke DI diesel engine

A COX BEng and **F J WALLACE** MSc, PhD, DSc, FEng, FIMechE
University of Bath, UK
G BIRD and **R HORROCKS**
Ford Motor Company Limited, UK

SYNOPSIS

A two cylinder 2-stroke DI diesel engine utilising the uniflow method of scavenging, has been designed and built using component parts from a donor four cylinder 4-stroke engine. The performance of the 2-stroke DI engine has been predicted using a simulation package and comparisons are drawn between this engine, its 2-stroke IDI sister and its 4-stroke development counterpart.

1. THE CASE FOR A SMALL AUTOMOTIVE 2-STROKE DIESEL ENGINE

The 2-stroke engine in its gasoline form dates back to the 1880's. Indeed, the engine in its simple ported crankcase compression form was first patented in Bath by Joseph Day in 1891. From its inception, the ported engine offered several advantages including:- fewer moving parts, a lightweight construction, and a higher specific power output than equivalent capacity 4-stroke engines. Its automotive heyday was probably the decade before the Second World War, when a two cylinder air cooled engine achieved wide spread acceptance in Germany in the DKW car. Post-war derivatives of this engine were fitted to Trabant, Wartburg and SAAB cars. However, modern emissions legislation has effectively eliminated the traditional crankcase scavenged engine from use in the passenger car, mainly because of its inherently higher hydrocarbon emissions associated with lube oil carry-over into the combustion chamber past the piston rings, and excessive smoke output due to poor scavenging. The last mass produced 2-stroke engine, fitted in the Eastern Bloc Trabant, fell victim to the collapse of the Berlin Wall, the re-unification of Germany and the adoption of Western emissions legislation.

In the past, medium speed / medium capacity 2-stroke diesel engines have been successfully used in many commercial vehicles, prime examples being:- the uniflow Foden FD6 engine, the Rootes TS3 opposed piston engine fitted to the Commer trucks of the 1950's, and the V6 Detroit Diesel uniflow engine used to power the trans-state Greyhound buses in North America. Additionally, almost all low speed / high capacity marine engines operate on the 2-stroke cycle, the Sulzer RTA being a good example. However, to date there have not been any successful applications of a high speed / small capacity 2-stroke diesel engine to passenger cars.

Nevertheless, there has been a distinct revival of interest in the 2-stroke engine in both its gasoline and diesel forms [1], perhaps the most well known being the Australian Orbital engine. This is a three cylinder, cross scavenged, crankcase compression, gasoline engine with fuel injection direct into the cylinder. Low emissions and good durability are claimed, and its high specific power output means that it is currently undergoing evaluation with a number of automotive manufacturers as a potential replacement for existing four cylinder 4-stroke engines fitted in small passenger cars.

Parallel investigations have been conducted by several automotive manufacturers to investigate the 2-stroke diesel engine to see if the advantages of higher thermal efficiency (and thus reduced fuel consumption) of diesel over gasoline, observed in 4-stroke operation, can be realised in 2-stroke form.

The very nature of operation of the 2-stroke diesel engine means that it must practically utilise some form of external compression charging; additionally, the fuel needs to be injected directly into the cylinder, thus overcoming the two main handicaps of the simple ported gasoline engine, albeit at the loss of simplicity. The merits of the DI vs. IDI combustion systems seem more clear cut in 2-stroke form. Popular opinion is that the IDI pre-chamber could be difficult to scavenge effectively, particularly at slow speeds, possibly making a 2-stroke IDI engine reluctant to start with a conventional 'Roots' type blower. Conversely, the 'open' DI combustion chamber should be easier to scavenge - providing the incoming air has an adequate swirling motion. The passage of the IDI pre-chamber also produces additional pumping and heat transfer losses, and thus the DI combustion chamber should also offer a higher thermal efficiency and improved BSFC.

The two most desirable features of a 2-stroke diesel engine, applicable to the passenger car designer are :-

Higher Specific Power Output
Due to the doubled firing frequency of the 2-stroke cycle, and allowing for inefficiencies, it is theoretically possible to replace a four cylinder 4-stroke engine with a three cylinder 2-stroke engine, offering a weight saving of up to 20 % and reduction in length of 25 %. The resultant smaller engine package could enable installation in a more aerodynamic and lighter bodyshell, with possible further gains in vehicle fuel economy.

Inherently Lower NOx Emissions
Experience has shown that 2-stroke diesel engines tend to produce lower NOx emissions than their 4-stroke counterparts due to their inherent EGR characteristics. This is a result of the retention of the residual exhaust gases in the cylinder, due to the inefficiency of the scavenging process. Thus, with careful design, it may be possible to produce a scavenging system to achieve the legislated level for NOx output without the need for additional EGR equipment. However, it is noted that the remaining exhaust emissions will have to be dealt with by similar means to those employed in current 4-stroke diesel engines, namely, electronic control of fuel injection, and use of carefully selected injectors to introduce finely atomised fuel at the correct rate and time, to improve mixing and thus combustion; and installation of oxidation catalysts to reduce excessive HC output.

2. PROJECT RATIONALE

It is against this backdrop that the University of Bath undertook a project on behalf of the Ford Motor Company, to design, manufacture, and test a 2-stroke diesel engine in both its IDI and DI forms. The subject of this paper is the design and development of the DI engine. The design, development and experimental testing of the IDI version of the engine is covered in some detail in an earlier paper [2], though reference to, and comparisons with this engine will be made where appropriate in this paper.

Time and financial constraints dictated that it was not feasible to build a prototype engine from scratch. Instead, it was decided to convert an existing small capacity automotive 4-stroke diesel engine, to operate on the 2-stroke cycle. The donor engine selected was in current production in its IDI form, and existed as a *development* prototype in its DI form. This meant that it was possible to utilise many of the existing sub-assemblies and component parts from the donor engine. However, it was realised that use of these components would place practical restrictions on both the design and the limits of operation of the engine. This compromise was considered acceptable, as any experimental results obtained from the *demonstrator* engine would be directly comparable with those from its 4-stroke counterpart and thus give an indication of the potential performance of the 2-stroke engine.

A preliminary design study, briefly detailed below, adopted the uniflow method of scavenging for the 2-stroke engine, whereby charge air enters through multiple ports in the base of the cylinder and the exhaust gas exits through poppet valves in the head. The fundamental fact that the 2-stroke cycle fires at twice the frequency of the 4-stroke cycle, meant two further limitations were applied to the engine design. Conversion of the existing four cylinder distributor type fuel injection pump to 2-stroke operation, meant that the engine could only have two firing cylinders - for general convenience and accessibility, cylinders 1 & 3 were chosen. It also transpired that the existing valve gear would have to operate at twice its normal design speed, therefore doubling the component stresses at any engine speed. Thus the maximum speed of the 2-stroke engine was limited to 2500 RPM, approximately half its rated 4-stroke speed. Finally, as initially there was no knowledge of the specific air requirements of this engine, it was decided that an external air source would be used to supply charge air, to ensure that sufficient scavenging was available at all speeds.

3. ENGINE DESIGN

From the outset of the project, it was decided that the method of scavenging, the port and valve timing, and the engine dimensions (bore & stroke), would be identical for both the IDI and DI engines. Additionally, the nominal compression ratio for each engine would be the same as the value optimised for use in the 4-stroke combustion system. It was intended that these constraints would enable a direct comparison to be made of the performance offered by each type of combustion system in both the 2-stroke and 4-stroke modes of operation.

The choice of scavenging method, the determination of suitable engine timing, and the ideal values for inlet and outlet flow areas were all determined using the commercially available *Simulation Program for Internal Combustion Engines (SPICE)* software modelling package. The existing 4-stroke engine was modelled first, and its performance was used as a benchmark against which the various 2-stroke engine parameters were optimised. The IDI engine was built first, since it was known that its tried and tested Ricardo 'Comet' pre-chamber design offered guaranteed mixing, despite being potentially difficult to scavenge. Successful combustion in a DI engine was more difficult to achieve, as thorough mixing required a strong swirling motion to be imparted to the incoming air. More time was required to optimise the inlet port design.

3.1 Choice of Scavenging Method

Efficient scavenging is the key to good performance in a 2-stroke engine. Because of the physical configuration of the existing production 4-stroke engine, scavenging by use of ports alone was discounted. The only practical solutions available for this design of engine were either the uniflow (inlet ports and exhaust poppet valves) or inverted loop (inlet and exhaust poppet valves) scavenging methods. It is well known that the uniflow method of scavenging, whereby the fresh charge is introduced through multiple ports in the base of the cylinder wall and combustion products are exhausted through poppet valves in the cylinder head, offers the highest charge purity for a given delivery ratio. An initial simulation study revealed that the existing 4-stroke cylinder head with two valves per cylinder, could not supply an adequate scavenging airflow through the single inlet poppet valve to sustain combustion above 2000 RPM. This naturally led to the adoption of the uniflow method of scavenging for this engine.

3.2 Determination of Port and Valve Timing

Specification of uniflow scavenging dictated that the inlet port timing must be symmetrical about BDC. However, use of a conventional camshaft and poppet valve arrangement, means it is possible, and indeed desirable, to have asymmetric exhaust valve timing. The angular timing of the exhaust valve opening (EVO) is vital; too early in the expansion stroke and vital work is lost, too late and there is the risk that the residual pressure in the cylinder will still be too high, and the exhaust gas will backflow into the inlet manifold when the inlet ports first open (IPO), thus reducing the effectiveness of scavenging process. Practical experience with uniflow 2-stroke engines has shown that a blowdown period (i.e. the interval between EVO and IPO) of approximately 30 degrees C.A. is necessary to ensure this will not happen.

A detailed simulation study determined that the timing schedule given below represented a good compromise between allowing an adequate time period for scavenging to occur without significantly penalising the power output of the engine at its rated speed. This schedule was used for both the IDI and DI 2-stroke engines.

EVO	95 deg ATDC	EVC	235 deg ATDC
IPO	125 deg ATDC	IPC	235 deg ATDC

3.3 Overriding Design Constraint

It was found to be impossible to resolve the twin conflicts of retaining the existing engine bore & stroke dimensions whilst overcoming the practical problem of how to incorporate 'directed' ports necessary to impart swirl to the incoming air, as required for a DI combustion system. A practical method could not be found to retro-fit directed ports into the base of a cylinder of a cast engine block with an integral water jacket. It was therefore decided to 'move' the working section of the cylinder to above the existing block deck line. This would be achieved by insertion of a 'sandwich' block between the cylinder head and block. The port belt could then be machined in the lower part of the block. The thickness of the block would have to be at least equal to the existing stroke of the engine, with a small allowance made for material below the ports. This modification would also necessitate increasing the combined piston / con-rod length by the same dimension. However, because the crankshaft throw would remain unaltered, the existing stroke would be maintained.

3.4 Inlet Port Design

Two conflicting requirements have to be taken into consideration when designing the ports for a DI combustion system, necessitating a trade off in the final solution. The DI 'combustion in bowl' system requires a swirling motion to be imparted to the incoming air to ensure thorough mixing with the fuel, and thus good combustion. Swirl is only achieved by inclining the ports at some tangent to the cylinder wall - the more oblique the angle of the port, the greater the swirl motion imparted to the incoming air. Unfortunately, this requirement conflicts directly with the desirability of maximising port area to reduce the pressure losses at inlet to a minimum, since the power demand from an engine driven 'Roots' blower at any given speed will be approximately proportional to the pressure drop across the engine (i.e. between the inlet ports and the exhaust valves). Additionally, for this particular engine, there was also a third practical design requirement to take into consideration due to the method by which the swirl ports were to be incorporated into the engine (detailed in the following section). Problems with accessing a clear path limited the number, width and thus total effective area of the ports (since the timing requirement dictated the port height). This consideration effectively out-weighed the other two and became the primary constraint upon which the final port design was based.

The port design was optimised using the commercially available *Otto Diesel Engine Simulation (ODES)* program. This software package has previously been developed by one of the co-authors of this paper primarily to accurately model the scavenge process of a 2-stroke engine. The program represents the actual in-cylinder scavenging process by use of a quasi single-zone two-phase model, whereby upon opening of the inlet port there is an initial displacement phase, followed by a combination of a mixing and short circuiting phase. A detailed explanation of the scavenging model is given in a previous paper [3]. The claimed advantage of this model is that it enables the notional displacement, mixing and short circuiting fractions to be varied to suit the particular type of scavenging system employed. The program also allows the exact specification of the port & valve geometry and specification of a notional blower, to individual airflow requirements. Detailed results are provided for power output, fuel consumption and heat loss. Additionally, a notional value for swirl of the trapped cylinder gases is calculated. This is based on the concepts of angular momentum transfer into and out of the cylinder, by the entering air and leaving gas mixture, assuming solid body rotation, and allowing for the effects of wall friction. The value of swirl at inlet port closure (IPC) gives a useful indication of the likely success of mixing of the cylinder contents and thus combustion. This value could not be considered absolute as no experimental verification had been performed. However it was a useful index by which the various port configurations could be judged.

Working within the practical constraints, several designs offering various combinations of number, width and orientation of ports were evaluated using the package. The chosen solution utilised four ports per cylinder, with two ports located either side of the block centreline. The design had two opposing *swirl* ports, inclined at 40 degrees to the local radial direction, and two opposing *directed* ports with a 15 degree orientation, as shown in Figure 1. This design was considered a good compromise between a complete 'swirl' system, which would have a high pressure drop at inlet, and a 'quiescent' system which would offer no swirling motion to the incoming air, and thus risk poor mixing and combustion. Simulations conducted at different speeds and loads indicated that this design offered a swirl ratio of approximately 2.0 (relative to engine rotational speed) at IPC.

4. MANUFACTURE OF THE ENGINE

Once the outline design of the engine was finalised, innovative thought was necessary to overcome the numerous manufacturing problems associated with converting the donor four cylinder 4-stroke engine to a two cylinder 2-stroke engine. Figure 2 highlights the main design features of the *demonstrator* engine.

4.1 Cylinder Block

It has previously been mentioned that because of the requirement for swirl ports, it was decided to move the working part of the stroke to above the level of the existing block deck, effectively increasing the length of the existing cylinder. Time and financial constraints dictated that it was impractical to cast a one piece sandwich block, therefore it was decided to subdivide the block into three unequal sections or 'sandwich plates' into which the swirl ports, and water jacket profile could be machined separately - as shown in Figure 3. To minimise the risk of differential rates of thermal expansion and thus distortion of the bores on the firing engine, the plates were machined from a grey cast iron, the same material used for the existing block and head. The total thickness of the three plates was equivalent to the existing stroke length plus a small allowance for material below the air-ports. The external dimensions of these plates were identical to those of the block.

The two firing cylinders were first bored in the doweled plate assembly. This was then separated and the swirl ports were machined in the lower sandwich plate, and a directed water jacket was machined in the middle and upper sandwich plates. Where appropriate, holes were drilled through the plates to accommodate head bolts, coolant entry & exit passages, and the oil supply & drain back to sump. The three plates were accurately aligned, located and clamped to the cylinder block by use of specially fabricated clamps picking up on the two redundant cylinders. The mating faces between the sandwich plates' and the top of the block were sealed using a 'zero thickness' jointing compound. The two firing cylinders (No.'s 1 and 3) were then sleeved with extra long liners, which were pushed through both the sandwich plate assembly and the cylinder block. Finally, the liners were spark-eroded from inside-out to meet the previously machined air-ports in the lower sandwich plate.

4.2 Coolant System

Since the 2-stroke engine fires on every revolution, at twice the frequency of its 4-stroke equivalent, the heat required to be dissipated by its cooling system is effectively doubled. However this *demonstrator* engine has only two firing cylinders compared with four cylinders for its 4-stroke counterpart, so the net heat output per engine revolution will be the same. With the sandwich plate assembly now becoming the working part of the cylinder, and thus subject to the bulk of the heat dissipation, provisions had to be made to supply adequate cooling to this section. Careful design enabled the existing coolant pump and circuit to be utilised for this task. An identical mirrored coolant jacket profile was cut in both the middle and upper sandwich plates. The profile was designed so that each firing cylinder was supplied with a dedicated and directed supply of coolant from a single entrance point. Two holes were drilled in the appropriate sandwich plates to allow water to pass from the existing block, through the lower (air) sandwich plate and into the machined jacket volume around the firing cylinders in the middle and upper plates. The coolant then exited the upper sandwich plate on the opposite side of the block through two holes into the appropriate positions in the cylinder head to provide a directed jet to cool the exhaust valve bridge of the two firing cylinders. The coolant jacket was of sufficient length to provide a heat flow path for the majority of the power stroke of the engine, from opposite the upper piston ring at TDC, down to approximately 95 deg ATDC, the point where the exhaust valves first open.

4.3 Con-Rods

Because of the increased cylinder length, it was necessary to manufacture longer con-rods for the 2-stroke DI engine. The increase in length was limited by the minimal clearance available between the lower part of the shank and the existing block casting. The external profile of the longer rod was similar to the standard item. It was forged and machined from EN24T steel, and shot peened for additional strength. Despite being 31% longer than the standard item, the 2-stroke DI con-rod were actually 19% lighter than its 4-stroke equivalent, this being due to removal of any unnecessary material. This weight-saving helped keep down the overall increase in the rotating and reciprocating masses for this engine to a minimum. Additionally, because there is a firing stroke every revolution in the 2-stroke cycle, the con-rod small end bearing is under a constant loading. To ensure adequate lubrication of this bearing, it was decided to provide it with a pressurised oil supply from the big end bearing, via an internal drilling through the shank of the con-rod.

4.4 Pistons and Rings

Special longer pistons had to be cast for this engine. These were similar in design to those used in the 4-stroke DI engine, except that the distance between the gudgeon pin and the piston crown had to be increased by 109 % over the standard length. This resulted in the 2-stroke DI piston being 64 % heavier than its 4-stroke counterpart, thus increasing the reciprocating mass of the assembly. Also, the skirt below the gudgeon pin was slightly lengthened to accommodate an extra compression ring. This was to ensure that charge air entering the engine through the inlet ports could not pressurise the crankcase. The piston profile was carefully designed to allow a controlled flow of lubricant past the lower ring to lubricate the bore wall and upper ring pack. The existing upper piston rings were pegged at 120 degree intervals to prevent rotation, and thus the risk of fouling on the lips of the air-ports. It was not considered necessary to peg the lower piston ring since this did not pass over the port belt. The volume of the 're-entrant' style piston bowl was adjusted to achieve the same *geometric* compression ratio as the 4-stroke DI engine.

4.5 Crankshaft

The existing 4-stroke crankshaft was utilised for this engine. Two collars were made to clamp around the redundant big end journals of cylinders 2 and 4. These served the dual purpose of preventing loss of oil pressure through the lubrication drillings, and additionally, to correct the out-of-balance forces evident because of the removal of the piston / con-rod assemblies from the two redundant cylinders. The mass of each collar was made equivalent to the rotating mass of one of the DI con-rod assemblies.

4.6 Cylinder Head, Valve Train and Camshaft

The DI cylinder head with its overhead camshaft arrangement was unchanged in design. It was decided to leave the original inlet valve swirl ports unchanged in shape, as the reverse direction flow coefficient was satisfactory, and thus would provide little obstruction to the passage of exhaust gas. The cylinder head was clamped to the block by use of extended length bolts that passed through the sandwich plate assembly and utilised the existing threads cut in the original block. Since both the inlet and exhaust valves were now to be used as exhaust valves, it was necessary to design and manufacture a special camshaft with two pairs of cams (phased 180 degrees apart), each cam with an identical profile, and each pair operating simultaneously. A single lobe camshaft design identical to that used on the 2-stroke IDI engine was utilised on this engine. The cam profile offered the same follower velocities, accelerations and lift as its 4-stroke equivalent. Choice of a single lobe design meant that the camshaft had to be driven at engine speed. Unfortunately, this meant that the rubbing speeds and subsequent forces on the valve train effectively doubled. For this reason, it was decided to be prudent to approximately halve the rated engine speed from 5000 RPM to 2500 RPM, to ensure longevity of the valve components. By doubling the size of the crankshaft pulley, and using a longer drivebelt, it was possible to retain the original 4-stroke overhead camshaft drive arrangement.

4.7 Inlet and Exhaust Systems

A controllable flow of charge air, supplied from an external compressor, was introduced to the ports via a large plenum and a pair of fabricated 'drainpipe' style manifolds attached to either side, and running along the length of the lower sandwich plate. It was decided to initially design an exhaust system that made no attempt at pulse tuning. The exhaust ports from the redundant cylinders were blocked off, and the working exhaust ports discharged into a fabricated plenum attached to the cylinder head. The volume of this plenum was calculated so that the exhaust pressure pulses were damped out to minimise resonance or wave effects.

4.8 Fuel Injection System

The 2-stroke cycle requires fuel to be injected into each cylinder at twice the frequency of the 4-stroke cycle. Cost limitations meant that the original distributor pump from the donor 4-stroke engine had to be utilised for the *demonstrator* 2-stroke engine. Cross-drilling of the internal distributor head meant that the pump, whilst still operating at half engine speed, could supply fuel at the correct frequency to two of its four H.P. outlet ports. This modification dictated that the engine could only have two firing cylinders. 'Slimtip' eight hole injectors were used to introduce a finely atomised well-dispersed fuel spray to the combustion chamber.

5. EXPERIMENTAL ENGINE TEST PROGRAMME

Time restrictions meant that it was only possible to run the 2-stroke DI engine on a steady-state test facility at Bath for a couple of hours, and no formal experimental testing as such, was conducted over this period. However, whilst running of the DI engine at low speeds and light loads, it was found to be necessary to increase the delivery ratio by about $^1/_3$ in order to achieve comparable smoke readings to those previously observed on the IDI engine operating at similar conditions. This was attributed to the fact the 2-stroke DI engine offered poorer scavenging and mixing for a given airflow rate, due to its compromised port design.

6. PREDICTED ENGINE PERFORMANCE

Due to the paucity of available experimental results, only predicted performance results for the 2-stroke DI engine can be given. These were obtained from running the ODES simulation programme with the model previously used to determine the port configuration. The simulation was run at several speeds and fuelling conditions, from 6 mm^3/shot up to a value of between 21 - 24 mm^3/shot, which corresponded to a *trapped* AFR of 20 : 1 - this was considered a suitable limiting value, as it was judged that visible smoke output would rise sharply at richer mixtures. The model was 'tweaked' to give identical airflows and exhaust back pressures to those values previously observed in the 2-stroke IDI engine experimental test programme. For those tests, notional blower characteristics were used, where the airflow was linearly increased with speed, resulting in a delivery ratio of 60 % in excess of unity at the lowest speed of 1000 RPM, reducing to unity at the rated speed of 2500 RPM.

6.1 DI vs. IDI 2-stroke Engine Predicted Results

Good agreement was found between the *predicted* ODES simulation results and the *actual* experimental performance results for the 2-stroke IDI engine. It was therefore decided that for the purposes of a fair comparison, results from ODES simulations would be used for both the DI and IDI versions of the 2-stroke engine. Identical assumptions for blower characteristics etc., were used for both the engine models, though the relevant port geometries, con-rod dimensions, compression ratio, start of injection timing etc. were changed to suit each engine. Figures 4(a-d) summarise the main predicted performance differences between the DI and IDI versions of the 2-stroke engine.

Figure 4a - Pressure Drop Across Engine

This figure compares the predicted pressure drop between the inlet and exhaust manifolds for both engines at maximum fuelling, over their operating speed range. The pressure drop increases with speed, and is between 25% to 39% worse for the DI engine. This was expected, as the *effective* port area of the DI engine is 24 % less than its sister IDI engine. A higher pressure drop means that more work will have to be done by the blower to deliver the same air flowrate, thus increasing its parasitic drain on engine power. It is predicted that the DI blower will absorb 0.91 kW at the rated engine speed of 2500 RPM, compared with 0.75 kW for the IDI blower.

Figure 4b - Net Brake Thermal Efficiency

This figure compares the brake thermal efficiency for both the DI and IDI engines, for increasing fuelling, at their rated speed of 2500 RPM. Despite a greater blower loss, the brake thermal efficiency for the DI engine is still almost 35 % at full load, approximately 3.5 % better than its IDI sister. Gains increase significantly at reduced fuelling levels, The DI engine being 43 % more efficient at minimum fuelling. These gains obviously translate to a greatly reduced fuel consumption for the DI engine at part load. Similar trends, though less significant, were observed at the lower engine speeds. The reduced thermal efficiency of the IDI engine can be directly attributed to the larger pumping loss suffered by the gas as it passes through the transfer passage or 'throat' connecting the pre-chamber to the main cylinder, and also an additional heat transfer loss due to the larger effective surface area of the combustion chamber in the IDI engine, because of the inclusion of the pre-chamber. For any given engine speed, these losses are independent of load. However, as fuelling reduces, the losses become a more significant proportion of the gross power produced, and thus increasingly reduce the part load efficiency of the IDI engine.

Figure 4c - BSFC

This figure illustrates how the gains in brake thermal efficiency described above, are practically translated into a reduction in fuel consumption for the DI engine. At full load, the minimum BSFC for the DI engine is 242 g/kWh, compared with 251 g/kWh for its IDI sister. However at minimum fuelling, the consumption for the IDI engine has increased to 566 g/kWh, whilst the DI engine only requires 404 g/kWh.

Figure 4d - Brake Power Output

This figure compares the maximum brake power available from each type of combustion system at the limiting *trapped* AFR of 20 : 1, for the speed range of the engines. It is observed that at the lower speeds there is virtually no difference between the power output of the two engines, however at the rated speed of 2500 RPM, and despite a greater parasitic blower loss, the DI engine produces a maximum output of slightly over 22 kW, just over 4% more power than its sister IDI engine.

Theoretical Improvements to the 2-Stroke DI Combustion System

One of the goals of the project was to compare the performance differences between DI and IDI combustion systems in the 2-stroke engine. However, it has previously been explained that the DI engine port design was compromised by several manufacturing practicalities. This led to two major differences in the geometric specification of the combustion systems on the existing 2-stroke engines, thus preventing a true 'back-to-back' comparison of the engines. Firstly, whilst the *geometric* compression ratio (C.R.) for the 2-stroke DI engine is identical to that of its 4-stroke counterpart, its *effective* C.R. (measured at port closure) is only 17.6 : 1, compared with 18.3 : 1 for its IDI sister. This is due to the much taller port required on the DI engine, necessary to maintain an identical inlet timing value for both engine designs. Secondly, restrictions on the positioning and orientation of the ports meant that the *effective* port area of the DI engine was actually 24% less than that available in its sister IDI engine. Additionally, the four ports are inclined at two different angles, and are located in pairs on opposite sides of the cylinder. In an *ideal* design, It would have been more desirable to have six swirl ports, all with a moderate inclination of say 30 degrees, and equi-spaced around the cylinder wall, as this would induce a more regular swirling motion to the incoming air.

In order to predict the *true* performance differences between the DI and IDI combustion systems, it was decided to repeat the 2-stroke DI engine simulations with the *effective* C.R. and *effective* port area suitably increased to identical values as used on the 2-stroke IDI engine. Additionally, the displacement fraction used in the scavenging model was increased from a conservative value of 0.4 to a more realistic value of 0.6, in view of the increased port area and improved swirl conditions offered by the *ideal* port design proposed above. Tabulated results showed that the effect of increasing the *effective* port area was to cause a reduced pressure drop across the engine, which in turn meant that the parasitic power loss to the blower was reduced. Additionally, the equi-spaced and orientated port design helped increase the swirl ratio by about 15%, from approximately 2.0 to 2.3 (relative to engine rotational speed). This led to improved scavenging and a higher charge purity, which in turn allowed marginally more fuel to be burnt. The small increase in compression ratio would also improve the thermal efficiency and thus reduce the fuel consumption.

The net result of all these improvements is seen as an overall increase in the net brake power output for the 2-stroke DI engine over its entire speed range, as indicated in Figure 4d. The greatest power increase is at the rated condition of the engine - 2500 RPM and a 20 : 1 *trapped* AFR. At this point the maximum net power output is 24.5 kW, this represents a 14 % improvement over the maximum power available from the 2-stroke IDI engine, with both engines now using similar combustion systems. Additionally, the minimum BSFC has improved from 242 g/kWh to 235 g/kWh at this condition, giving the *optimised* DI engine 6 % less fuel consumption that its IDI sister.

6.2 Comparison of 2-Stroke vs. 4-Stroke DI Diesel Engine Results

It is also of interest to compare the performance of the 2-stroke DI engine against its 4-stroke counterpart. However, since the 2-stroke engine fires every cycle, it would be incorrect to compare the two engines in terms of absolute power output. Additionally, the 2-stroke *demonstrator* engine has only half the capacity of its 4-stroke counterpart. Thus, for an unbiased comparison, it would only be fair to normalise the performance for each engine by dividing the power output at any speed by the total swept capacity of the cylinders. The results can then be presented on common axes, with specific power in kW/Litre, as the ordinate and engine speed in RPM, as the abscissa.

Figure 5 - Net Specific Brake Power Output

The lower three curves on his figure compare the predicted maximum specific power output of both the 2-stroke DI and IDI *demonstrator* engines against experimental results obtained from recently performed tests conducted on a turbocharged and intercooled *development* version of the 4-stroke DI engine. Since, in practice, a 2-stroke automotive engine would be required to operate at speeds in excess of 2500 RPM in order to develop sufficient power and improve its driveability, it was therefore decided to run the simulation up to a theoretical maximum engine speed of 4000 RPM. Previously used values for airflow and pressure drop at the lower engine speeds were linearly extrapolated to give notional values at the higher speeds.

The figure indicates that at the lowest speed of 1000 RPM, both the DI and IDI versions of the 2-stroke engine are predicted to produce approximately $^2/_3$ more specific power than the 4-stroke DI engine. At this speed, the fuel quantities injected per stroke, and the inlet manifold pressures for the three engines are all broadly similar. It is therefore suggested that the increase in specific power merely reflects the doubled firing frequency of the 2-stroke engine. As the speed increases, the turbocharger starts to work, and the boost pressure in the inlet manifold for the 4-stroke engine soon exceeds the values obtainable from the mechanically blown 2-stroke engine, enabling more fuel to be injected per stroke, and thus proportionally reducing the specific power advantage offered by both the 2-stroke engines at low speed.

The relevant performance results for the three engines operating at 2500 RPM are summarised in Table 1. The simulation predicts that in its mechanically blown form, the 2-stroke DI engine will produce an almost equal specific power output as its turbocharged and intercooled 4-stroke counterpart. Of course, it is technically feasible to turbocharge the 2-stroke engine, however, a mechanical blower would still be needed for start up and low speed running. Assuming it were possible to boost the inlet air to the equivalent pressure of the 4-stroke engine, then by simple extrapolation, the theoretical specific power output of the 2-stroke DI engine would increase to 57 % higher than its 4-stroke counterpart. This implies that a three cylinder 2-stroke engine, of an identical swept volume per cylinder, could at least match and possibly exceed the power output of a four cylinder 4-stroke engine at this speed. The 2-stroke IDI engine produces a similar output when turbocharged.

Extrapolated values for the theoretical maximum specific power output for both the 2-stroke engines, when operating at the equivalent boost levels of the 4-stroke engine, are shown as the upper two curves of the figure. At engine speeds above 2500 RPM, it is observed that the predicted specific power output from the 2-stroke DI engine increasingly diverges from its IDI sister. It is suggested that this due to the transfer passage pumping losses becoming more significant for the IDI engine, and additionally, poorer scavenging of the pre-chamber, both contributing to a reduction in net specific power.

Table 2 summarises the engines' performance at the theoretical maximum speed of 4000 RPM. It is observed that whilst the 4-stroke turbocharger produces almost two times more boost to the incoming air, enabling the 4-stroke DI engine to burn proportionally more fuel; this is adequately compensated for by the doubled firing frequency of both the mechanically blown 2-stroke engines, which are predicted to produce a comparable specific power output. When turbocharged, both engines show similar gains to those observed at 2500 RPM, however the DI engine now produces 11 % more specific power than its IDI sister, because its reduced pumping losses and superior scavenging mean it can burn slightly more fuel at the limiting *trapped* AFR of 20 : 1.

7. CONCLUSIONS

The ODES simulation package predicted that the two cylinder 2-stroke DI *demonstrator* engine will produce a slightly higher power output and reduced fuel consumption when compared to its sister IDI engine. At its rated speed of 2500 RPM, it is predicted that the specific power output from the mechanically blown 2-stroke DI engine will almost equal that from its turbocharged and intercooled 4-stroke counterpart. It was also shown that with suitable turbocharging, a three cylinder 2-stroke engine produces sufficient power to be a potential alternative to a turbocharged four cylinder 4-stroke engine.

However, it must be acknowledged that in the developed world, the emissions output from the engine clearly has to be considered in parallel with performance - it has not been discussed in this paper due to absence of relevant data at this stage. Experimental tests conducted on the 2-stroke IDI indicate that this engine does not appear to be at a significant disadvantage relative to its 4-stroke counterpart, and it may be suggested that the situation would be similar for the DI engine, though this has yet to be proved.

It is anticipated that a full set of experimental tests will be conducted on the DI engine in the future, and these results will form the basis of a future paper comparing the *actual* performance of the 2-stroke DI engine with its IDI sister and 4-stroke counterpart.

ACKNOWLEDGEMENTS

The authors are grateful to the Ford Motor Company for its interest and sponsorship of this project, for the assistance from its engineers in help with design of some of the engine components, and their permission to publish the findings of the research work undertaken to date. Additionally, thanks are due to the postgraduate student, Mark Mathieson [4], who assisted in the design of the engine; and to the technical staff, Martin Tanner and Andy Green, who turned sketches into components. A final mention is due to Dr. S.J. Charlton who conceived the project and who was its inspiration in the early days.

REFERENCES

[1] Bulmer, C. ***'Two-Stroke Renaissance ? (Parts 1 & 2)'***.
Car Design and Technology Yearbook 1992, Brackland Publishing PLC.

[2] Cox, A., Wallace, F.J. and Bird, G.L. ***'Initial Development of an Experimental Uniflow IDI 2-stroke Diesel Engine'***. EAEC 5th International Congress, Strasbourg, 1995, (SIA 9506A10).

[3] Dang, D. and Wallace, F.J. ***'Some Single Zone Scavenging models for Two-Stroke Engines'***.
International Journal of Mechanical sciences, Vol. 34, No. 8, 1992.

[4] Mathieson, M.F. ***'Two Stroke Diesel Engine Design'***.
M.Phil Thesis, University of Bath, 1994.

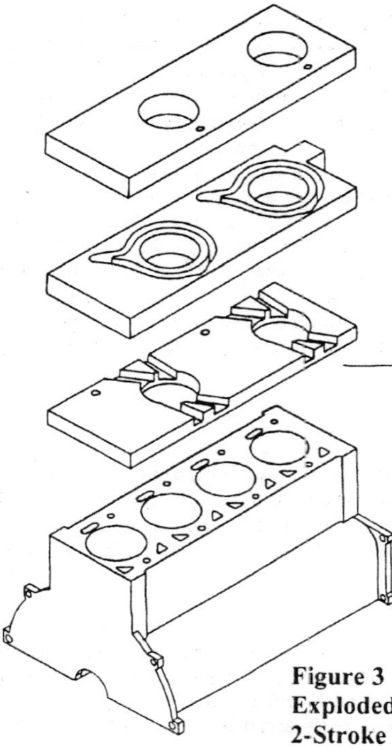

Figure 3
Exploded View of the 2-Stroke DI Block

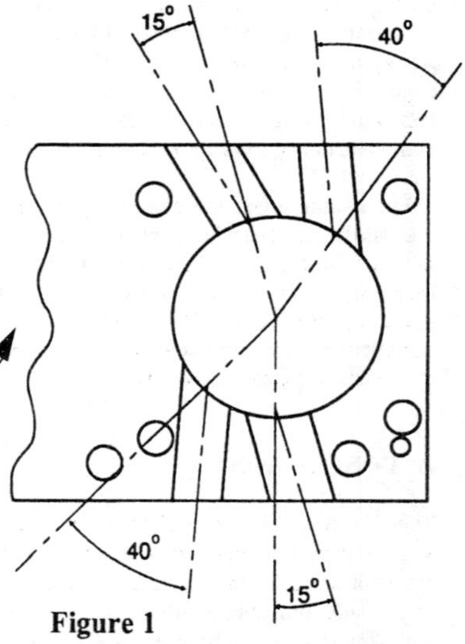

Figure 1
Inlet Port Design for the 2-Stroke DI Engine

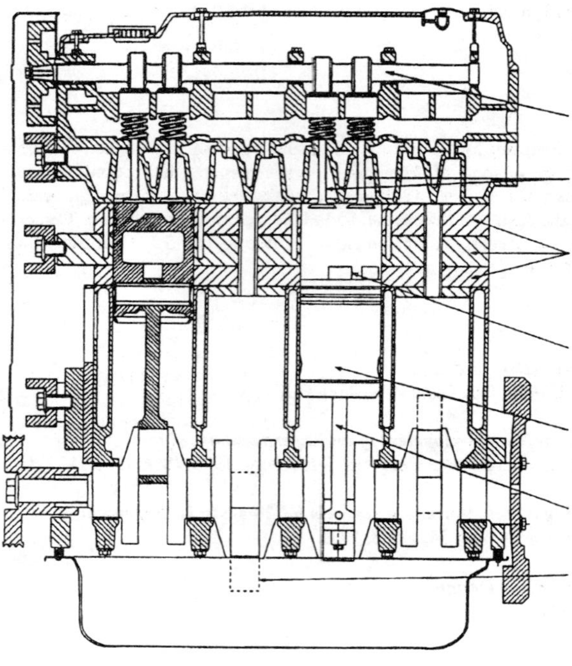

Figure 2
Design Features of the 2-Stroke DI Engine

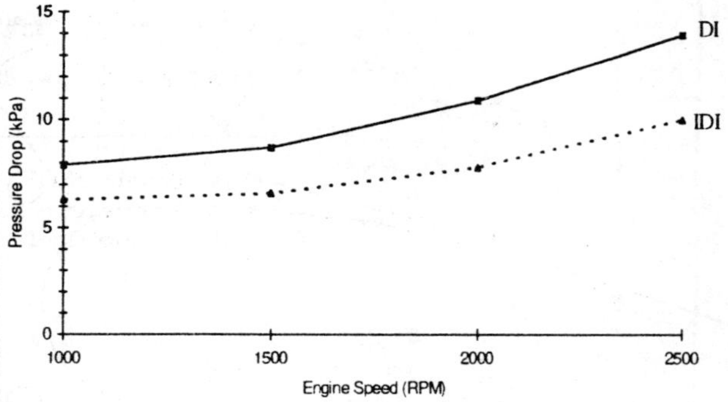

Figure 4a
Predicted Pressure Drop Across Engine at Maximum Fuelling - *Trapped* AFR = 20 : 1

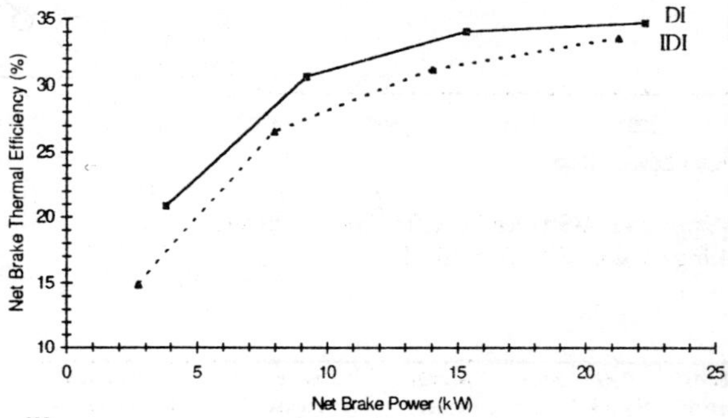

Figure 4b
Predicted Net Brake Thermal Efficiency at 2500 RPM (Includes 'Blower' Losses)

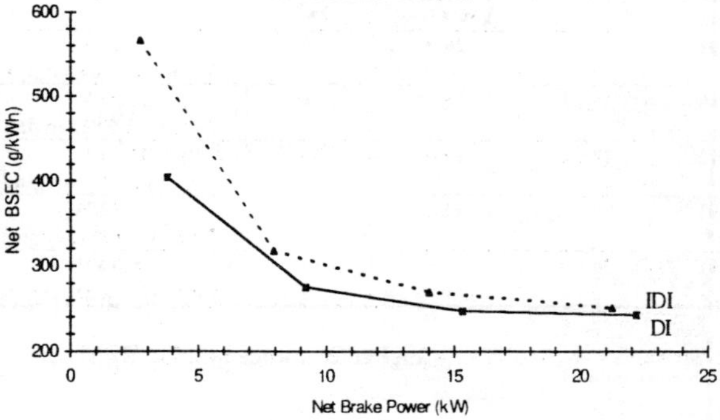

Figure 4c
Predicted Net Brake Specific Fuel Consumption at 2500 RPM (Includes 'Blower' Losses)

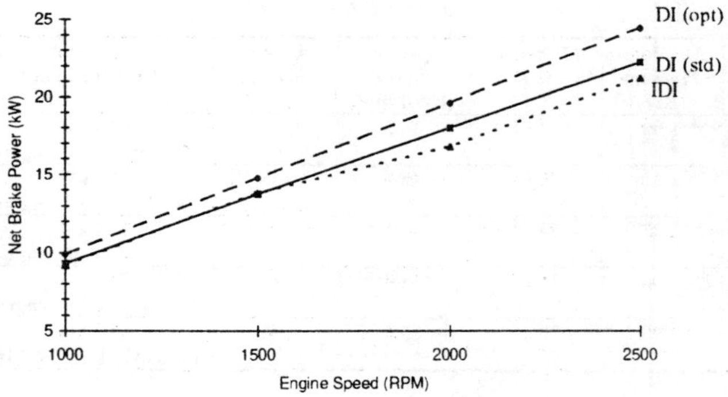

Figure 4d
Predicted Net Brake Power at Maximum Fuelling - *Trapped* AFR = 20 : 1 (Includes 'Blower' Losses)

Figure 4(a-d) Comparison of 2-Stroke DI vs. IDI Predicted Engine Performance

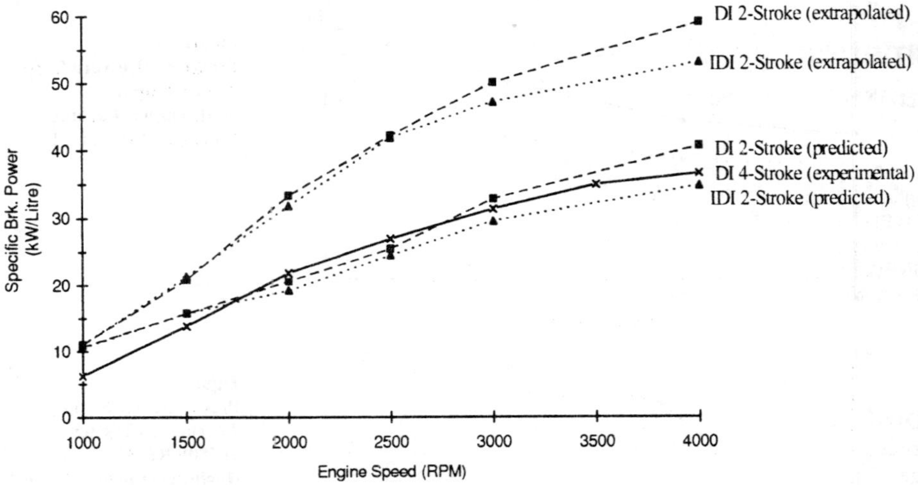

Figure 5 Comparison of 2-Stroke vs. 4-Stroke Specific Power Output at Maximum Fuelling (*Trapped* AFR = 20 : 1)

Engine Type	Build Level (results)	Abs.Boost Pressure (Bar)	*Trapped* AFR	Specific Power (kW/Litre)	Power Increase (%)	Practical Interpretation
4-Stroke DI TCi	Development (as tested)	1.91	24 : 1	26.9	0	47.1 kW for 4 cyl / 1.8L engine
2-Stroke DI Blower	Existing (predicted)	1.15	20 : 1	25.4	-6	22.2 kW for 2 cyl / 0.9L engine
2-Stroke IDI Blower	Existing (predicted)	1.12	20 : 1	24.4	-9	21.4 kW for 2 cyl / 0.9L engine
2-Stroke DI TCi + Blower	Turbocharged (extrapolated)	1.91	20 : 1	42.1	+57	55.3 kW for 3 cyl /1.35L engine
2-Stroke IDI TCi + Blower	Turbocharged (extrapolated)	1.91	20 : 1	41.8	+56	55.0 kW for 3 cyl / 1.35L engine

Table 1 Comparison of 2-Stroke vs. 4-Stroke Engine Performance at 2500 RPM

Engine Type	Build Level (results)	Abs.Boost Pressure (Bar)	*Trapped* AFR	Specific Power (kW/Litre)	Power Increase (%)	Practical Interpretation
4-Stroke DI TCi	Development (as tested)	1.90	23 : 1	36.5	0	64.0 kW for 4 cyl / 1.8L engine
2-Stroke DI Blower	Existing (predicted)	1.31	20 : 1	40.4	+11	35.4 kW for 2 cyl / 0.9L engine
2-Stroke IDI Blower	Existing (predicted)	1.25	20 : 1	34.7	-5	30.4 kW for 2 cyl / 0.9L engine
2-Stroke DI TCi + Blower	Turbocharged (extrapolated)	1.90	20 : 1	58.9	+61	77.4 kW for 3 cyl / 1.35L engine
2-Stroke IDI TCi + Blower	Turbocharged (extrapolated)	1.90	20 : 1	53.0	+45	69.7 kW for 3 cyl / 1.35 L engine

Table 2 Comparison of 2-Stroke vs. 4-Stroke Engine Performance at 4000 RPM

C498/9/040/95

Numerical prediction of the acceleration performance of a rotary engine

J MA PhD
Shanghai Jiatong University, People's Republic of China
R BOWEN MPhil and **R ALLEN** PhD
University of Southampton, UK
B AGNEW PhD
University of Newcastle-upon-Tyne, UK

SYNOPSIS A comprehensive computer simulation has been developed to predict the transient performance of a rotary engine. An adiabatic throttling process has been assumed to account for crevice leakage. The overlapping double Wiebe curve has been introduced to represent the combustion process. The simulation has been tuned to fit experimental data. The potential of the model as an aid to engine design and development is illustrated by means of predictions of the effects of changes in some of the main design parameters.

1 INTRODUCTION

The rotary engine has a small size, low weight, high specific power density and a construction far less complicated than a reciprocating piston engine. The transient performance of the engine is extremely important for application on for example racing motorcycles and aircraft drones.

Mathematical simulation now forms the basis of many design and development projects. It provides a convenient, reliable and low-cost opportunity to undertake studies of the effects of changes on engine performance and can highlight sensitivities in the response to particular parameters.

The construction of the rotary engine provides an ideal candidate for simulation since changes in rotor geometry, for example, are extremely difficult to assess in practice without first having a firm basis for the nature of the changes to be made and their likely effect on performance. The leakage of the side and apex seals, the influence of port area and cross section and the cycle work of the three chambers related to the rotor produce analysis problems which are different from those posed by reciprocating piston engines.

This paper discusses a quasi-steady homogeneous model that has been used to simulate the transient performance of a rotary engine. The parameters within the model were fitted using experimental data obtained from tests carried out on a Norton NR 601 rotary engine. The engine test facility is introduced and the results of the tests on the engine are presented and compared with the simulation predictions.

NOTATION

a	ellipse's major axis
A	area
A_m	maximum area
b	ellipse's minor axis
Cv	specific heat at constant volume
d	distance defined in text
$dx/d\theta$	combustion velocity
e	rotor's eccentricity
E	internal energy
H_u	fuel calorific value
h	enthalpy
J_e	engine moment of inertia
J_l	load moment of inertia
k	ratio of specific heats, trochoid constant
M	mass of working substance
m	combustion quality index
M_e	engine torque
m_f	fuel feed
n	shaft speed
N_e	shaft power
P	pressure in chamber
P_e	mean effective pressure
P_i	mean indicated pressure
P_m	mean mechanical loss pressure
P_o	ambient pressure
Q	heat
Q_f	heat release from combustion
R	rotor's generating radius, gas constant
T	temperature
V	volume of chamber
v	specific volume
W	rotor's width
x	mass fraction burned
Z	number of rotors
α_{max}	rotor maximum angle of oscillation
μ	flow coefficient
μ_m	maximal flow coefficient
μf	equivalent cross sectional area of the flow
θ	shaft angle
θ_z	period of sustained combustion
θ_0	shaft angle at the beginning of the combustion

Subscripts:

a	air
cr	through crevice on side seal
d	down stream
exh	exhaust gas
g	gas
in	incoming
leak	leakage through apex seal
out	outgoing
p	port
rec	rotor recess
u	upstream
wall	side wall of chamber

2 SIMULATION MODEL

The simulation is based upon a single zone thermodynamic model of the engine cycle. An aim of the work was to investigate the potential of fuel injection as a means of improving the low speed and transient performance of the engine. For the baseline studies, however, a carburettor is being used and has been assumed in the model at this stage. A block diagram of the model is presented in figure 1 to show the relationships between the various effects that have been described:

Atmospheric air is inducted into the chamber through the filter, the inlet and the carburettor. The fuel flow rate is determined by the opening of the throttle. The processes in the chamber are governed by thermodynamic and mechanics considerations and include induction, compression, combustion, expansion and exhaust. Leakage from the combustion chamber of the rotary engine has been considered in two parts, the permeation through the apex seal to neighbouring chambers and the leakage through the gap between the apex of the seal and the housing. The effects of engine dynamics, that is the inertia and load, are also included in the simulation. The friction losses and the heat transfer are considered at the same time.

Combustion is modelled as a single zone process described by the Wiebe function. Although the transient response is being modelled, during each complete cycle the speed is assumed to remain constant. The single zone thermodynamic model has been shown to be capable of simulating the low speed and transient performance of the rotary engine effectively. At the end of each cycle the main parameters are calculated together with the acceleration.

3 PROCESSES IN THE CHAMBER

The main processes in the cycle take place in the combustion chamber. In order to study the effect of the processes on engine performance the conservation equations of energy and mass are used to describe changes in the working substances.

$$\frac{dE}{d\theta} = \frac{dQ_{in}}{d\theta} - \frac{dQ_{out}}{d\theta} - p\frac{dv}{d\theta}$$

$$\frac{dm}{d\theta} = \frac{dm_{in}}{d\theta} - \frac{dm_{exh}}{d\theta} - m_f \frac{dX}{d\theta} - \frac{dm_{cr}}{d\theta} - \frac{dm_{leak}}{d\theta}$$

where Q_{in} is the incoming heat into the chamber and is expressed as follows:

$$\frac{dQ_{in}}{d\theta} = \frac{dQ_f}{d\theta} + \frac{dm_{in}}{d\theta} h_{in}$$

The two terms on the right are the change rates of the combustion heat release and the enthalpy brought in by the incoming fluid, both with respect to the angle of rotation. Q_{out} is the heat loss which is described by the following equation:

$$\frac{dQ_{out}}{d\theta} = \frac{dQ_w}{d\theta} + \frac{dm_{exh}}{d\theta} h_{exh} + \frac{dm_{cr}}{d\theta} h_{cr} + \frac{dm_{leak}}{d\theta} h_{leak}$$

In the above equation the terms on the right hand side are the rates of change of heat transfer loss, enthalpy of exhaust, leaking enthalpy through the crevice on the side seal and leaking enthalpy through the apex seal.

The internal energy E of the working substance in the chamber can be represented as:

$$\frac{dE}{d\theta} = \frac{d(Cv\ M\ T)}{d\theta} = \bar{Cv}\ \frac{d(M\ T)}{d\theta} = \bar{Cv}\ T\frac{dM}{d\theta} + \bar{Cv}\ M\ \frac{dT}{d\varphi}$$

where, T is the temperature of the working substance in the chamber, Cv is the mean specific heat capacity at constant volume, hereafter written as Cv for convenience.
The energy equation for can then be rearranged to produce an expression that describes the rate change with crank angle of the mean fluid temperature.

$$\frac{dT}{d\theta} = \left[\frac{dQ_f}{d\theta} + \frac{dM_{in}}{d\theta} . h_{in} - \frac{dQ_w}{d\theta} - \frac{dM_{exh}}{d\theta} h_{exh} - \frac{dM_{cr}}{d\theta} h_{cr} - \frac{dM_{leak}}{d\theta} h_{leak} - P \frac{dV}{d\theta} - Cv\, T \frac{dm}{d\theta} \right] \frac{1}{M\, C}$$

4 HEAT TRANSFER

In the above equation the heat transfer through the cylinder walls and the rotor is represented as Qw. This is modelled as a convective process with the heat transfer coefficient h_w being taken from Blair and Fleck [1].The heat transfer process then takes the form

$$\frac{dQ_w}{d\theta} = h_w \left[A_{rot} (T - T_{rot}) + (A_{wall} + A_{mar}) (T - T_{wall}) \right] \frac{1}{1080\ 60\ n}$$

where, A_{rot} represents one side area of the rotor and A_{mar} the corresponding chamber wall area. These areas are determined from the basic geometry of the epitrochoid [2]:

The heat transfer coefficient h_w is as below [2]:

$$h_w = \frac{590}{861 \times 4.1868} \left(\frac{n}{1000} \frac{p}{0.9807} \right)^{0.786} - 100\ T^{-0.525} \quad \frac{kJ}{m^2\ s\ K}$$

5 THE COMBUSTION PROCESS

The combustion process in the chamber can be simulated by an overlapping double Wiebe function.In a typical indicator diagram of a petrol engine, it is possible to distinguish the stages in the combustion process. The first stage is mainly a chemical process, which corresponds to the time for the formation of the self-propagating nucleus of the flame and the second stage is both a chemical and a physical process, which corresponds to the propagation of the flame. There is a big difference between the combustion quality index of two stages so that the double overlapping Wiebe curves is introduced. The analysis of heat release provides the necessary parameters for the simulation.

The mass fraction burned at any angle is given by [3],[4]:

$$x = 1 - e^{6.908 \left[\frac{\theta - \theta_o}{\theta_z} \right]^{m+1}}$$

and the rate of the burn by its derivative [4]

$$\frac{dx}{d\theta} = 6.908 \times \frac{m+1}{\theta_z} \left[\frac{\theta - \theta_0}{\theta_z} \right]^m e^{-6.908 \left[\frac{\theta - \theta_0}{\theta_z} \right]^{m+1}}$$

where, m is the combustion quality index, θ_z is the shaft angle corresponding to the period of the sustained combustion, $dx/d\theta$ is the combustion velocity, θ_0 is the crank angle at the beginning of the combustion. The rate of heat release can then be determined :

$$\frac{dQ_f}{d\theta} = H_u \, m_f \, \frac{dx}{d\theta}$$

in which H_u and M_f represent the fuel calorific value and fuel feed which is proscribed by the opening of the throttle and the characteristic map of the carburettor [5].

6 INTAKE AND EXHAUST

The flow rates during the intake and exhaust processes depend on whether the flow is choked. The flow paths in the engine are complicated in shape and at any time the outline of the active port is the intersection of the port and the trochoidal cylinder. The ports are of elliptical cross section with major and minor axis designated by a and b. When a port is partially masked by the rotor, the effective sectional area must first be calculated. This area is the projection of the opened section of the port in the direction vertical to the passage centre line.

$$A_p = \frac{\pi a b}{2} + a\, b \arcsin \left[\frac{d}{b} \right] + d\, a \sqrt{1 - \frac{d^2}{b^2}}$$

where d is the distance between the major axis and the edge of the flow area. The calculation of the flow coefficient μ uses the formula fitted to the flow through the port of the two-stroke diesel engine taken from [6].

$$\mu = \mu_m \left[1 + a \left[e^{-\left[b \frac{A_p}{A_m} \right]} - e^{-b} \right] \right]$$

In this μ_m is the maximum flow coefficient and A_p/A_m the ratio of the area of the opening of the port to the area of the fully open port.

7 CREVICE SIZE AND LEAKAGE

Leakage from the combustion chamber of the rotary engine is composed of two parts, the permeation through the apex seal to neighbouring chambers and the leakage through the gap between the apex of the seal and the housing. The leakage through the apex seal arises from the crevice between the neighbouring chambers and the pressure difference between the chambers. One dimensional isentropic conditions are assumed to be appropriate for calculating the leakage flow.

In the subcritical case,

$$\frac{dM_{leak}}{d\theta} = \frac{1}{18n} \frac{A_{cr} P_u}{\sqrt{R T_u}} \left(\frac{P_d}{P_u} \right)^{1/k} \left[\frac{2k}{k-1} \left(1 - \frac{P_d}{P_u} \right)^{\frac{k-1}{k}} \right]^{\frac{1}{2}}$$

and in the critical case

$$\frac{dM_{leak}}{d\theta} = \frac{1}{18n} \frac{A_{cr} P_u}{\sqrt{R T_u}} k^{\frac{1}{2}} \left[\frac{2}{k+1} \right]^{\frac{k+1}{2(k-1)}}$$

where A_{cr} represents the area of the crevice in the apex seal and subscripts u and d represents the upstream and downstream conditions.

The leakage through the crevice on the rotor side seal is wholly determined by the state of the working substance. This leakage process can be divided into two stages.

In the first stage, the working substance in the chamber enters the groove through the side seal's crevice where the cross sectional area of the passage contracts abruptly and gives rise to a pressure drop across the seal. It can be treated as an adiabatic throttling process and the velocity is assumed constant before and after the seal. The temperature of the working substance in the groove is equal to that in the chamber. The mass rate of the working substance in groove is given by

$$M_{cr} = \frac{P_{cr} V_{cr}}{RT}$$

where P_{er} represents the pressure in the groove and V_{cr} the volume of the groove.

Along with the change of pressure and temperature, the mass of the working substance in the grove changing continuously. The mass in the groove can be expected to behave approximately as described by the following equation:

$$\frac{dM_{cr}}{d\theta} = M_{cr}\left[\frac{1}{P_{cr}}\frac{dP_{cr}}{d\theta} - \frac{1}{T}\frac{dT}{d\theta}\right] = \frac{P_{cr}\,V_{cr}}{RT}\left[\frac{1}{P_{cr}}\frac{dP_{cr}}{d\theta} - \frac{1}{T}\frac{dT}{d\theta}\right]$$

In the second stage of this process the working substance goes through the side seal once again and escapes to the atmosphere outside. The side seal is assumed to behave as an adiabatic air-bleed from a rigid container to the outside atmosphere. At the end, the pressure and temperature of the working substance will be the same as the atmosphere environment (Po, To).

$$P_{cr} = P_0\left[\frac{T}{T_0}\right]^{\frac{k}{k-1}}$$

Therefore,

$$\frac{dP_{cr}}{d\theta} = \frac{k}{k-1}\,P_0\left[\frac{T}{T_0}\right]^{\frac{k}{k-1}}\frac{dT}{d\theta}$$

which gives rise to the following equation,

$$\frac{dM_{cr}}{d\theta} = \frac{1}{k-1}\,\frac{P_0\,V_{cr}}{R\,T}\left[\frac{T^{(2-k)}}{T_0{}^{k}}\right]^{\frac{1}{k-1}}\frac{dT}{d\theta}$$

Thus at any crank angle the mass of fluid in the crevices and the fluid exiting the combustion chambe and its associated enthalpy can be determined.

8 ENGINE OUTPUT

The mean indicated pressure can be solved by the integration of work done over the entire cycle whic in this case occupies three revolutions of the output shaft (1080°):

$$P_i = \frac{1}{V}\int_{\theta = 0}^{\theta = 1080} P\,dV$$

The mean effective pressure is given by:

$$P_e = P_i - P_m$$

where P_m represents the mean mechanical loss pressure. It can be expressed as a function of the engine speed n, and determined from a curve fit to the test results [2].

$$P_m = 1.403811 \times 10^5 - 16.13745\ n + 1.2984516 n^2 - 8.264925 \times 10^{-7}\ n^3 \quad (Pa)$$

The shaft power calculation is determined from:

$$N_e = P_e \frac{V\ n\ Z}{60000} \qquad (kW)$$

and the shaft torque:

$$M_e = \frac{60000\ N_e}{2\ \pi\ n} \qquad Nm$$

In calculating the engine speed, it is assumed that the speed change during each working cycle is determined by the motive torque during the last cycle.

$$\frac{dn}{d\varphi} = \frac{5}{3} \frac{1}{\pi\ n} \frac{M_e}{J_e + J_l}$$

where, J_e, J_l represent the engine and load moments of inertia.

9 CYCLE CALCULATIONS

The change in the volume of the combustion chamber can be represented by the following equation [2]:

$$V = V_{min} + V_{rec} + \frac{3\sqrt{3}}{2} e\ R\ b \left[1 - \sin\left(\frac{2}{3}\theta + \frac{\pi}{6} \right) \right]$$

where, V_{min}, is the minimal volume of the chamber and is usually expressed as below.

$$V_{min} = \left[\frac{\pi}{3}e^2 + 2\ e\ R\ \cos\theta_{max} + \left(\frac{2}{9}\ R^2 + 4e^2 \right) \theta_{max} - \frac{3\sqrt{3}}{2} e\ R \right] W$$

The recess volume in the rotor is designated as V_{rec} in the above and is determined by measurement. V_{min} and V_{rec} are constant so that the period of the chamber volume change occupies 3π whilst the whole cycle of rotor rotation corresponds to 6π of rotation angle of the rotor shaft.

At the beginning of the simulation it is necessary to input certain data. At this point the working cycle number is set to zero, and every parameter is set to the initial value at the very beginning of the transient process. At the starting point of compression the calculation begins and passes through combustion, expansion, exhaust and induction in succession until the end of the cycle. If the engine speed has not reached a steady state the calculation is continued until a steady speed is obtained. The speed of the next cycle is then determined from the calculated excess torque available to accelerate the rotor over the time interval under consideration.

Since the rate of change of many of the variables differs during the various processes during the cycle, different size time steps have been used for some processes. The longest time step (2 SAD--shaft angle degrees) used is during the compression process while the shortest time step (0.25 SAD) is during the port overlap period. The remaining processes use a medium time step of (1 SAD). This process is shown schematically in the appendix

10 ENGINE TESTING AND SIMULATION PREDICTION

A computerised test facility has been created around a Norton NR601 rotary engine. It is a water cooled single rotor engine with a chamber of 294 cc. A block diagram of the rig is shown in figure 2 which indicates the interaction of the various components. A detailed description of the rig is presented by Harding et al. [7] together with the background to its development.

An acceleration test was performed on an unloaded engine in order to provide initial data of rotor shaft inertia for tuning the mathematical model. The speed response of the engine is shown plotted against time in figure 3, together with the response of the model following tuning. A close correlation between the two responses is apparent.

A series of tests have been carried out on the model to determine the sensitivity of the performance to various engine parameters. For comparison purposes all model responses are shown in terms of the output shaft acceleration and each of the different events examined depicts the measured response of the engine in the standard build configuration and in the new configuration.

10.1 Effect of Groove volume in the Crevice

The results of this simulation are shown in figure 4. When the volume of the crevice is enlarged the leakage is increased with a corresponding deterioration in the transient performance of the engine. The reverse is the case if the volume decreases. The sensitivity of the performance to groove volume is however slight as there is no difference in the acceleration performance for a variation in groove volume between one half and one tenth of the initial volume. That is, the groove volume in the crevice can be reduced to comply with other criteria without significantly adversely effecting the engine performance.

10.2 Leakage Through the Apex Seal

Aside from the crevice volume the leakage through the tip apex seal has also been studied as shown in Figure 5. The cross sectional area of the crevice in the apex seal is far less than the crevice in the side seal, so that leakage of the apex seal is small by comparison. Changes in the size of this gap have only a small effect on transient performance as indicated.

10.3 Change in the Recess Volume

The significance of the recess volume in this simulation is in the effect it has on the rotor compression ratio so it is of vital importance to the engine thermal efficiency. If the volume of the recess is increased, the compression ratio will reduce and visa-versa. There are however other effects such as leakage and heat transfer that are associated with compression ratio thus it is possible that an optimal value of compression ratio can be found. The simulation shows this effect with the acceleration performance showing the fastest response for the standard build case, figure 6. Increasing the compression ratio (0.25 standard recess volume) reduces the response and increasing the volume also has the same effect but not to the same extent. The performance is less sensitive to a reduction in compression ratio through an increase in recess volume than it is to a reduction in the recess volume.

10.4 Change in the Port Area

The enlargement of the port cross section is beneficial to the improvement of the gas exchange process and a high degree of port overlap is known to be beneficial at high rotor speeds on a wide open throttle. However, excessive enlargement has a deleterious effect at the beginning of the transient process when the port overlap cannot be used to advantage. This is clearly illustrated in figure 7 in which the large port at 1.5 x standard port shows an initial slow engine response followed by a rapid acceleration after a rotor speed of 2000RPM. The 2 x standard port is clearly unsuitable for the engine. The 0.66 x standard port shows an initial response similar to the standard port but the gas exchange process cannot be maintained at the high rotor speeds.

10.5 Change in the Port Position

The position of the inlet (exhaust) port is defined in terms of the included angles between the lower (upper) edge of the port and the horizontal axis of the rotor. The numbers in the legend of figure 8 refer to these angles. If the distance between the two ports is too far, the scavenging process will be shortened resulting amongst other things in an increase in the residual exhaust gas and generally poor scavenging. If the distance is too small the effective expansion ratio of the engine will be shortened resulting in a low power output. There is an optimum configuration of the port position that will produce an optimised output for the engine design conditions. This can be seen in figure 8. The trends associated with port position are not easily deduced from this figure as the interaction of the different effects described are very complicated. It can be seen however that apart from the case of very small included angles (8-13) the sensitivity of the transient response to port site is much less than that to port size.

10.6 Change in Ignition Advance

As with all spark ignition engines the point of the ignition depends on several factors including engine speed, cylinder pressure at ignition and fuel type. For a given engine an optimal ignition advance is chosen for the acceleration process. As the simulation of the combustion process used in this work was very simplistic the results shown in figure 9 should be considered only as indications of trends. Again an optimum value is indicated around which earlier or later timing will adversely affect the engine acceleration. The simulation calculation results provide some guidance for the selection of an optimum spark ignition timing.

10.7 Effect of Change in Eccentricity

The geometry of the rotor is determined by two main parameters. These are the generating radius which is normally fixed for a given rotor size and the rotor eccentricity. The eccentricity which is expressed in terms of the trochoid constant can be treated as a variable for the purpose of the simulation. It can be seen in figure 10 that increasing the eccentricity and reducing the trochoid constant will tend to improve the engine performance.

10.8 Change in Moment of Inertia

As might be expected, the moment of inertia of the rotor and other rotating parts should be reduced as much as possible, figure 11. In practice there will be limits on how much this can be done due to design and material constraints. However, the prediction of the simulation is still meaningful as it shows the correct trends that would be asscciated with changes in the rotor inertia.

10.9 Overall Effects

Of the variations in the design or engine modifications that have been studied and depicted in the figures they can be divided into two sub groups. The first is associated with the basic geometry of the engine and would include aspects such as port design, rotor eccentricity and seal leakage while the second group can be more closely identified with phenomena that are common with reciprocating engines. The second group would include ignition advance, port area, recess volume (through compression ratio) Referring to the second group the effect of these parameters on the engine performance are as would be expected from a reciprocating engine which leads one to have some confidence in the simulation The sensitivity of the model to the parameters in this group is larger than that for the parameters in the first group which tends to indicate that the design of the engine can be such to accommodate mechanical considerations without unduly restricting the thermodynamic processes.

11 CONCLUSION

A quasi-steady homogeneous model of a rotary engine has been shown, by the limited data mentioned above, to be capable of simulating the transient performance of the engine effectively.

The resulting simulation has been used to predict the effect of changes in engine design parameters on the transient performance of an engine.

Examples have been provided to illustrate the use of the model in identifying sensitive parameters and their effects on the transient performance of the engine.

There is still a great deal of work to be done in refining the model to include progressive combustion and emissions formation. Work is also taking place to simulate a turbocharged and intercooled engine It is hoped that these improvements coupled with the simulation of an electronic fuel injection system will improve the accuracy and application of the model.

ACKNOWLEDGEMENTS

The authors would like to thank the following: Mr David Garside and Mr Chris Biddulph of Alvis UAV UK Ltd for kindly supplying the test engine. Mr Brian Kenny of Tillotson Limited who kindly provided the carburettor performance map for this project. The engine rig is being developed by Mr Ian Harding and Mr Geoff Spencer of the Department of Mechanical Engineering, University of Southampton, and the authors are very grateful for all their help in carrying out the engine tests.

REFERENCES

[1] Blair, G. O., and Fleck, P., The Unsteady Gas Flow Behaviour in a Charge Cooled Rotary Piston Engine, SAE Off-Highway Vehicle Meeting and Exhibition, Milwaukee, Wis., SAE Paper No. 770763, 1977.

[2] Yamamoto, K., Rotary Engine, Sankaido, Tokyo, 1981.

[3] Ramos, J.I., Internal Combustion Engine Modelling, Hemisphere Publishing Corporation, 1989

[4] Wiebe, I., Halbempirische Formel durch die Verbrennungsgeschwindigkeit, in Kraftstoffaufbereitung und Verbrennung bei Dieselmotoren, Springer-Verlag, 1964

[5] Ma, J., Allen, R., and Bowen, R., Mathematical Simulation of the Transient Performance of a Petrol Engine, SAE International Congress and Exposition, Detroit, Mich, SAE Paper No. 930855, 1993.

[6] Gu, H., Simulation Calculation of Thermal Process of Turbo-charged Diesel Engine, Shanghai Jiaotong University Press, 1985.

[7] Harding, I., Spencer, G., Bowen, R. and Allen, R. The development of a diagnostic test rig for the assessment of the low speed performance of a rotary Wankel engine. Proc. COMADEM '93, 21-23 July, 1993, Bristol.

APPENDIX

SIMULATION FLOW CHART

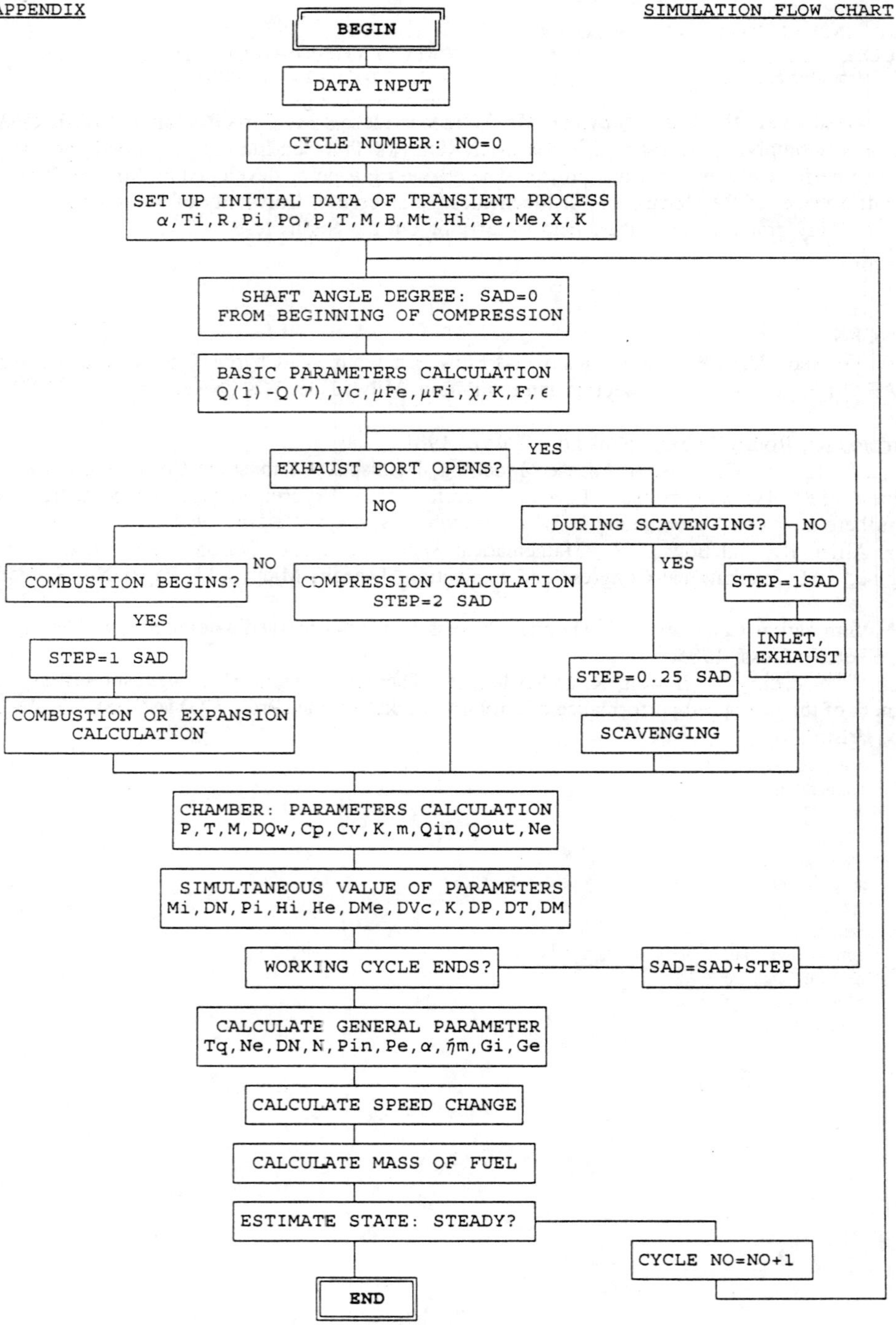

INPUT: INITIAL DATA OF TRANSIENT PROCESS

α:	excessive coefficient
Ti:	temperature in inlet
R:	gas constant
Pi:	pressure in inlet
Po:	atmosphere pressure
P:	pressure in chamber
T:	temperature in chamber
M:	mass of gas in chamber
B:	fuel feed
Mt:	gas mass transferred into at normal state
Hi:	enthalpy in inlet
Pe:	pressure in exhaust
Me:	mass of gas in exhaust
X:	fraction of burned fuel
K:	isentropic index

BASIC PARAMETERS

Q(1) -Q(7):	port time
Vc:	chamber volume
μFe:	equivalent cross sectional area of exhaust port
μFi:.	equivalent cross sectional area of inlet port
χ:	angle of oscillation
K:	trochoid constant
F:	side area of chamber
ϵ:	compression ratio

CALCULATION RESULTS: PARAMETERS IN CHAMBER

P:	pressure
T:	temperature
M:	mass of working substance
DQw:	change rate of heat transfer of chamber wall
Cp:	specific heat capacity at constant pressure
Cv:	specific heat capacity at constant volume
K:	isentropic index
m:	combustion quality index
Qin:	heat into chamber
Qout:	heat out of chamber
Ne:	engine power

CALCULATION RESULTS: SIMULTANEOUS VALUE OF PARAMETERS

Mi:	mass of gas in inlet
DN:	change rate of engine speed
Pi:	pressure in inlet
Hi:	enthalpy in inlet
He:	enthalpy in exhaust
DMe:	change rate of mass in exhaust
DVc:	change rate of chamber volume
K:	isentropic index
DP:	change rate of pressure
DT:	change rate of temperature
DM:	change rate of mass of gas

CALCULATION RESULTS: GENERAL PARAMETERS

Tq:	engine torque
Ne:	engine power
DN:	change rate of engine speed
N:	engine speed
Pin:	mean indicated pressure
Pe:	mean effective pressure
α:	excessive air coefficient
ηm:	mechanical efficiency
Gi:	indicated fuel consumption
Ge:	effective fuel consumption

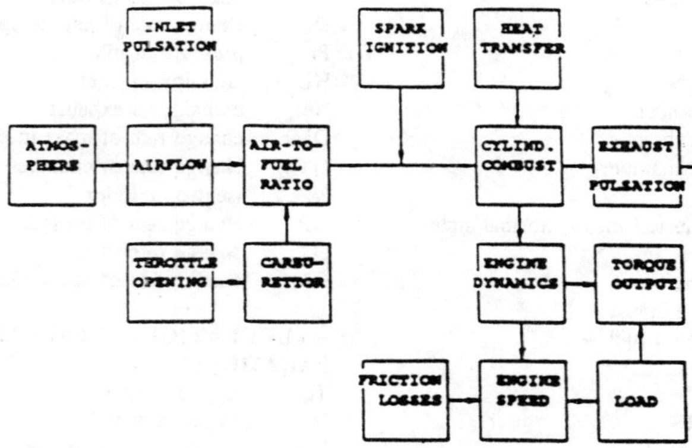

FIG.1 BLOCK DIAGRAM OF SIMULATION

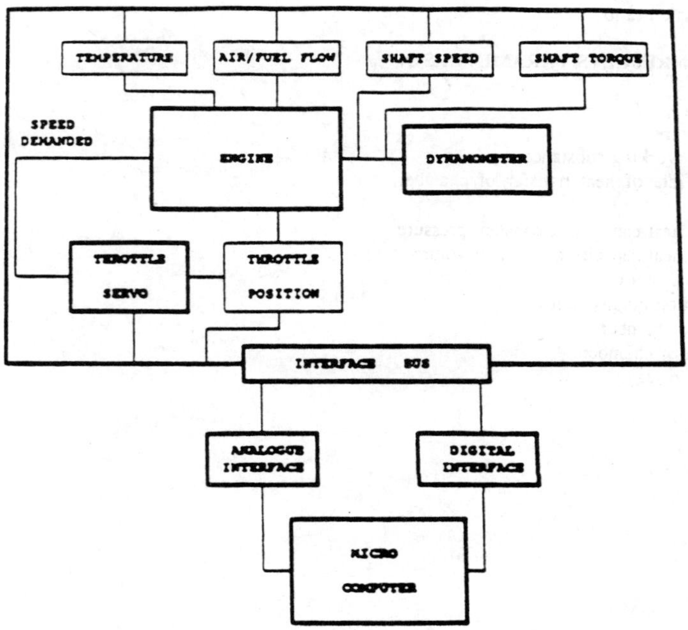

FIG.2 OUTLINE OF TEST FACILITY

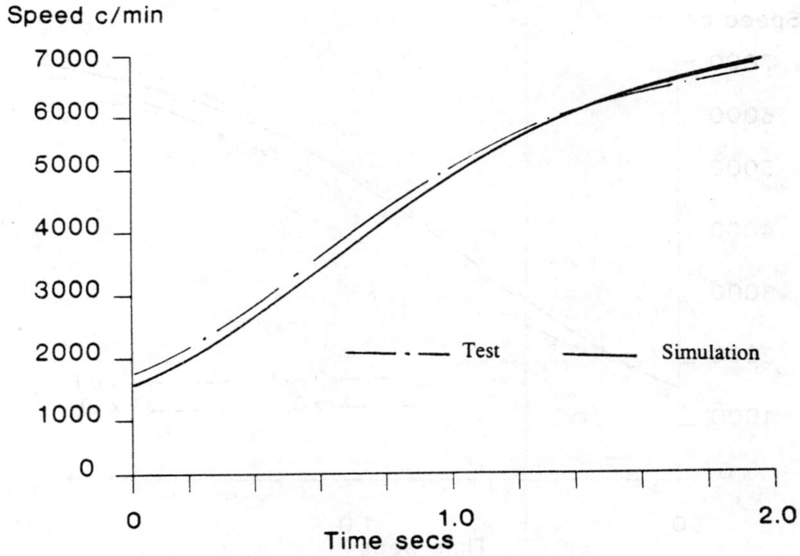

Figure 3 Comparison Between Test and Simulation

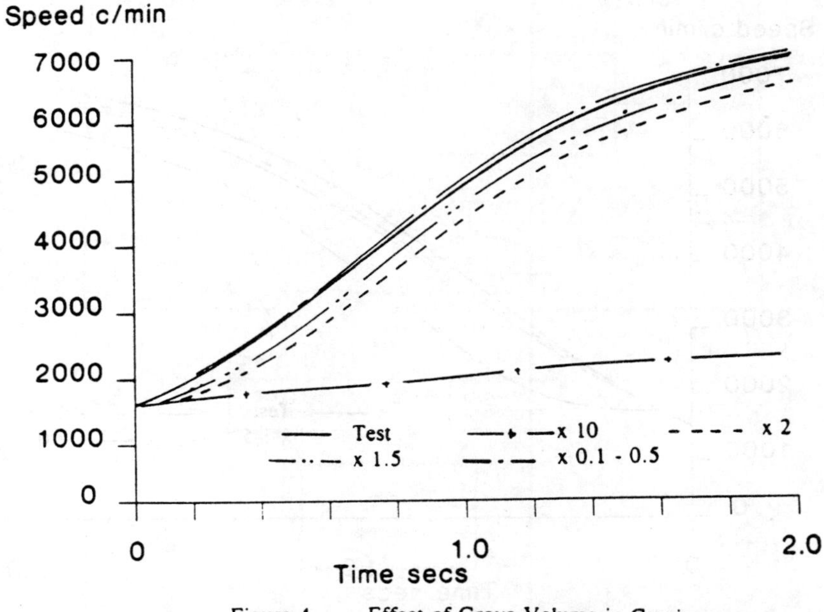

Figure 4 Effect of Grove Volume in Crevice

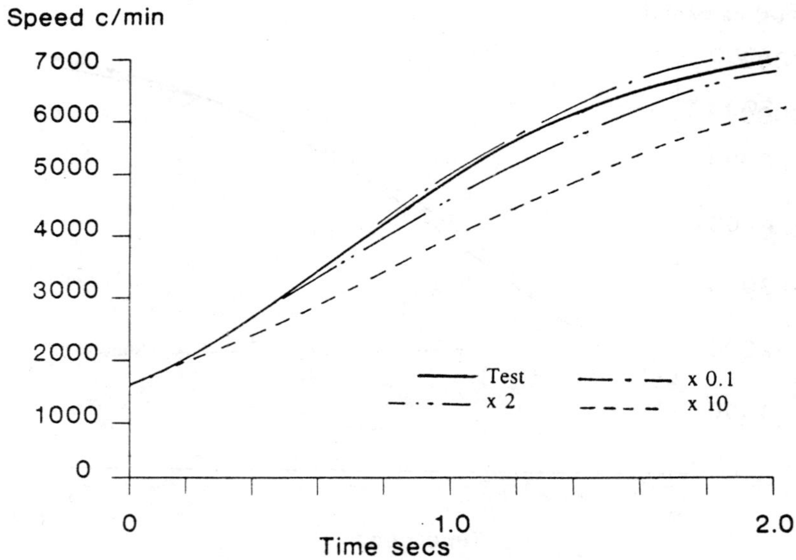

Figure 5 Effect of Leakage Through Apex Seal

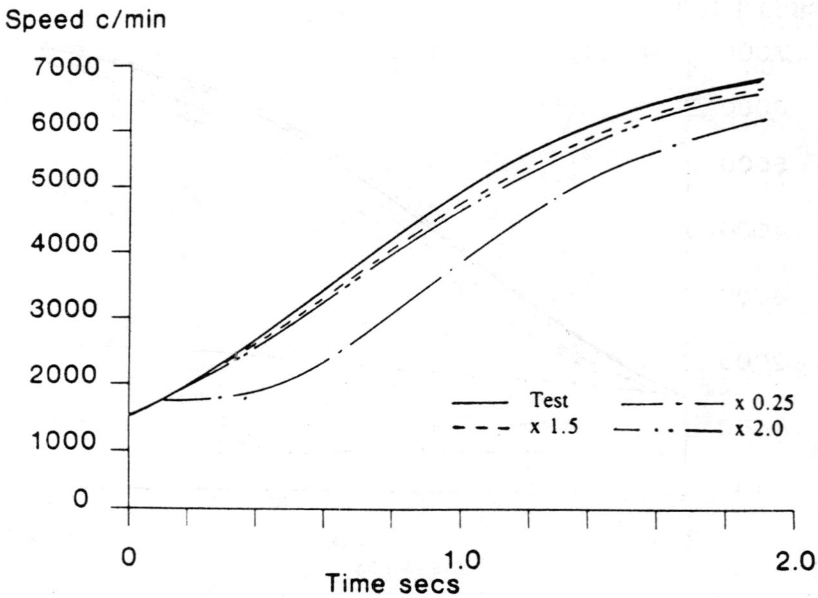

Figure 6 Effect of Recess Volume

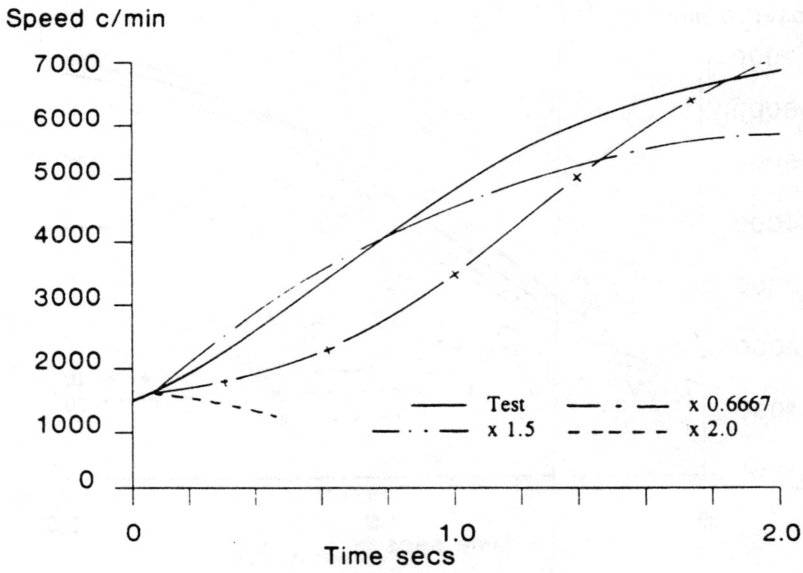

Figure 7 Effect of Port Area

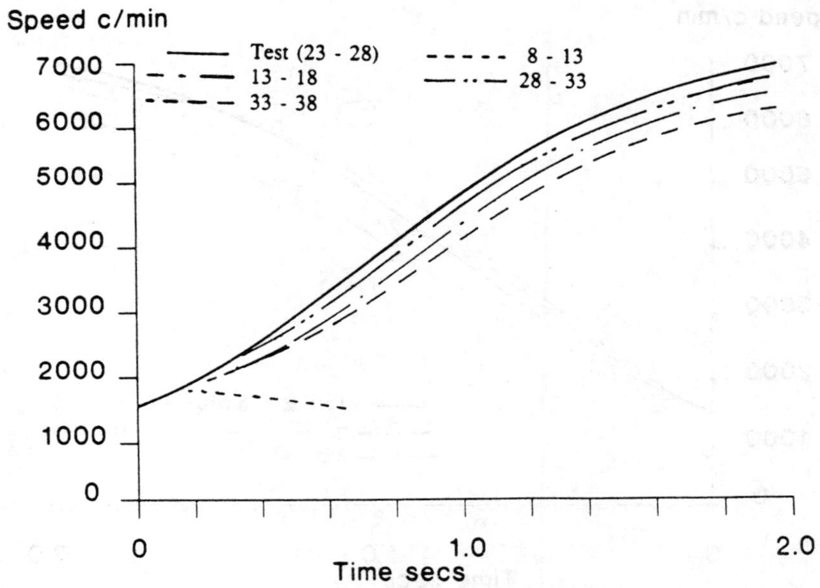

Figure 8 Effect of Port Site

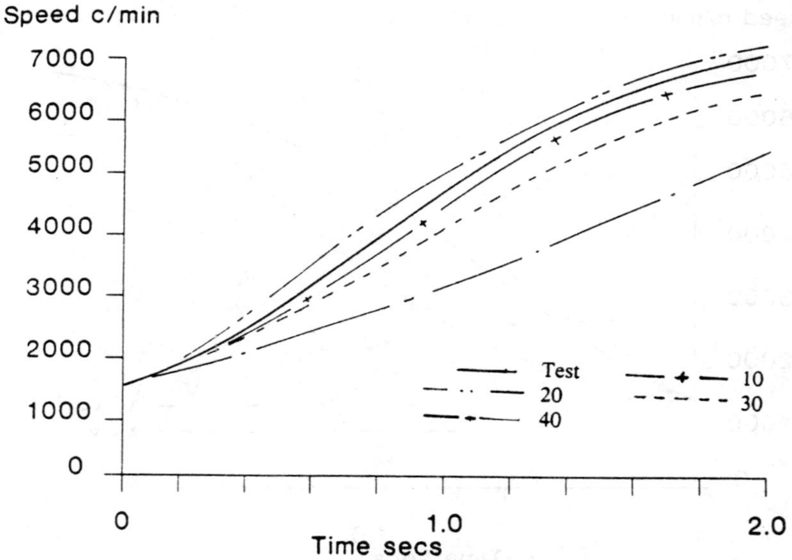

Figure 9 Effect of Ignition Advance

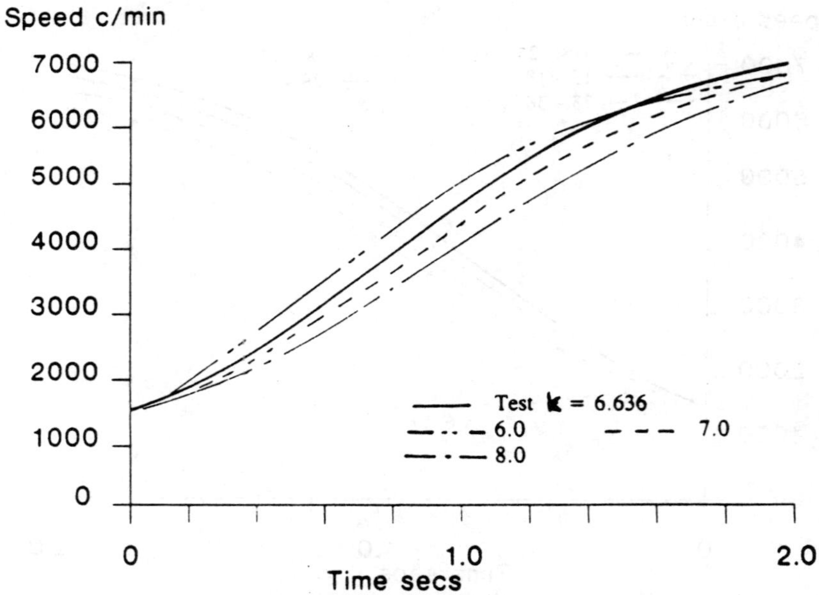

Figure 10 Effect of Change in Eccentricity

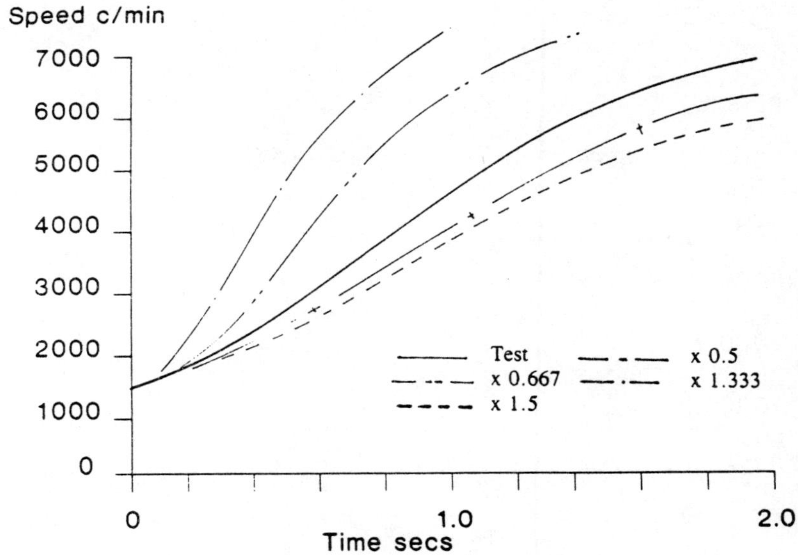

Figure 11 Effect of Change of Inertia

C498/9/228/95

The development of diesel fuel injection equipment utilizing classical and computational analysis techniques

D J HOPLEY BEng and **A M GREANEY** BSc
Lucas Diesel Systems, UK

SYNOPSIS

This paper describes an approach to product development in which classical and computational analysis techniques are used in the design process. The value of this approach for complex products, such as those of the automotive industry, is discussed. The development of a mechanical governor for a rotary distributor diesel pump has been used as an example. By using classical as well as computational analysis techniques, improvements in performance and durability have been achieved. This approach has used less resource than previous work using trial and error methods and has provided a greater understanding of the problem.

1. INTRODUCTION

In the modern day automotive industry the market demands continual improvements in the performance, reliability and durability of complex products, such as diesel fuel injection equipment. Consequently, the design and development process is of great importance for business success. This suggests a greater emphasis should be placed on optimising the design and development process.

This paper concentrates on the analytical techniques used in the design of product performance. Some consideration of experimental methods for validation is also made. Other factors such as cost, manufacture and assembly strongly influence product design but have not been addressed specifically.

2. OBJECTIVES

The objective of any design and development process is to meet or exceed the market requirements for a product. The product specification must account for any ambiguity in customer requirements, and it can be expected to change during the design and development process.

At the point when a new design is released for production an adequate degree of confidence in the change must be achieved. This level of confidence is determined by the understanding gained from the analysis, supported by the results obtained from validation work. The overall aim must be to reach this point in the most time and cost efficient manner. This requires a considered approach to solving the problem and careful selection of analytical and experimental methods.

In the design and development process there are numerous methods of problem solving. Frequently a trial and error approach is adopted which involves a lot of testing and requires considerable resource. Alternatively, the use of Computer Aided Engineering methods allows in-depth analysis which can provide more information than is required to produce a solution. For any specific problem there is an optimum depth of analysis. This provides sufficient information to identify the root cause of the problem and derive a solution to eliminate it. These ideas are illustrated in Fig 1, where the resource required to provide a solution varies with the approach to solving the problem.

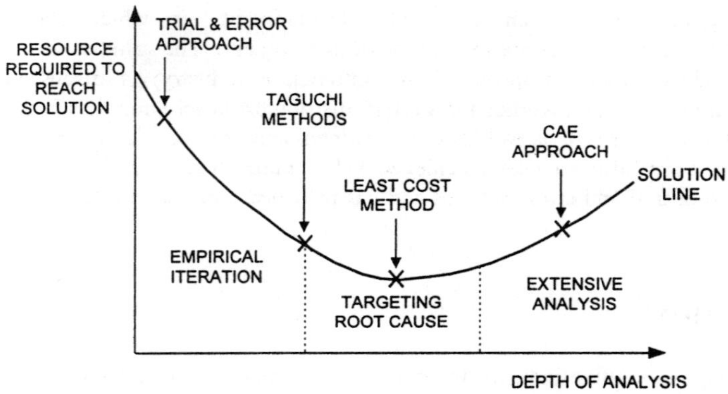

Fig 1 Problem analysis model

3. LEVELS OF ANALYSIS

Irrespective of the overall depth, four stages of analysis can be identified. The model in Fig 2 suggests that in comparison with computational methods, a classical approach requires less resource and provides a greater understanding of a product. However, computation provides specific knowledge which cannot be obtained by classical methods.

3.1. Simple linearised analysis

When designing performance into a product assumptions have to be made at different levels of the analysis. The first stage of analysis should include assumptions that simplify and linearise the problem to isolate only the dominant performance characteristics. This form of analysis lends itself to classical techniques, an example being the investigation of differential equations describing the behaviour of a governor within a diesel fuel pump. This approach can quickly provide a broad understanding of the system. For relatively simple devices this level may provide sufficient understanding to devise a solution to the problem. Thorough investigation at this level can reveal alternative design approaches.

3.2. Linearised analysis

This stage of analysis is more focused and attempts to introduce secondary effects which influence performance, such as vibrations, temperature effects and wear sensitivity. Although the analysis is still linear in form, the inclusion of such characteristics complicates the mathematical description of the product and limits the techniques available for analysis. However, in many cases these secondary effects can have a significant influence on product behaviour.

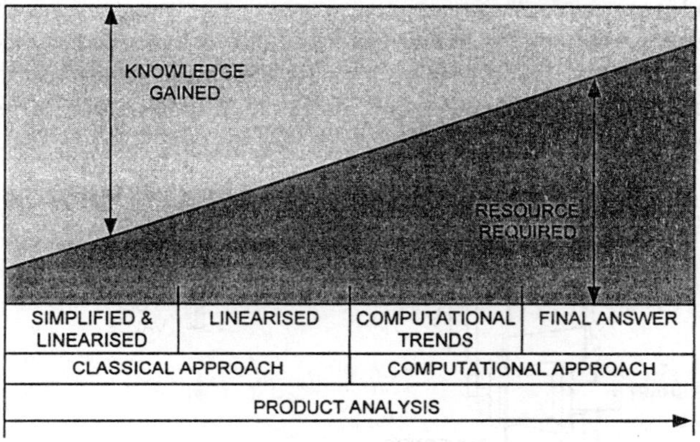

Fig 2 Comparing the knowledge gained and resource required for the various analysis stages.

3.3. Computational trend analysis

Having gained a comprehensive understanding of the linearised behaviour of the product, computational methods can be adopted to incorporate non-linear effects. This allows step functions to be used and removes the constraint of performing analysis about a nominal operating condition. In the case of the mechanical governor the effect of varying friction was analysed as discussed in section 5. A higher level of investment is required to gain a smaller amount of understanding about the product but for some applications this may be essential.

3.4. Final answer analysis

The highest level of analysis uses computational methods to provide an answer to a specific set of conditions. In the case of the mechanical governor this would involve comprehensively modelling the governor, pump and engine. This method is often a necessity for complex problems and is a useful tool for verifying the performance of an optimum design. However, it requires considerable resource and little knowledge is gained about the robustness of the design.

4. CLASSICAL ANALYSIS METHODS

Classical analysis methods have been used in the development of a mechanical governor within a diesel fuel injection pump. The governor maintains an approximately constant engine speed by increasing the fuel delivery when a load is applied to the engine. As shown in Fig 3, centrifugal weights act against a spring via a pivoting arm. The fuel delivery is determined by the position of a fuel metering valve which is linked to the governor arm. When a load is removed from the engine the speed initially increases causing the weights to move out. This reduces the fuel delivery and lowers the engine speed. Traditionally, to maintain engine stability, mechanical governors have been designed to give an increase in speed when a load is removed. This can be termed positive governing and is shown in Fig 4. Development was required to allow tighter governing (where the speed change with load is smaller) without causing instability. Problems concerning durability also had to be addressed as wear of the components resulted in rapid deterioration of governor performance when close governing was specified.

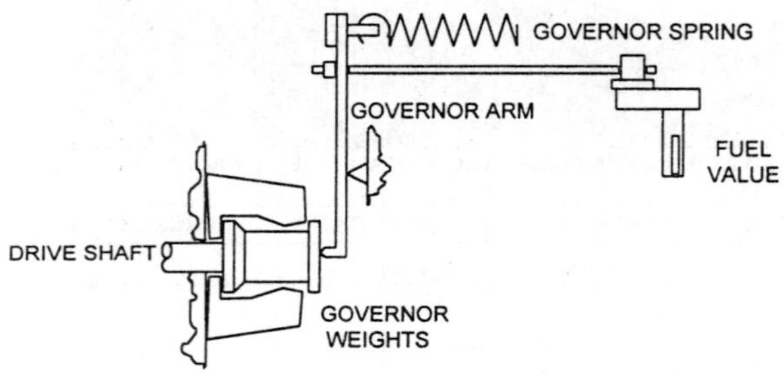

Fig 3 The conventional mechanical governor.

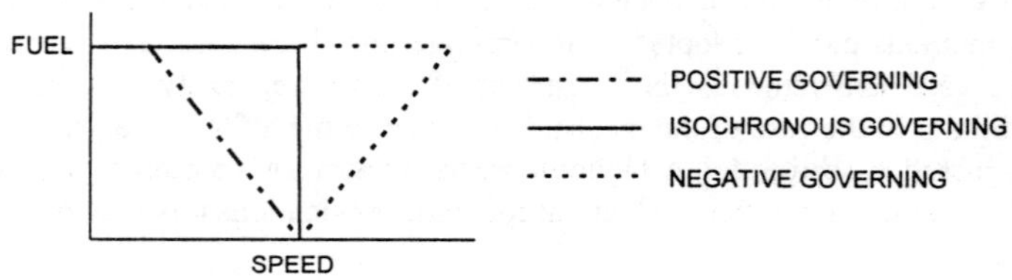

Fig 4 Governor characteristics showing positive, isochronous and negative governing.

In the example of the mechanical governor, a simple linear analysis was used to optimise the performance of an additional stabilising device. In order to do this the mechanics were simplified by grouping terms that were kinetically linked. This removed geometric relationships which were not of immediate interest. Once the transfer function was obtained, classical control techniques were used to analyse the system. Stability was assessed using the Routh-Hurwitz stability criterion (1) to define an envelope within which the pump was stable. The analysis showed that negative governing, shown in Fig 4, is possible. This condition was previously considered to be inherently unstable and experiments were devised to demonstrate a stable negative pull-off. As a result an isochronous governor (no change in engine speed with load) was designed with a large stability margin.

To demonstrate the increased stability a bode plot was produced comparing two governors of equal pull-off, one standard and one with the stabiliser. The modifications to the governor have reduced the systems gain to less than unity at a phase shift of 180° and is therefore stable. It has also moved the natural frequency of the governor away from the running speed of the pump. Using a classical approach design improvements were identified which would have been missed if trial and error or purely computational methods were adopted.

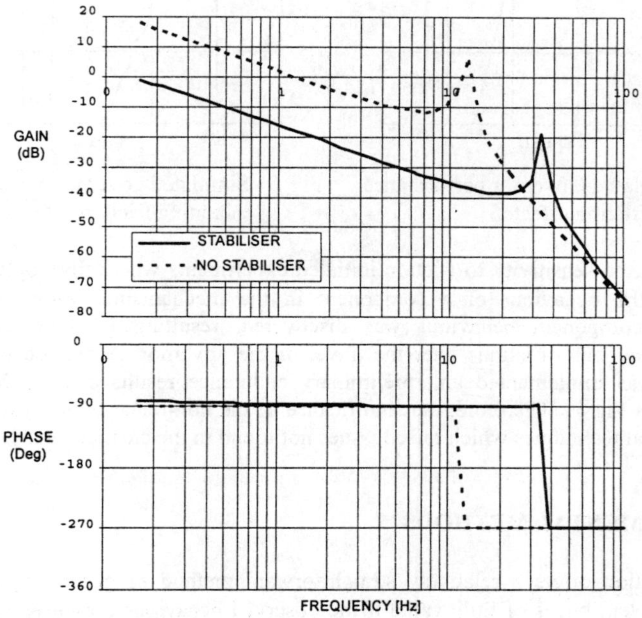

Fig 4 Bode plots illustrating improved governor stability

5. COMPUTATIONAL ANALYSIS METHODS

A more thorough examination of the mechanical governor required computational analysis techniques where the non-linear effects of static friction within the system were examined. The results show that small friction variations within the mechanism produce large changes in performance. Fig 5 shows the governor performance as defined by the linear analysis levels. Fig 6 illustrates the governor curve including the effect of friction variations. These results were produced using the Advanced Continuous Simulation Language (ACSL).

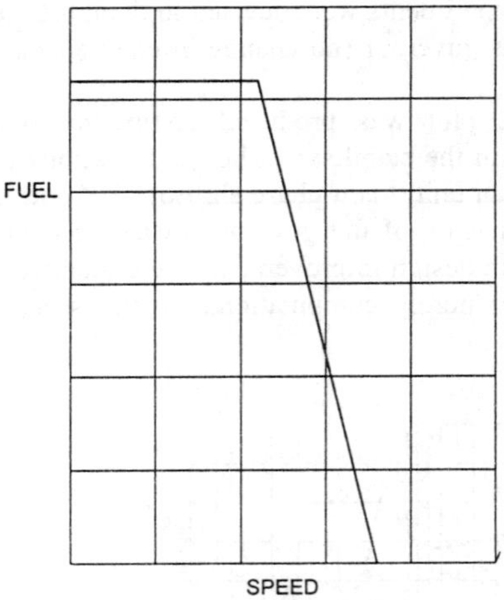

Fig 5 Simulated Governor performance without friction effects.

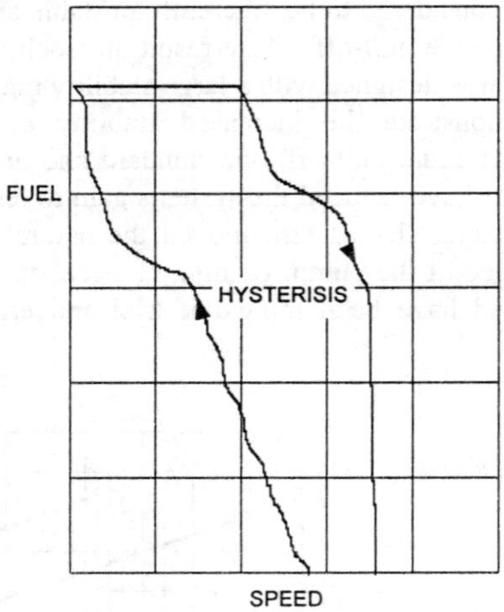

Fig 6 Simulated governor performance including friction effects.

This unexpected sensitivity to friction initiated experiments where fibre optics were used to investigate the behaviour of a component in the mechanism. From this work some unforeseen component behaviour was discovered, resulting in increased wear, which produced frictional variations over the travel of the governor. A method of correcting the behaviour was implemented and preliminary endurance results show a reduction in the distortions in Fig 6. Therefore, the significance of the non-linear effects justified the use of computational techniques which raised issues not found in the classical analysis.

6. EXPERIMENTAL METHODS

Experimentation offers a relatively straightforward method of generating an answer to a specific problem but is of little value if the observed behaviour cannot be explained. Thus comprehensive analysis increases the efficiency and value of experimental work by producing a fuller picture of the problem. This approach minimises risk at the development stage, hence avoiding high cost penalties in production.

6.1. Fractional Factorial Design

In the case of diesel fuel injection pumps, validating design to meet specific customer requirements is hindered by the large number of factors involved and the complex relationships between them. In the example of the mechanical governor, several design features were changed simultaneously and Fractional Factorial Design methods (as applied by Taguchi) were used in the validation. Such statistical techniques provide a lot of information from little testing.

In the example of the diesel pump mechanical governor, six proposed design changes were tested at two levels; comparing the current design with the proposed modifications. The six changes, or factors, were tested in an L8 matrix (2); eight tests by seven possible factors at two levels. Reference (3) describes the derivation of two level Fractional Factorial Design matrices and gives examples of their use. In our example, to cover all possible combinations sixty-four tests would have been required for a full factorial array. By performing the eight tests sufficient data was produced to judge the relative merits of each factor. The L8 matrix showed that three out of the six proposed modifications made little improvement to governor performance and as a result these features where not pursued to endurance testing. The aforementioned stabilising device showed improvements in steady state stability with a larger stability margin, as predicted by the analysis. Other design modifications showed reduced hysterisis and improved consistency.

7. CONCLUSIONS

Both classical and computational analysis techniques have a place in the design and development process, but it can be concluded that a greater understanding can be attained from a classical approach. Computational methods are most useful when modelling complex non-linear mechanics. With this in mind it is important to select methods of analysis carefully. Empirical results should be used to support analytical data, not as stand-alone evidence of a product's performance. By using this approach to design and development a higher degree of performance from the mechanical governor was achieved which was supported by a higher level of confidence in the robustness of the design solutions.

8. ACKNOWLEDGEMENTS

The authors thank the directors of Lucas for giving their permission for the publication of this paper. The authors also thank their colleagues at Lucas Diesel Systems Gillingham in all marketing, manufacturing, quality, purchasing and engineering functions who have contributed to this work.

9. REFERENCES

(1) SCHWARZENBACH, J. and GILL, K.F. System modelling and control, 1992, (Edward Arnold) pp.143-145.

(2) TAGUCHI, G. Introduction to quality engineering, 1986, (APO) pp.182.

(3) BOX, G.E.P., HUNTER, W.G. and HUNTER, J.S. Statistics for experimenters, 1978, (Wiley) pp. 374-417.

C498/9/165/95

The influence of piston design and cylinder liner finish on total unburned hydrocarbon emissions from a direct injection diesel engine

P W YATES BEng and **P W FOSS** BSc, MSc, PhD, CEng, MRAeS
Sheffield Hallam University, UK
D H TIDMARSH BSc, PhD, CEng, FIMechE, MBIM, MIMgt
University of Central England, UK
M WILLCOCK BEng, PhD
AE Goetze Limited, Bradford, UK

SYNOPSIS

A major source of exhausted unburned hydrocarbons is in the crevice volumes around the piston, including the lands and piston ring grooves. The chemical mechanism of adsorption and desorption between the fuel and the lubricating oil is shown to be greatly influenced by the oil's characteristics and cylinder liner finish. Experimental results are compared with those from a computer model and are shown to be in close agreement, illustrating that it is possible to model the effects of small design changes in piston and cylinder design.

NOTATION

A_{ref}	=	second land reference area
A_{sf}	=	constant for soot formation
A_{sc}	=	constant for soot consumption
A^*	=	non-dimensional second land puddle area
Cd	=	discharge coefficient
d_2	=	distance between second land and cylinder
d_n	=	nozzle diameter
dms	=	incremental mass of soot

dmsf = incremental mass of soot formed
dmsc = incremental mass of soot consumed
F/A = fuel air ratio
ga = ring gap width
h = specific enthalpy
h_i = second land oil film thickness before top ring reversal
h* oil = non-dimensional change in film thickness
m = mass
mfv = mass of fuel vapour .
mi = mass of fuel injected
ms = mass of soot
mu = mass of fuel yet to be injected
M = mass per mole
N = engine speed (RPM)
Pcyl = cylinder pressure
PPO2 = oxygen partial pressure
Q = gross heat release
Qb_{max} = max flow rate of blowby gas through top ring gap
Q_{ht} = heat transfer
R = gas constant
r_o = radius of puddle of oil consumed
$\mathcal{R}$ = universal gas constant
S = spray penetration
t = time
t_{max} = time for Qb to reach max value
T = temperature
Ta = Taylor number = $\mu U / \sigma$
Tw = wall temperature
u = specific internal energy
Ua = gas flow velocity over puddle
V = volume
γ = ratio of specific heats
μ = oil dynamic viscosity
μa = air dynamic viscosity
$\rho_{s,ntp}$ = soot density
ρ_g = gas density
ρ_o = oil density
Φ = equivalence ratio.
Θ = angle

All units are SI unless otherwise stated.

INTRODUCTION

Continuing tightening of emission regulations has encouraged extensive research into fuel spray vaporisation and combustion. The developments from this research have significantly reduced total emissions, but reduction in fuel spray derived emissions has increased the piston and liner contribution to the total emitted matter. The research described in this paper expands the low emission model developed by Willcock[1], to evaluate the effect upon emissions of small but significant design changes to the piston-liner interface. Crevice mixture oxidation, flow into the piston lands and adsorption/desorption are based upon the same theories. The effect of adsorption and desorption of the fuel into the lubricating oil is by the same mechanism in diesel engines and spark ignition engines but the impact of this will be different.

MATHEMATICAL MODEL

Because of the complexity of the combustion process within an engine, certain assumptions have been made:-

1) The piston is completely circular and positioned centrally in the circular liner.
2) Flow is isothermal.
3) Gas in the crevice volume is at the piston temperature.
4) Gas in the crevice volume is unburnt mixture.
5) Gas entering the crevice is air until the fuel spray reaches the liner wall.
6) Gas entering the crevice volume with fuel entrained has the same equivalence ratio as the cylinder.
7) Piston rings remain on the base of the ring groove.
8) The gas does not flow past the 2nd ring.
9) Cylinder contents are at a uniform temperature.
10) Fuel is a single hydrocarbon species.

These assumptions, despite their obvious simplicity, enable numerical results to be predicted with a high degree of accuracy, and correlate with observed trends due to changes in detail design.

THE DERIVATION OF THE HEAT RELEASE EQUATION

The cylinder contents are assumed to be a single open system. The only mass flowing across the boundaries while all the valves are closed are the fuel and crevice gases. Gross heat release with convective heat transfer(2) to the cylinder wall is

$$\frac{dQ_{ht}}{dt} = Ah_c(T - T_w) \tag{1}$$

Crevice volumes are included in the standard heat release equation by adding a term containing V_{CR}

$$\frac{dQ}{dt} = \frac{\gamma}{\gamma-1}p\frac{dV}{dt} + \frac{1}{\gamma-1}V\frac{dV}{dt} + VCR\left[\frac{T'}{TW} + \frac{T}{TW(\gamma-1)} + \frac{1}{bTW}\ln\left(\frac{\gamma-1}{\gamma-1}\right)\right]\frac{dp}{dt} + \frac{dQht}{dt} \tag{2}$$

Radiation heat transfer is ignored.

FUEL MASS BURNING RATE

The routine used was developed by Krieger and Borman[3] and is as follows:-

u=u(T,p,Φ) R=R(t,pΦ)

also

$$\Phi = \Phi_0 + \left(\frac{m}{m_0} - 1\right)\frac{1+(F/A)_0}{(F/A)_s} \qquad (3)$$

and $\frac{d\Phi}{dt} = \frac{1+(F/A)_0}{(F/A)_s m_o}\frac{dm}{dt}$

Suffix 'o' denotes initial value prior to fuel injection ,and 's' is the stoichiometric value. It then follows that

$$\frac{1}{m}\frac{dm}{dt} = \frac{-(RT/V)(dV/dt)-(\partial u/\partial p)(dp/dt)+(1/m)(dQ/dt)-CB}{u - hf + D(\partial u/\partial\Phi)-C[1+(D/R)(\partial R/\partial\Phi)]}$$

where

$$B = \frac{1}{p}\frac{dp}{dt} - \frac{1}{R}\frac{\partial R}{\partial p}\frac{dp}{dt} + \frac{1}{V}\frac{dV}{dt} \qquad C = \frac{T(\partial u/\partial T)}{1+(T/R)(\partial R/\partial T)} \qquad D = \frac{[1+(F/A)_o]m}{(F/A)_s m_o}$$

FUEL INJECTION

The accurate prediction of the fuel behaviour within the injection system requires a sophisticated model. For this model the fuel injected is based on quasi steady, one dimensional and incompressible flow through a nozzle.

Mass of fuel injected (mf) is $m_f = C_D A_n \sqrt{2\rho\Delta p}\,\frac{\Delta\Theta}{360N}$ (4)

where Θ is nozzle open period in degrees, N is engine speed in revolutions per minute. The duration of injection and fuel line pressure was determined from pressure traces taken from the test engine.

FUEL SPRAY PENETRATION

Fuel penetration into quiescent air in the combustion chamber (S), uses the formula developed by Dent[4]. Lower penetration rates produce increased quantities of soot from insufficient air utilisation, whilst higher penetration rates give higher values of unburnt hydrocarbons from the crevices and wall impingement.

$$S = 3.07\left(\frac{\Delta p}{\rho_g}\right)^{1/4}(td_n)\left(\frac{294}{T_g}\right)^{1/4} \qquad (5)$$

FLOW INTO THE 2ND LAND AREA

The flow into the second land is based upon the conservation of energy for a fluid flowing through an orifice; the formula is as follows:

$$V_2 = Cd\sqrt{\frac{2\Delta P}{\left(\left(\frac{A_2}{A_1}\right)^2 - 1\right)\rho}} \qquad (6)$$

THE MOLECULAR WEIGHT OF THE FUEL AIR MIXTURE

Due to the fuel being injected during the cycle the properties of the contained gas change. To determine the mass of hydrocarbons in any one area the number of moles of each compound flowing in or out needs to be obtained.

The ideal gas law is used to determine the number of moles of fuel and air present at each program step, normally 1 degree.

$$m = \frac{pVM}{RT} \qquad (7)$$

For use in the crevice flow component of the model equation 7 is differentiated to give,

$$\frac{dm}{d\Theta} = \frac{VM}{RT}\frac{dp}{d\Theta} \qquad (8)$$

The integral of equation 8 is

$$dm = \int_{p2}^{p1} VMdp$$

thus the incremental mass change per degree is

$$am = \frac{VM}{RT}(p_1 - p_2)$$

to relate this to time the engine speed is included

$$dm = \frac{VM(6N)}{R\ T}(p_1 - p_2)$$

POLLUTANT MODELLING

Combustion has been modelled using the approach suggested by Whitehouse[5]. This model is comprised of two parts: the first based on the Arrhenius equation predicts the fuel burning rate and the second predicts the fuel preparation rate.

$$FBR = \frac{K}{N} \times \frac{PPO2}{\sqrt{T}} \exp\left(\frac{-Ea}{T}\right) (FPR - FBR')d\Theta \qquad (9)$$

where

$$FPR = K' \times mi^{(1-x)} \times mu^{x} \times (PPO2)^{z} \qquad (10)$$

The model utilises calibration values for the fuel burn and preparation rates within those discussed by Benson [(6)].

$$0.01 < x < 1$$
$$z = 0.4$$
$$0.008 < K < 0.020 \ (bar^{-2})$$
$$1.2 \times 10^{10} < K' < 65 \times 10^{10} \ (K^{1/2}/bar\ s)$$
$$Ea = 1.5 \times 10^{4} \ (K)$$

Within this equation the term (FPR - FBR)dΘ is equal to the quantity of fuel within the cylinder that has been prepared but not yet burnt. For each step the value of FBR' is calculated using the value of Σ(FBR - FPR)dΘ

SOOT, SOLUBLE FRACTION AND PARTICULATES

The full formation mechanism of the soot produced within the combustion chamber of diesel engines is not fully understood but the controlling factors have been identified; these are gas temperature and oxygen availability. There is no generally accepted method for the calculation of in cylinder soot levels. Due to the structural similarities between soot and pyrograhites, the oxidation rates should be similar. Studies based on the Nagle and Strickland-Constable[7] oxidation mechanism have shown a reasonable correlation for temperatures between 1100 and 2500 °K, provided that the oxygen partial pressure is below one atmosphere. The operating fundamentals of this mechanism are based on the theory that as oxidation progresses, the sites available on the carbon for oxygenation change from the reactive type A sites to the less reactive type B sites. Conditions found within the combustion chamber of a diesel engine are not within the range that has been proved accurate for this mechanism, where the partial pressure of oxygen is in the region of several atmospheres and the temperature attains a peak of approximately 2800°K.

The soot formation mechanism used within the model has been developed by Nishida[8], and is as follows.

$$\frac{dms}{d\Theta} = \frac{dmsf}{d\Theta} - \frac{dmsc}{d\Theta} \tag{11}$$

where

$$\frac{dmsf}{d\Theta} = Asf \times mfv \times P_{cyl}^{0.5} \times \exp(-6313/T) \tag{12}$$

and

$$\frac{dmsc}{d\Theta} = ASC \times ms \times PPO2 \times P_{cyl}^{1.8} \times \exp(-7070/T) \tag{13}$$

The results from this equation produce a soot concentration curve for the complete cylinder, which has the same form as a soot concentration curve using the global average for a three dimensional model. Soot formation is a direct function of fuel vapour within the cylinder and the injection timing and duration can influence the total quantity considerably.

The soot content that enters the atmosphere is assumed to be the same as the concentration within the cylinder when the exhaust valve opens. This soot mass concentration can be converted to other units such as Bosch smoke units. The correlation used within the model was derived by Matsui[9]:

$$BSU = 40.03231\rho_s - 221.9747\rho_s^2 + 819.641\rho_s^3 - 1758.072\rho_s^4 + 2195.714\rho_s^5 - 1574.006\rho_s^6 + 599.589\rho_s^7 - 93.9501\rho_s^8 \tag{14}$$

The soluble organic fraction is related to the unburnt hydrocarbon(uHC) mass concentration within the cylinder when the exhaust valve opens, and includes :-

1) Fuel from regions where the concentration was not sufficient to promote combustion.
2) Fuel from regions where the temperature was not sufficient to allow combustion to complete.
3) Fuel from locally over rich areas.
4) Fuel adsorbed and desorbed from the lubricating oil film.
5) Fuel entering the crevice volumes and emerging after flame extinction.
6) Lubricating oil consumption.

The contribution to the uHC from items 1,2 and 3 are derived from the combustion model. The adsorbed/desorbed fuel has been obtained by a method developed by Korematsu[10] : Figure 1 shows the change in adsorbed/desorbed matter at 1600 RPM, 5.72 bar BMEP caused by changing the oil film thickness from 5μm to 4μm. Fuel starts to be injected at 12° BTDC. It is absorbed into the oil film, reaching a maximum of 3 to 4 μg depending on the thickness of the oil layer at the top of the liner. The fuel is then desorbed from the oil and released into the combustion chamber. It is unlikely to be burned and so contributes to the uHC emitted. Figure 2 shows the effect of engine speed and cylinder pressure.

OIL CONSUMPTION

The oil consumption of the test engine was measured by the drain and weigh method after running at several steady state test points. These results were compared to the values from the model, which uses the puddle theory for oil consumption developed by Hoult[(11)]. The majority of the oil entering the combustion chamber is not fully oxidised and contributes to the SOF.

$$\text{Oil consumption} = 3\times10^{4} \times RPM \times \rho h_i h^{*} A_{ref} A^{*} \quad \text{(g/hr)} \qquad (15)$$

Where

$$A^{*} = 15.28(Ta)^{-2/3}$$

$$h^{*} = 1.30(U_a t\max)(Ta)1/3\left(\frac{\mu a}{\mu}\right)+0.61$$

Also

$$U'_a = \frac{Qb_{\max}}{\pi d_2 (r_o - g/2)} \ln\left(\frac{r_o}{ga/2}\right)$$

RESULTS

In cylinder hydrocarbon mass

The quantity of hydrocarbon remaining within the cylinder boundary when the exhaust valve opens is assumed to be that detected in the exhaust system. The concentrations for 1600 and 2500 RPM / 8.59 bar BMEP are shown in figure 3; the fuel injection duration remained constant (15°) but the peak injection pressure increased from 250 to 350 atmospheres. The calculated delay period before combustion did not change by greater than 0.5°, the resolution of the model. The increase in concentration after 8° ATDC (peak cylinder pressure) is primarily due to the crevice and oil effects. The concentration of fuel molecules in the cylinder crevices for 2500 RPM 8.52 bar BMEP are shown in figure 4. Fuel molecules that pass the first ring and enter the 2nd land remain at a higher pressure longer than that of the other crevices, and start to return to the cylinder later in the cycle when the cylinder gas temperature is lower, and has a greater possibility of not being oxidised. C.I. engines have large temperature gradients within the combustion chamber and the gases from the crevice and oil enter the cylinder in the relatively cool areas. End of in cylinder oxidation is determined by the temperature of sudden freezing[12] which is a function of cylinder boundary and the gas temperature.

Crevice mass

Doubling the 2nd land volume reduced the observed pressure by one half, but this reduction in pressure allowed the gases to flow into this region for a longer period at the same mass flow rate. This increase in time for the flow allowed a greater mass of gas to accumulate in this region, before being released later in the cycle than the gas from a standard piston. The entrained mass was slightly more than double that of the standard piston.

In cylinder soot

The main formation and degradation processes occur within a period slightly longer in duration than the injection period and the peak soot concentration occurs approximately at the end of injection. The burn up rate for the in cylinder soot is dependent upon the gas temperature and the available oxygen. The higher injection rate that occurs at 2500 RPM produces a lower peak level of carbon compared to that emitted into the exhaust system (fig 5). The peak in cylinder soot concentration is approximately double that of 1600 RPM 8.59 bar but the emitted matter is a factor of four greater.

Conclusions

The model results compare favourably with those obtained experimentally (fig 6), the model being suitably sensitive to all engine and operating parameters and predicting all trends correctly to present theories relating to engine combustion and pollutant formation mechanisms. The experimental results were collected for differing speed and load conditions using a FID for the gas phase hydrocarbons and a dilution tunnel for the particulate matter, CO and NOx were also monitored during the testing. It is suitably flexible in allowing other combustion chamber configurations to be analysed. The model shows that the majority of unburned hydrocarbon emissions are derived from the fuel spray and oil consumption. Crevice volumes and oil film thickness contribute to a lesser degree.

ACKNOWLEDGEMENTS

The authors of this report would like to thank AEGoetze for the funding and technical assistance contributed during this research program.

REFERENCES

(1) Willcock, M., Foss P.W. and Tidmarsh D.H.: "Modelling the optimal design for a low emission piston system". ISATA 1994 paper 94EN019, pp243-250.

(2) Heywood J.B.:"Internal CombustionEngine Fundamentals".McGraw-Hill.ISBN 0-07-100499-8.

(3) Krieger, R.B., and Borman, G.L.: "Thecomputation of apparent heat release for internal combustion engines," ASME paper 66-WA/DGP-4, in Proc. Diesel gas Power, ASME, 1966.

(4) Dent J.C.: Basis for the comparison ofvarious experimental methods for studying spray penetration". SAE paper 710571,SAE Trans., Vol 80, 1971.

(5) Whitehouse N.D. and Way R.: "Rate of heat release in diesel engines and its correlation with fuel injection data", Proc Instn Mech. Engrs, vol 184, pt 3,Paper 1, pp 17 - 27 1969-1970.

6) Benson R.S. and Whitehouse N.D. (1979). "Internal Combustion Engines", vols 1 and 2, Pergamon Oxford.

(7) Park C., and Appleton J. P.: "Shock tube measurements of soot oxidation rates." Combustion Flame, vol 20, pp 369-379, 1973.

(8) Nishida K. and Hiroyasu H. "Simplified three dimensional modelling of mixture formation and combustion in a DI diesel engine". SAE 890269, 1989.

(9) Matsui Y., Kanimoto T. and Matsuoka S. " Formation and oxidation processes of soot particles in a DI diesel engine - An experimental study via the two colour method". SAE 820464, SAE Trans., vol 91, 1982.

(10) Korematsu,K: "Effects of fuel absorbed in oil film on unburnt hydrocarbon emissions from spark ignition engines". JapaneseSociety of Mechanical Engineers. 1990.Vol. 33. No.3 pp 606-614.

(11) Hoult D.P. and Shaw B.T.: "The puddle theory of oil consumption".S.T.L.E. Tribology Transactions, 1994, Vol 37, pp 75-82.

Predicted Adsorption / Desorption
Standard Piston, Injection starts @348°
The effect of oil film thickness change

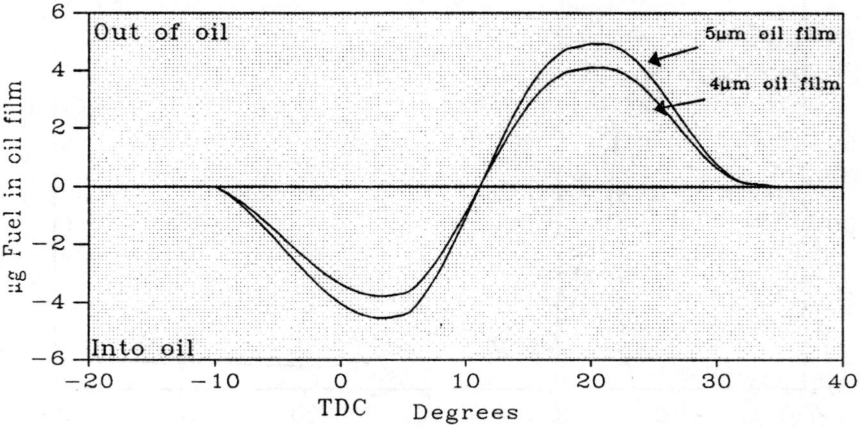

Perkins 4 litre direct injection diesel   Figure 1

Predicted Adsorption / Desorption
Injection starts @ 12°BTDC

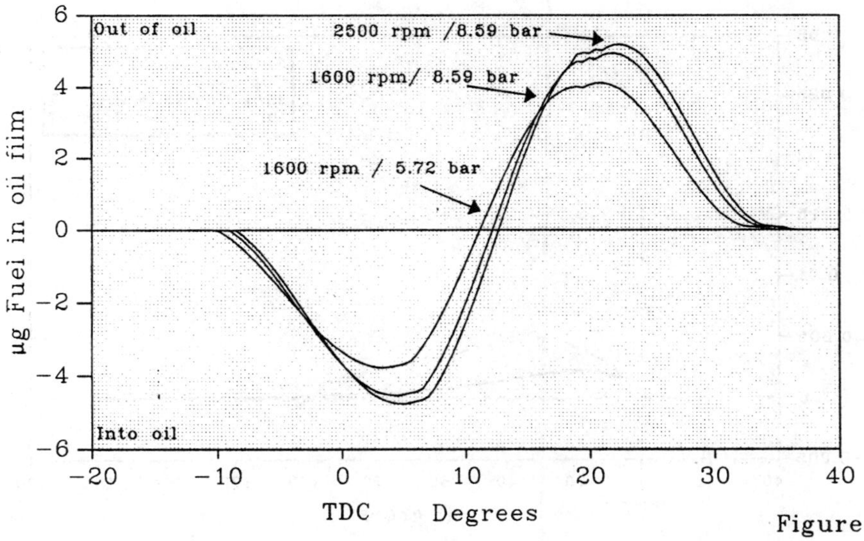

Figure 2

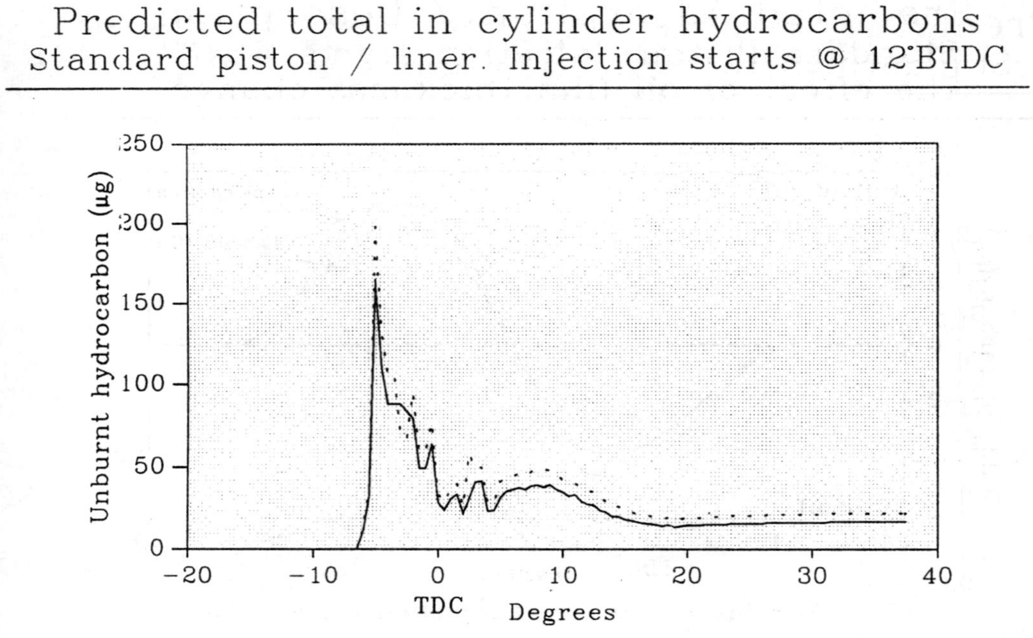

Figure 3

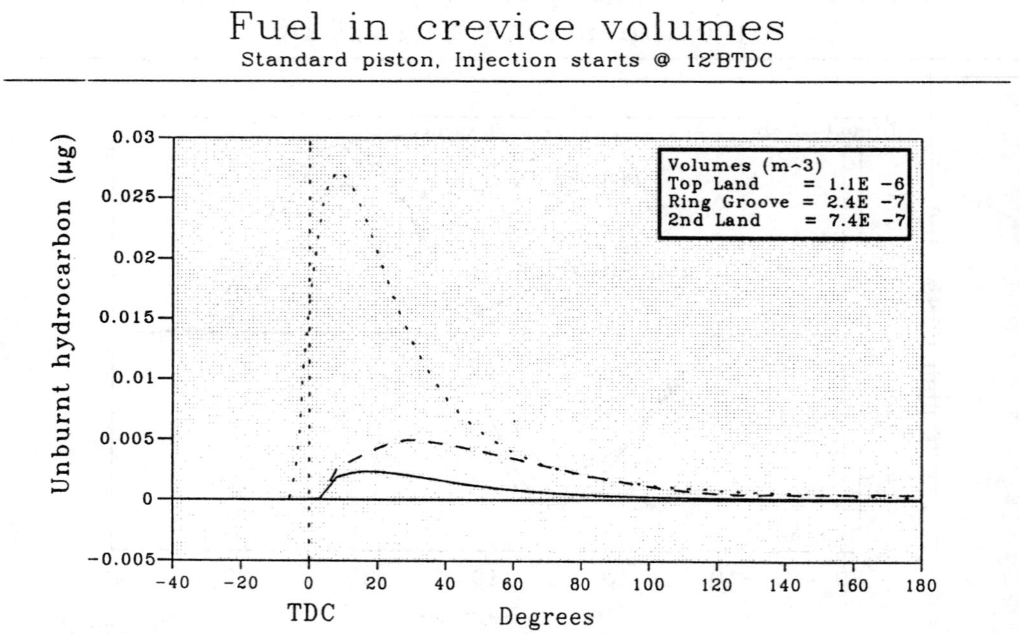

Figure 4

Predicted in cylinder soot concentration
Standard piston. Injection starts @ 12°BTDC

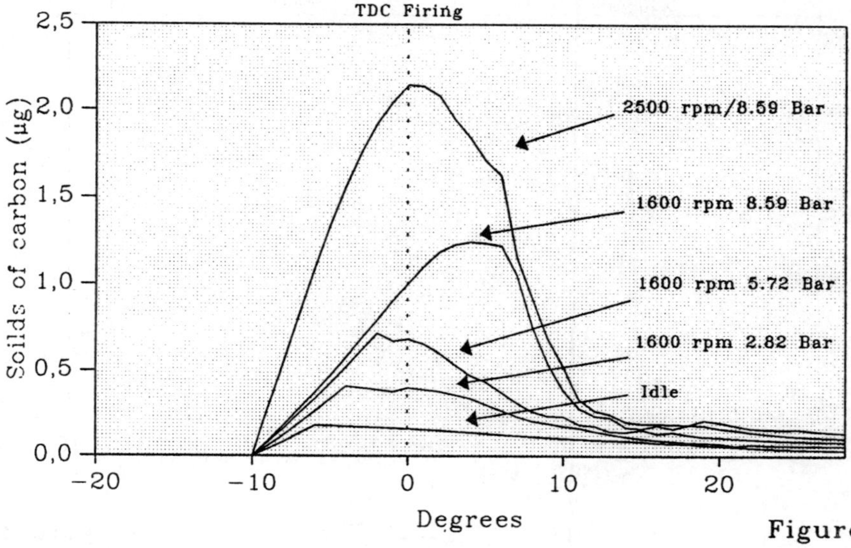

Figure 5

uHC Emissions

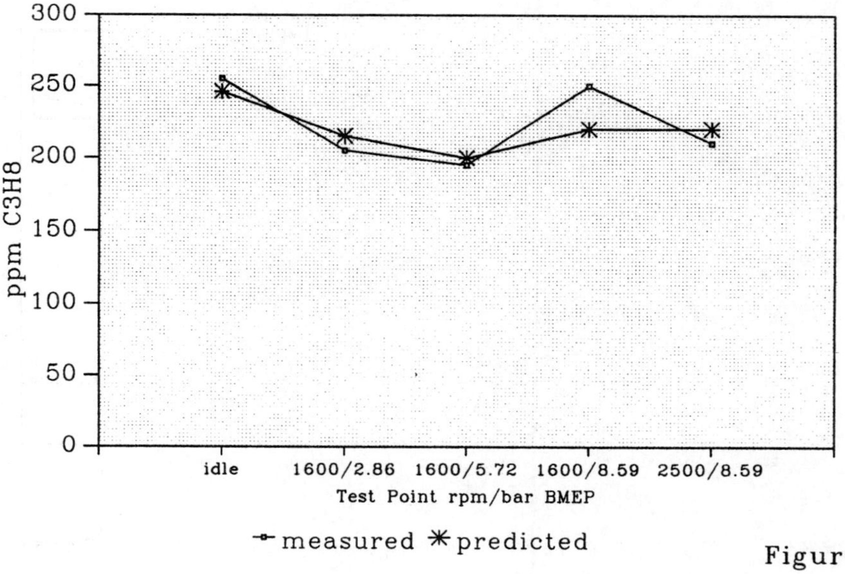

Figure 6

Engine Developments

C498/15/169/95

Combustion stability improvements through the use of port mounted barrel throttles

R F DUCKWORTH and **D PATES**
Cosworth Engineering, Northampton, UK

SYNOPSIS

Using barrel throttles in the intake port of a spark ignition engine, high overlap valve events have been employed to give improved performance through increased volumetric efficiency whilst maintaining excellent combustion stability at low loads. Mixture motion generated by the throttles plays an important role in achieving excellent idle stability and part load burn rates.

1 INTRODUCTION

In order to extract high performance from a naturally aspirated four stroke spark ignition engine, maximisation of the air charge is required. One way of achieving this is to use increased duration valve events. However for road engine applications this can be problematical because the increased intake valve opening duration leads to high overlap between inlet opening and exhaust closing, which in turn leads to charge dilution, especially at low engine loads. This dilution of the intake charge by the exhaust residuals leads to poor idle stability. A well known method of reducing these dilution effects is to use individual port throttling in combination with high overlap valve events.

2 PORT THROTTLING

As described in Newman et al.[1], in a conventional plenum throttled engine at part load the low port pressure at inlet valve opening means that gas from the cylinder will be drawn into the port and manifold and as the exhaust valve is still open, and the exhaust port at atmospheric pressure, exhaust gas is drawn back into the cylinder thereby diluting the following inlet air charge. It is this effect on a conventional engine that makes the combustion stability deteriorate rapidly with increasing valve event overlap.

With port throttles, due to the small volume between valve and throttle when the intake valve closes, the port pressure recovers to atmospheric pressure so that when the intake valve opens again there is no initial pressure gradient to create the reverse flows.

2.1 Barrel throttles

When Cosworth designed and built its own concept engine, a 2.5 l V6 designated the MBA, high specific performance was required and hence the consideration of port throttles.

The use of conventional butterfly throttles was considered difficult due to the requirement for small volume between the throttle and intake valves. For the preferred port design the distance from the two intake valves to a possible circular cross-section would not meet the low port volume requirements. With butterfly throttles each plate must be assembled individually to the shaft in situ. leading to difficulty in balancing individual port flows.

The barrel throttles employed allow placement in the oval section of the intake ports giving low throttle to valve volume. By the use of a single rotating barrel for all cylinders on each bank, with machined, matching port and throttle detail, there is no need for adjustment between individual cylinder throttles. Another advantage over the butterfly throttles is that when fully open there is no pressure drop incurred across the throttle which is likely due to the high speed nature of the flow in the intake port. Figure 1 shows a section through the port and barrel throttle.

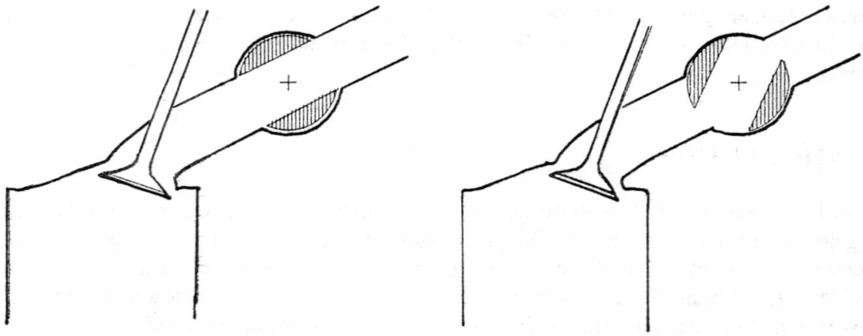

Figure 1 - Barrel throttle layout in port

3 IDLE / LOW LOAD STABILITY

Initial tests were conducted on the engine using the valve events indicated in Figure 2. The event overlap of 60 crankshaft degrees is 40 degrees greater than is commonly used on conventional road car engines.

The engine was run at a range of low speed, low load conditions to represent a section of the possible idle conditions seen in a vehicle where an automatic transmission can put the engine under variable load depending on mode and temperature.

The results of these tests are shown in Figure 3 against a similar range of loads for a conventional plenum throttled V6 engine, fitted with standard camshafts employing the more usual 20 degrees of event overlap. The marked improvement in standard deviation of i.m.e.p. (sdimep) of the port throttled engine is clearly demonstrated despite the far greater overlap period.

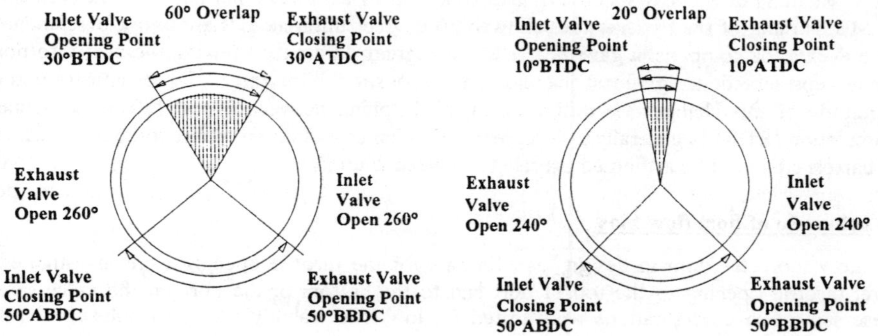

Figure 2 - Long duration valve events shown against conventional valve timing.

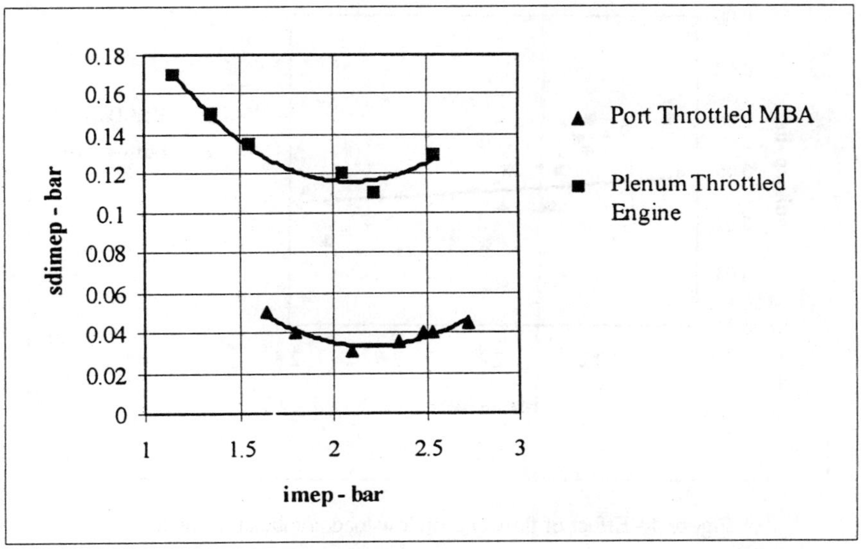

Figure 3 - Idle stability at 700revs/min

The results shown above, despite the increased overlap over that of the standard engine, may have been due to the port throttling reducing dilution of the inlet charge to a level below the standard case but another mechanism was proffered and later proved; the barrel throttles induce greater mixture motion in the cylinder than that present in the original port.

Initial work on the MBA engine compared high overlap with barrels against low overlap with plenum throttling but the latter was achieved using the same camshafts and therefore intake valve closing point was 20 degrees later and exhaust opening 20 degrees earlier. [2]

Running at the low speed and load condition of 1500 revs/min and 2.62 bar b.m.e.p. the MBT timing of the two versions showed 20 degrees difference. Later work on standard valve events for comparison gave nearer to 10 degrees difference showing the port throttled engine represented a significant increase in combustion burn rate. To demonstrate the magnitude of this change, the difference in MBT timing between 0 and 18% exhaust gas recirculation (EGR) is generally 3-4 degrees. This indicates that the effect, on burn rates, of the barrels could not be attributed entirely to reduced dilution.

3.1 Influence of port flow bias

To allow more freedom in design, two formats for the throttle opening were investigated, barrel throttle opening to the top of port and to the bottom of the port on the valve side. These same two configurations were tested for low load stability and the results shown in Figure 4 show that there is very little difference between the wall attached flow to the bottom or top of the port.

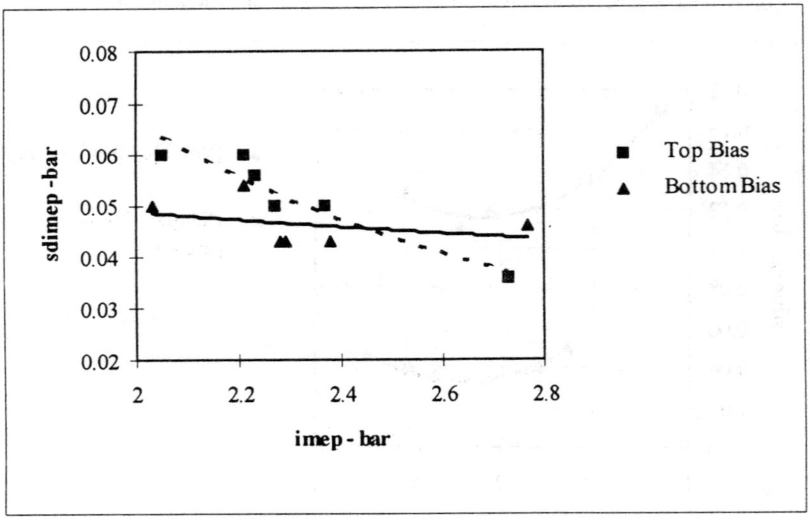

Figure 4 - Effect of flow bias on low load combustion stability.

To understand why there is little difference further work was undertaken to measure what mixture motion was created by the various configurations.

4 FUEL ECONOMY

Running the barrel throttled engine at the 1500 revs/min 2.62 bar b.m.e.p. condition although showing increased burn rates and consequential decrease in MBT spark advance, showed only a small (1.5%) improvement in fuel consumption over the standard port and valve event configuration (350 against 355 g/kWhr). However the benefit of the port throttling was clearly shown at the lower load condition of 2 bar i.m.e.p. at the same speed. At this condition MBT spark advance was again 12 degrees different but at this point fuel consumption was improved from 642 to 601 g/kWhr, a 6.5% improvement. It can be expected, due to reduced pumping work, that the lower the plenum pressure is on the conventional plenum throttled engine the greater fuel economy benefit that will be seen with the port throttles. These improvements are not of the same magnitude as those seen by Baker and Nightingale [3] despite the port to cylinder volume being 16% which is sufficiently low to enable full pressure recovery in the ports (see Newman et al. [1]).

5 AIR MOTION TESTS

The increased burn rates associated with the barrel throttles was thought to be due to the high speed jet created on the port side of the throttle. This jet attaches to the port wall and is maintained until it passes over the valve, in the case of the four valve combustion system this aids the tumbling motion created during the induction stroke.

Tests were conducted to measure tumble motion on a steady-state airflow rig and this work compared to a simple two dimensional model of the system to visualise the effects.

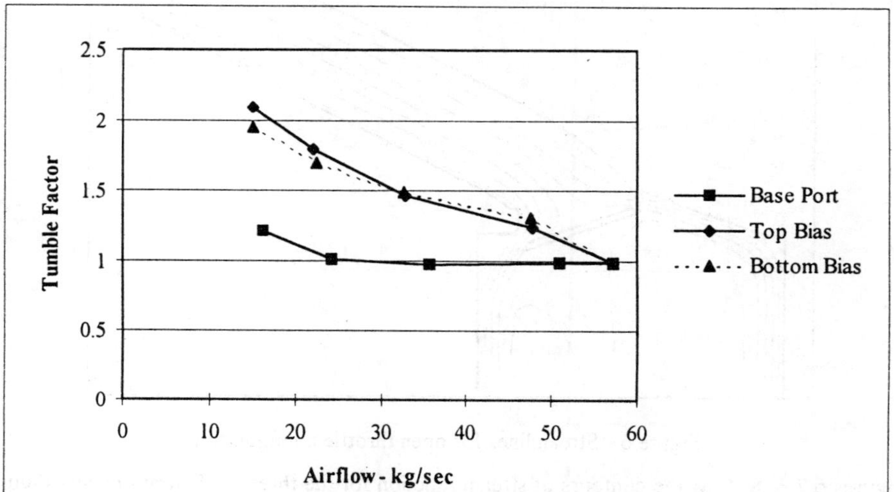

Figure 5 - Influence of airflow bias on non-dimensional rig tumble.

To compare the air motions created in the cylinder the cylinder head is mounted on a flow bench with a tumble adapter. With fixed full valve lift the airflow is increased on the bare port by increasing pressure drop, and on the throttled port by opening the barrels at a fixed pressure drop. The results in Figure 5 show that at low flows the tumble created by the barrel throttles is approximately twice that of the bare port but as the throttles are opened further, to achieve higher flows, the difference becomes less. Also shown is that there is little or no difference between bottom and top biased throttle opening which explains the similarity, in combustion stability, of these two configurations .
[Tumble Factor is a value derived from dividing tumble rig torque by mass flow times volume flow]

6 TWO-DIMENSIONAL FLOW SIMULATION

To gain a better understanding of how the air motion in the intake port is affected by the barrel throttle mechanism a two dimensional model has been used. Using the boundary conditions of a known mass flow into the port and pressure upstream of the barrel being atmospheric three different situations were modelled, conventional port (or barrel fully open), throttle opening to the bottom of the port and to the top. To set the boundary conditions for the conventional port situation the same mass flow was used but the in-cylinder conditions set from the results of the port throttled simulation. The simulation has been done at mass flows equating to low speed and load - 1500 rpm and 2.5 bar b.m.e.p

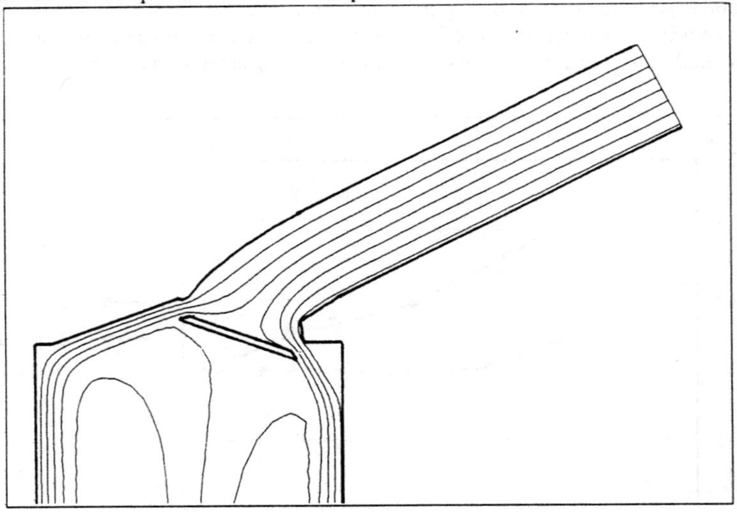

Figure 6 - Streamlines for open throttle configuration

Figures 6,7 & 8 show the contours of stream function for the three configurations and Figure 9 shows the velocity distribution across the cylinder bore at a plane 30mm below the cylinder head face. This distribution gives an indication of how the tumble motion is formed. The flow bench and tumble adapter can only measure the tumble as a single rotating mass air movement so the tumble of this type formed by the open throttle configuration is limited because the velocity distribution across the cylinder is almost even thereby cancelling out the bulk tumbling motion.

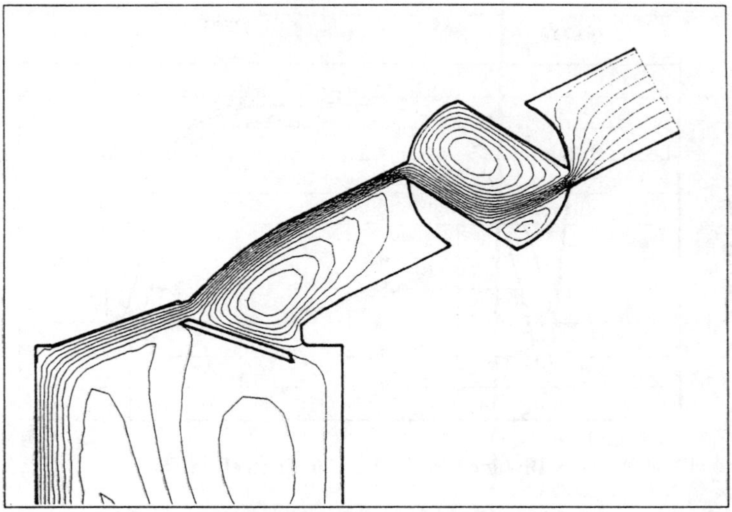

Figure 7 - Streamlines for top bias throttle

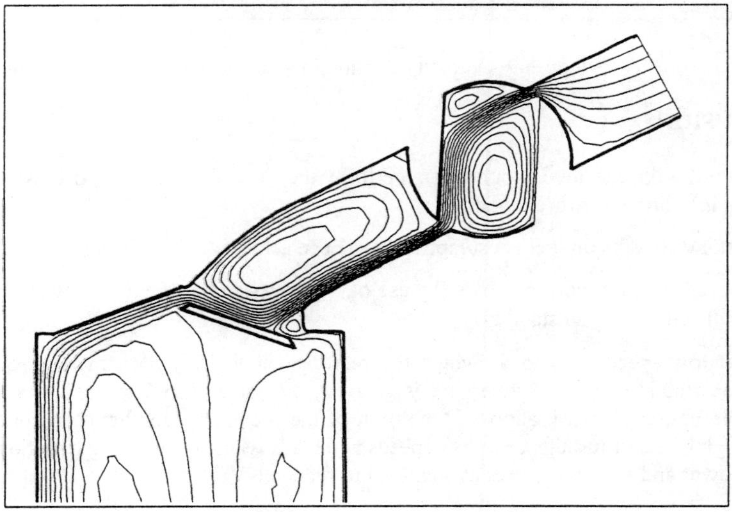

Figure 8 - Streamlines for bottom bias throttle

The configurations of top and bottom opening bias show only small differences but have a greater velocity, and thereby mass flow, down the cylinder wall on the exhaust side of the cylinder than down the intake side which will result in a bulk rotational tumbling motion. The boundary conditions set are similar to the lowest range of the flowbench test data where the barrel throttles show twice the tumble of the open port, the wall attached flow is clearly demonstrated and the effect that the high speed jet has on the resultant in-cylinder motion is clear.

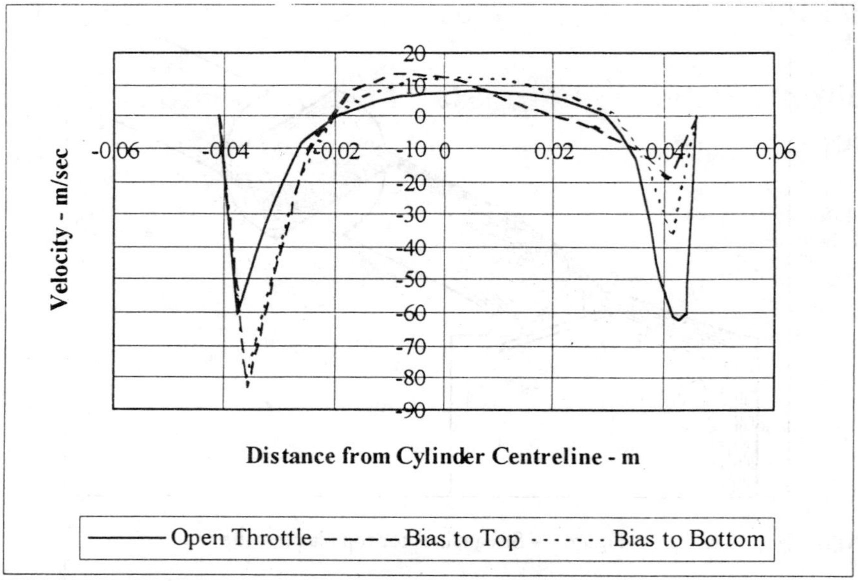

Figure 9 - Velocity distribution across cylinder

CONCLUSIONS

Barrel throttles in the intake ports can perform the dual function of port throttling and improving in cylinder mixture motion.

Port throttling benefits on fuel consumption have been achieved with this design.

The improved mixture motion allows the use of high overlap valve events whilst maintaining excellent idle and low load stability.

As it is at low speeds and loads where the requirement for fast burn rates to give good combustion stability and EGR tolerance is greatest, the barrel throttles fitted to a high flow, low tumble design port will allow optimisation of the previously conflicting requirements to provide high levels of turbulence at low speeds and loads whilst maintaining high flow for high specific power and slow burn rates at high load to reduce NVH.

REFERENCES

[1] NEWMAN, C.E., STEIN, R.A., WARREN, C.C. and DAVIS, C.G. The Effects of Load Control with Port Throttling at Idle - Measurements and Analyses, SAE Paper 890679

[2] POOLE, C.G., HANCOCK, R.J. and CAIRNS, D.C. Cosworth MBA Engine, SAE Paper 920849

[3] BAKER, T.G. and NIGHTINGALE, C.J.E. Port Throttles Applied to a High Performance Four-Valve S.I. Engine, IMechE Paper C448/032

C498/15/207/95

Requirements and potentials of future valve train concepts

K-H OEHLING, R TEICHMANN, and **H UNGER**
BMW AG, Germany

SYNOPSIS

Apart from future emission standards of general nature in the USA and the European Union, the reduction of CO_2 emissions represents a great challenge for the years to come. Focusing on the various options for improving engine efficiency, we see that the valve train offers a great potential for reducing both fuel consumption and CO_2 emissions.
This report presents and discusses current developments and future prospects offered by variable valve train systems. It also describes the potential improvements which may be made in terms of torque, exhaust emissions, and fuel consumption, finally providing a cost:benefit analysis.
An analysis of losses suffered with the spark-ignition engine enables us to quantify the individual loss shares. Friction and pumping losses as well as thermodynamically induced losses represent a substantial potential for improvement.

One of the general development aims with modern valve train systems is to reduce mechanical losses by way of low masses or low spring forces, or by way of roll friction. The first step into variable valve control is the use of phase adjusters (VANOS systems) determining the volume of exhaust gas recirculation and improving full load running conditions. Maximum variability is provided by systems with unthrottled load control substantially reducing pumping losses.

1. Introduction

Increasing environmental awareness, legal standards, customer demands, and voluntary promises made by the German automobile industry to reduce fuel consumption, all add up in making even greater requirements of the automobile. Apart from even stricter general emission standards to be applied in future both in the USA and the European Union, the reduction in carbon dioxide emissions represents a great challenge.

Stricter legal standards are currently being prepared in many of the world's major markets, the objective being to considerably reduce carbon dioxide emissions and, as a result, fuel consumption. Fig 1 shows how carbon dioxide emissions, the mileage covered, and fuel consumption have developed in German motor transport over the years [1].

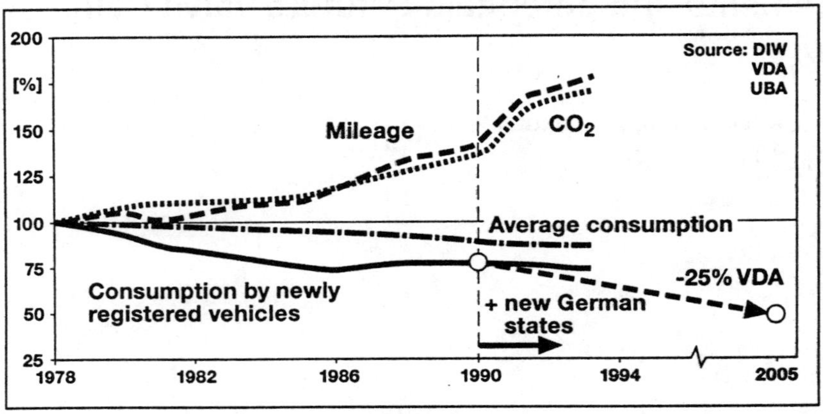

Fig 1: CO_2 emissions, mileage, and fuel consumption

The average mileage covered by motorists in Germany is increasing by approximately 4 per cent each year. The result is a steady increase in CO_2 emissions from road transport, although average fleet consumption is constantly decreasing.
In 1990 VDA, the German Association of Automobile Manufacturers, promised voluntarily to reduce fuel consumption up to the year 2005 by 25 per cent. In March of this year this commitment was then expressed again by the automobile industry in a statement to the Federal Ministry of the Environment and Transport.
Vehicle-specific measures alone are however not sufficient to compensate for the increase in CO_2 caused by increasing mileage in transport today. Accordingly, other measures such as traffic management, the use of alternative fuels, or fiscal incentives are essential.

2. Basic potentials for the further reduction of fuel consumption

An analysis of the process losses sustained with the spark-ignition engine reveals the potentials available to improve this process in its entirety. Referring to an existing four-valve power unit, Fig 2 presents an overview of both energy and loss shares with relation to the fuel consumed.

The valve train may be modified to directly influence the losses suffered in gas exchange and due to friction in the cylinder head, while indirectly - via the quality of the combustion process - influencing the amount of heat lost through the cylinder walls. In all, these losses represent a significant factor in comparison with the energy consumed. In other words, any reduction of such losses translates directly into an improvement in efficiency, meaning that for this reason alone it is highly worthwhile to focus on these items in engine development.

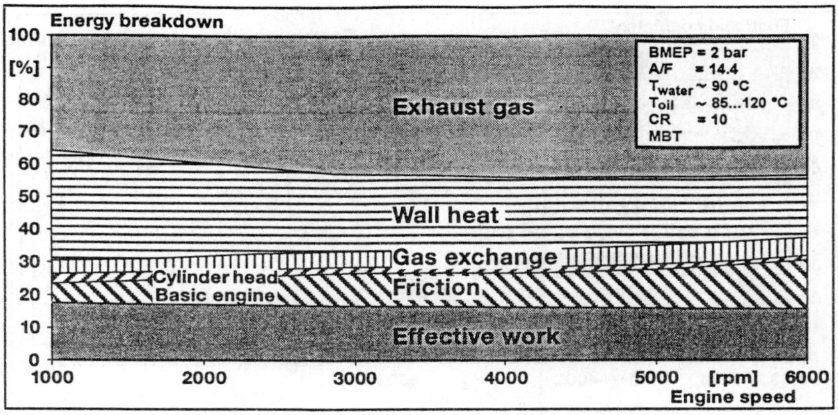

Fig 2: Energy breakdown with a four-valve power unit

3. Possibilities and potentials of future valve train concepts

The valve train concept and the degree of valve train variability are crucial to the properties of an engine. Through the introduction of multi-valve technology, the customer benefitted initially from an increase in power and performance, four-valve technology now being the usual standard in the upper performance segment with many engine manufacturers and thus providing the basis for ongoing development.

3.1 Conventional valve train systems

The primary objective when designing a valve train system is to ensure optimum gas exchange by making the entire configuration as stiff and stable as possible, in this way generating a powerful and stable valve lift curve. Further requirements are to make the system compact in design and inexpensive in production. Precisely these features are offered by bucket tappet valve trains which, on a four-valve engine, keep the combustion chamber compact and the valve angle small.

First-generation four-valve power units involved greater mechanical losses than a comparable two-valve engine. The energy breakdown of a four-valve engine presented in Fig 2 shows that cylinder head friction accounts for 2-3 per cent of the total fuel energy applied. Introducing a number of optimisation measures, for example by optimising valve train mass and spring forces, we were able on our four-valve engines to reduce frictional loss in the cylinder head by an average of 30 per cent (Fig 3 on the left), in this way reaching the same level as with a two-valve power unit.

The use of roller-type rocker arms converting hydrodynamic friction or mixed friction into roll friction (Fig 3 on the left) then serves to reduce valve train friction versus the first generation of four-valve power units by approximately 70 per cent. In all, this means a reduction in fuel consumption in the ECE combined cycle by up to 3.5 per cent. This virtually exhausts the potential reduction of friction on the cams, other factors being negligible.

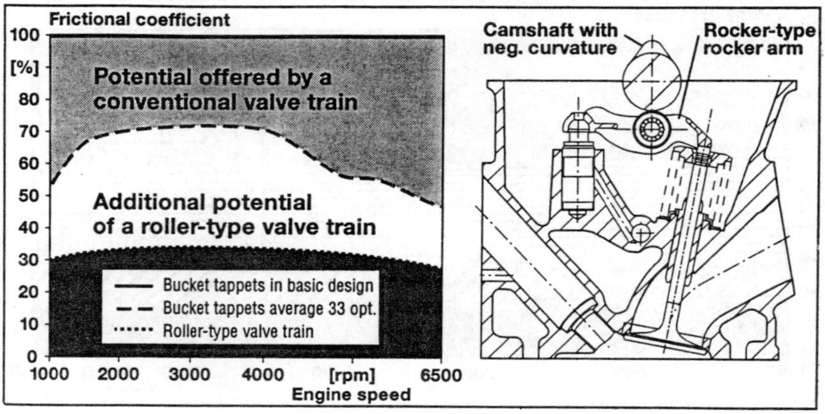

Fig 3: Potential valve train friction, valve train design and configuration

The roller-type valve train nevertheless requires extra space inside the cylinder head due to the built-in rollers. The best way to meet this requirement is to use rocker-type roller arms, a further advantage in this case being that the "hollow" cam contour can be reduced by the lever transmission. The roll friction philosophy therefore proves most worthwhile from a cost/benefit perspective if all cost factors are consistently optimised.
The objective in designing new valve train concepts, as shown in Fig 3 on the right, should therefore be to introduce a roller-type valve train as on our M43 four-cylinder and M73 12-cylinder.

3.2 Variable valve train

Fig 4 presents the basic function principles of some variable valve train control concepts examined by BMW. Starting with the function principle as such, it then describes the specific function obtained and its main effect on engine running conditions.

	Function principle	Function	Main effects
1		Variable intake and/or outlet cam spread	Emissions, full load, fuel economy, running smoothness
2	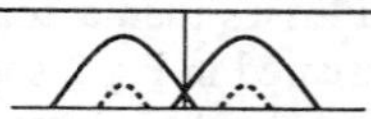	Switchable intake/outlet valve lift, cylinder deactiv.	Fuel economy (full load)
3	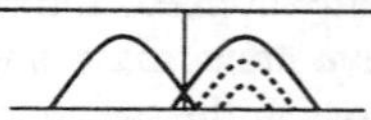	Variable valve lift (variable duration)	Fuel economy (full load)
4	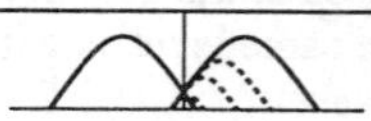	Variable intake closing (lost motion)	Fuel economy full load
5	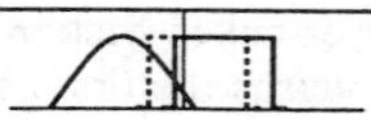	Variable intake closing, variable intake opening	Fuel economy full load

Fig 4: Basic function principles of variable valve train concepts

As a function of its phase position, VANOS valve timing on the intake and/or outlet side serves to control the engine's fuel economy, emissions, full load, and running smoothness.

Technical potentials and thermodynamic conditions

- VANOS systems

The first step towards variable valve train is a VANOS system able to adjust the phase setting of the camshaft versus the crankshaft. Simple systems of this kind allow two-point adjustment while more sophisticated concepts are able to provide infinite adjustment of the camshaft.

In principle all systems used so far are very similar in design, Fig 5 on the left presenting BMW's six-cylinder power unit with VANOS adjustment on the intake side. On the right, Fig 5 shows the overall range of adjustment with VANOS.

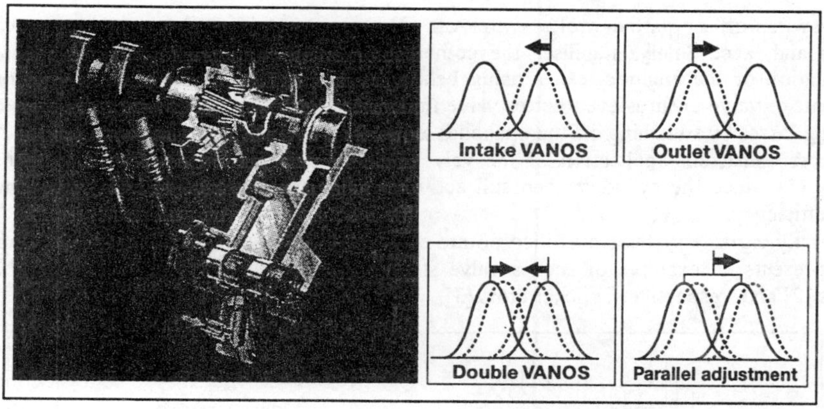

Fig 5: VANOS system on a BMW six-cylinder power unit and the various adjustment options

The potential of phase adjustment on the intake and/or outlet side depends largely on the optimum amount of residual gas to be obtained at the various operating points, the objective therefore being to find a compromise between a reduction in fuel consumption and the reduction of hydrocarbon emissions. The reduction of NOx emissions, in turn, depends on the level of residual gas at which the car may still be run.
An improvement in fuel economy will result primarily from a reduction of pumping losses and heat loss through the cylinder walls. Any deterioration of thermodynamic combustion conditions suffered in this case will be more than compensated for by the improvement in fuel economy. Contrary to the optimum configuration under part load, the objective under full load is to minimise the residual gas content in order to maximise the cylinder charge. Enhanced exhaust emission dynamics with large valve overlap will also improve the dynamic intake behaviour, a further advantage of the intake camshaft closing earlier being that air reflow losses back into the intake manifold are reduced accordingly.
With this full load configuration our power plants are able to develop a high specific torque particularly at low and medium engine speeds.

- Partly variable valve train

Another way to vary valve train is to opt for valve lift switchover, Fig 6 presenting the basic potentials of such a concept and its effect on engine running conditions in the case of a power unit with two intake valves.

	Function principle	Function	Main effects
1	E2 / E1	Partial lift on one intake valve	Faster combustion by charge motion
2	E2 / E1	Partial lift on both intake valves	Charge motion, reduction of pumping losses
3		Valve switch-off (cylinder deactivation)	Engine operating point shifted by cylinder deactivation

Fig 6: Basic principles and functions with partly variable valve train

Valve switch-off or partial-stroke switch-off (1) on one intake valve activates the charge motion and, accordingly, stabilises the combustion process and may improve exhaust gas recirculation or the engine's lean running behaviour. Additional partial valve lift (2) on the other intake valve ensures even better valve timing under part load and, accordingly, reduces pumping losses by switching off the intake lift at an earlier point.

The most consistent application of the valve switch-off concept is to deactivate cylinders entirely (3), since the cylinders then still active can be run at optimum operating points for better efficiency.

Such partly variable systems may incorporate either switchable rocker arms or bucket tappets. Fig 7 presents an example of such a valve stroke switchover concept incorporating bucket tappets [2] and, respectively, rocker arms [3].

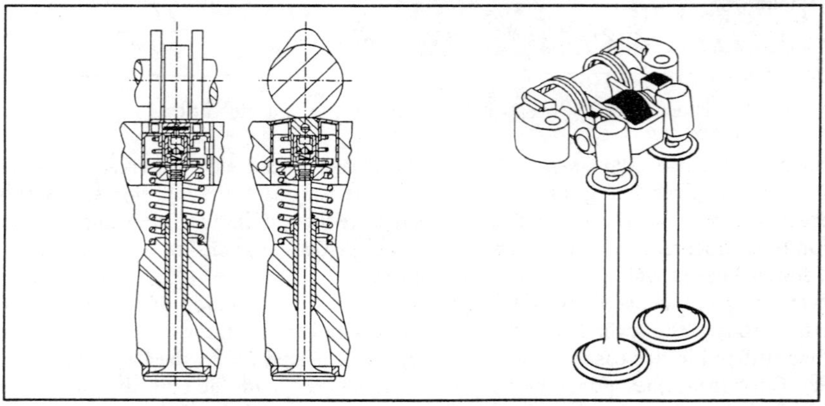

Fig 7: Switchable bucket tappets, switchable rocker arms

With the switchable rocker-type rocker arms shown in Fig 7 on the right each cam has its own arm and is switched over by a lock acting on the individual bolts and controlled by oil pressure. The switchable bucket tappets shown in Fig 7 on the left, on the other hand, feature a small tappet integrated within the bucket also incorporating valve lash compensation. This tappet is operated by a cam with a small valve lift, the outer tappet resting concentrically on the inner unit and being operated by two identical cams with conventional valve lift, positioned on either side of the small cam. Depending on whether the two tappets are locked or not, the valve will open in accordance with the position of the inner or outer cam.

The bucket tappet systems generally in use today differ mainly through the way in which the two tappets are locked to one another. On the version shown in Fig 7 this interlocking is provided by two bolts pressed apart by oil pressure.

The switchable valve train involves two fundamental problems: The first point is that the switch mechanism as such has to meet very demanding standards. The lock mechanism must operate reliably and, in particular, very quickly under all engine running conditions relevant to customer use and test purposes, meaning that following the switchover signal the valve lift must be modified very quickly indeed on all the cylinders involved. Should both intake valves operate with partial valve lift the switchover process will instantly change the cylinder charge and, accordingly, the torque generated by the engine, thus causing problems in terms of emission control and driving features. Such undesirable behaviour when the car is moving calls for special counter-measures and/or the use of EML control, which in turn eliminates part of the improvement potential offered by the system.

The second point is that since both of these systems may also be combined with VANOS control, the fundamental factors already mentioned remain in existence. The additional improvement of fuel economy compared with the sole use of a VANOS system alone results in the case of valve switch-off mainly from faster combustion on account of the improved charge motion. Thanks to better exhaust gas recirculation both pumping and heat transfer losses through the cylinder walls can be reduced. If, furthermore, both intake valves operate with part load control timing the pumping loop and, accordingly, pumping losses may be reduced by earlier closing of the intake valve.

- Fully variable valve train

Considering the problems presented by the switchable valve train, as described above, the actual target in engine development is to make the valve train system infinitely variable. Once such infinite adjustment covers all engine operating points from idling all the way to full load, we have - at least in principle - an unlimited option to use the engine's full potential for improvement.
The primary objective in this case is to minimise gas exchange losses, which can be done best by closing the intake stroke at an early point.

The fundamental functions of fully variable systems are presented in Fig 4 with items (3), (4), and (5). Fig 8, in turn, presents some typical examples of mechanical, hydraulic, and electromagnetic valve train.

A large number of basic concepts and options for an infinitely adjustable valve train with mechanical control are presented in patent literature [4]. Some of these concepts have already been developed to prototype stage and have proven their basic functionality, as is the case with the option shown in Fig 8 on the left [5].

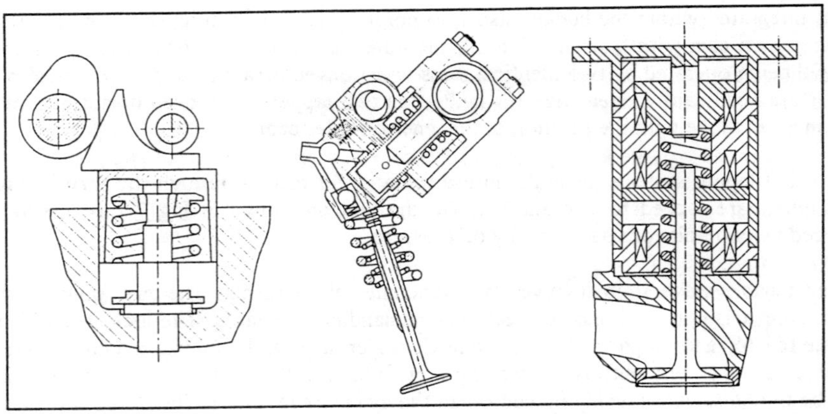

Fig 8: Various types of fully variable valve train

While such mechanical systems do not involve any general, fundamental drawbacks, there may be specific problems encountered in each case. Keeping the opening period shorter, mechanical systems are also able to provide the requisite valve lift, in this way improving charge motion in the combustion chamber as a result of the higher flow rate.

With the hydraulic valve train shown in Fig 8 in the middle [6], most existing systems incorporate a conventional camshaft acting on a hydraulic tappet the effective length of which may be modified by the controllable run-off cross-section. Accordingly, the valve may be closed as soon as the volume required at each respective load point has flowed into the cylinder, the theoretical pumping loop potential being achieved by closing the intake valve at an earlier point. Since, however, the energy required for opening the valves cannot be recycled when the valves are closed, as is the case with a conventional camshaft, the potential of such a system is considerably reduced by the high amount of power taken up by the camshaft. A further point is that with systems of this kind the closing of the valves is not verified by a cam, but rather by a hydraulic valve brake with the big disadvantage of being dependent of the viscosity of the oil. The result is undesirable variation in the closing behaviour of the valves, depending on the type of oil used and on oil temperature.

In the case of the electromagnetic valve train shown in Fig 8 on the right [7] the valves are driven by two springs, each valve being connected to an oscillating armature in the middle. Electric magnets then serve to compensate for friction and set the valve timing, as desired. In this context electronic control allows a very high standard of flexibility, valve opening and closing times being varied independently of each other and from one work cycle to the next. Further options are to deactivate individual valves completely or to use exotic solutions such as a six-stroke mode of operation.
Thanks to this high standard of flexibility the electromagnetic valve train provides a great number of different functions and applications able to create the right thermodynamic conditions for unthrottled load control. The potential which may be achieved in this process depends largely on the extent to which energy supply to the valve train system can be optimised. The energy required to control the valves is fundamentally the same as with a conventional valve train, but has to be generated via the alternator efficiency chain.

3.3 Assessment of variable valve control systems

The decisive factor in assessing the technical potentials offered by the valve train is the cost/benefit ratio (this applies in particular to measures taken in order to reduce fuel consumption): The larger the degree of variability required, the more modifications have to be made to accommodate the valve train concept and re-design the engine as such. Apart from the reduction of fuel consumption in the combined ECE cycle, the most important criteria for assessing variable valve train systems are the potential improvement of full load behaviour and the possible increase in the cost of production.
The starting point in any such assessment is that hydrocarbon and nitric oxide emissions must be just as good or better than with the initial configuration without variable valve control. On some systems unable to control their own internal exhaust gas recirculation, this makes it imperative to include at least an intake and/or an outlet VANOS.

An intake VANOS is able to generate 50-70 per cent of the possible full load potential in the lower to medium speed range, any further improvement under full load provided by a fully variable valve train then being reduced on account of this good starting point. The factors crucial to valve train variability, in turn, are the potential improvement of fuel economy and the additional expenditure and design modifications required for this purpose. Fig 9 provides such an overall assessment with a view to fuel economy and the cost of manufacturing a variable control system in the case of a four-cylinder power unit, without considering the further potential of cylinder deactivation assessed on an 8-cylinder engine.

Function principle	Function	Fuel economy potential*	Benefits/ costs
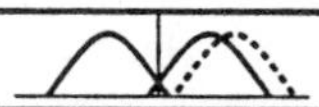	Intake VANOS	2%	O (Basis)
	Intake/outlet VANOS	3%	O
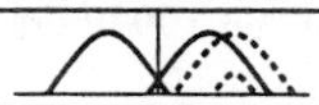	Intake/outlet VANOS, 1x intake valve lift switchover	4%	O
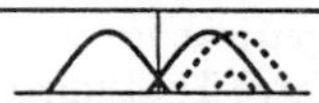	Intake/outlet VANOS, 2x intake valve lift switchover	5%	+
	Intake VANOS, cylinder deactivation	7 - 10%	+
	Outlet VANOS, variable intake close (lost motion)	7 - 9%	-
	Intake/outlet VANOS, variable valve shift	10 - 12%	+
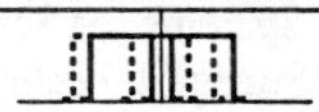	electromagnetic valve train	10 - 12%	+

* calculated/ estimated from measured results

Fig 9: Potential improvement of fuel economy and benefits/cost ratio

Looking at the benefits/cost ratio, we see first of all that cylinder deactivation, both intake valve lift switchover and fully variable systems achieve a result better than the basic system. The lost motion system, on the other hand, clearly does not qualify as a technical measure for reducing fuel consumption.
Depending on the need to reduce fuel consumption, therefore, the development engineer is able to choose a suitable system for variable valve timing in order to reach his specific target.

Such variability must of course also be considered with a view to the modifications required and any re-engineering of the basic power unit as such. Since these criteria differ substantially from one type of engine to the next, the engineer may possibly opt for a variable control system less positive at first sight, since it requires fewer design modifications.

4. Outlook

The frictional coefficient of conventional valve train systems is already reduced today by the increasing use of roller systems involving roll friction. VANOS systems, in turn, are being used increasingly by a growing number of engine manufacturers, since they have already been enhanced to production quality and offer a positive cost/benefit ratio.

Which specific concept of partial variability provided by switchable valve lift will ultimately be introduced depends on the technical potentials offered by each type of valve train system. A further point is that the specific control strategy chosen as well as the instationary behaviour of the engine determine the potential which may be achieved and, accordingly, the cost/benefit ratio.

In the long term infinite valve train adjustment will become the general technology of choice, since it offers the greatest potential for improvements and reflects the growing need to minimise fuel consumption. The general conclusion is that the valve train will take on increasing significance in the process of ongoing development. This alone provides a great potential for meeting the requirements engineers and manufacturers will face in future.

Literature

[1] Deutsches Institut für Wirtschaftsforschung (DIW), Verkehr in Zahlen 1993 (Transport in Figures 1993)

[2] Dr Ing P Tenberge, Verstellsysteme für kurven-gesteuerte Ventilgetriebe in Hubkolbenmotoren (Adjustment Systems for Curve-Controlled Valve Transmissions in Spark Ignition Engines), VDI-EKV Annual 1995

[3] W Ruffing, U Hoyer, Schwinghebel-Baugruppe mit miteinander verbindbaren Armen (A Swinging-Lever Assembly with Interconnectable Arms), patent document WO 95 / 04873

[4] W Hannibal, Vergleichende Untersuchung verschiedener variabler Ventilsteuerungen für Serien-Ottomotoren (Comparative Study of Variable Valve Train Systems for Standard Production Reciprocating-Piston Petrol Engines), doctor's thesis, Stuttgart Technical University, June 1993

[5] P Kuhn, H Schön, Vorrichtung zur Betätigung der Ventile an Verbrennungsmotoren mit veränderlicher Ventilerhebungskurve (Device for Operating Valves on Combustion Engines with a Variable Valve Lift Curve), patent document DE 3833540

[6] Q-H Vo, K-H Oehling, Untersuchungen an hydraulischen variablen Ventilsteuerungen (Examinations on Hydraulically Variable Valve Control Systems), MTZ 52 (1991), pp 622-630

[7] F Pischinger, P Kreuter, Elektromagnetisch arbeitende Stelleinrichtung (An Electro-magnetically Driven Adjustment System), patent document DE 3024109

C498/22/174/95

Development of a five valve per cylinder variant of the Ford Duratec 2.5 litre V6 engine

R SYKES
Tickford Limited, UK

SYNOPSIS

A 24 valve Vee 6 engine has been converted to a patented five valve per cylinder layout. The additional inlet valve area compensates for the inherent flow losses in very high tumble inlet ports. Efficient combustion and good EGR tolerance has been achieved without an inlet port de-activation device as used on the standard engine. The five valve head in conjunction with the freedom allowed in the inlet manifold has contributed to a high volumetric efficiency. This has resulted in an engine particularly suitable for passenger cars which require low fuel consumption and exhaust emissions with a good performance.

1. INTRODUCTION

Two valves per cylinder were the norm for many years but as piston speeds have risen the limited valve area has become restrictive on volumetric efficiency. The four valve layout has gradually been developed to the point where, in conjunction with long period cams, it can be made to provide adequate airflow even for the most highly rated engines.

So initially there seems to be no point in trying to increase the percentage of the bore occupied by valves by adoption of the three inlet, two exhaust layout which gives the largest valve area and best inlet/exhaust area ratio of all possible layouts.

However road going cars need attributes other than maximum power. They should have a good spread of torque to aid driveability, they must be as fuel efficient as possible and pass exhaust emission regulations. They should provide this with the lowest manufacturing cost and superb reliability. When such things are considered there are good reasons why an increase of valve area may be desirable.

Thermal efficiency is improved by minimising the heat loss to the metal walls surrounding the combustion gas in the cylinder. Large bores with short strokes are undesirable especially at low speeds where there is most time for the heat to escape. Thus it may be that the best all round bore and stroke combination does not provide sufficient valve area for the desired maximum engine speed and power.

At lower engine speeds Torque, Fuel efficiency and Smooth running can all be improved by employing short valve opening periods yet typically these limit the high speed breathing capability. It is well established that apparently excessive valve area can compensate and such a combination of short cam periods and big valves gives a good volumetric efficiency over the widest possible speed range without resorting to variable cam timing devices.

It is also well established that a strong motion of the incoming charge can vastly improve the combustion, particularly with regard to burning lean mixtures or those containing high percentages of recirculated exhaust gas. Whether axial swirl or tumble is employed it normally results in some reduction in flow efficiency.

So there are several reasons why a road-going engine benefits from the extra valve area provided by a five valve layout. In fact there are more good reasons for having 5 valves on a road engine than there are on a pure racing engine.

This paper presents the design and testing of a five valve per cylinder conversion of a Ford Duratec 2.544 litre Vee6 engine. The programme has shown that the desired combustion can be achieved by high tumble instead of Axial Swirl created by an Inlet Manifold Runner Control (IMRC) and the benefits in volumetric efficiency have made a significant improvement in performance.

2 BACKGROUND RESEARCH ON THE AIRFLOW BENCH.

It was known from many years of flow bench tests on a wide variety of cylinder heads and flow blocks that both the inlet port and combustion chamber shapes can significantly affect the flow and tumble.

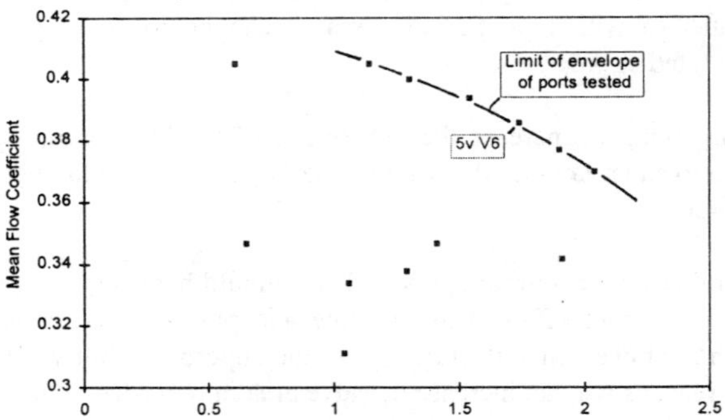

Figure 1. Chart of Flow and Tumble characteristics

The mean flow and mean tumble numbers for every port are entered on to a continually updated chart to judge the performance in relation to others. A selection of results is shown in Figure 1. where it can be seen that there are many cylinder head designs which were nowhere near the optimum on either scale. However the upper edge of the envelope confirms the logical characteristic that is as tumble increases the maximum airflow potential decreases.

A program of work was undertaken to determine some fundamental guidelines for the design of inlet ports to generate high tumble ratios with the maximum possible airflow. A single port model was made in which the valve to port angle, the intersection height and the floor to seat inner radius could be varied. The main port diameter and the valve seat remained constant with a typical area relationship between them. One of the many interesting relationships discovered was that a rapid increase in tumble occurred as the port to valve angle increased from 40 to 50 degrees.

This helped to explain why, all other things being equal, the Tickford design of 5 valve head generally has a higher tumble ratio than the more common Yamaha layout. This was an unexpected bonus from the design which started out as a means of providing a good combustion chamber shape.

3. DESIGN OF THE 5 VALVE / CYLINDER CONVERSION

A 2.54 litre Ford Duratec Vee6 engine was chosen as being representative of the new generation of compact, lightweight yet powerful engines which could power mid sized passenger cars to achieve better fuel consumption than the 3 to 4 litre engines currently in use. These new multivalve engines achieve good maximum power but inevitably have difficulty in matching the low speed torque of a larger swept volume engine.

The objectives of the design were to produce a 5 valve per cylinder variant of the engine with superior torque and equal or greater power than the standard 4 valve engine at no more than 6500 rev/min. The combustion efficiency and EGR tolerance at selected low load conditions was to be similar to that achieved by the standard engine by its use of Inlet Manifold Runner Control (IMRC).

As a rather separate requirement it was requested that the height of the cam covers above the crankshaft centreline should be reduced by approximately 40mm. The initial layouts showed that such a compact design could not be achieved whilst retaining roller finger cam followers.

Eventually it was decided to use an unusual mix of valvegear types. Mechanically adjusted inverted bucket direct attack tappets were used on four out of the five valves with a finger follower pivoted on a hydraulically adjusted post for the central inlet valve on each cylinder. This unconventional solution was adopted for this experimental engine partly to limit any start up lubrication problems and partly to provide a direct comparison between the two valve gear layouts and thus help to settle another current point of discussion.

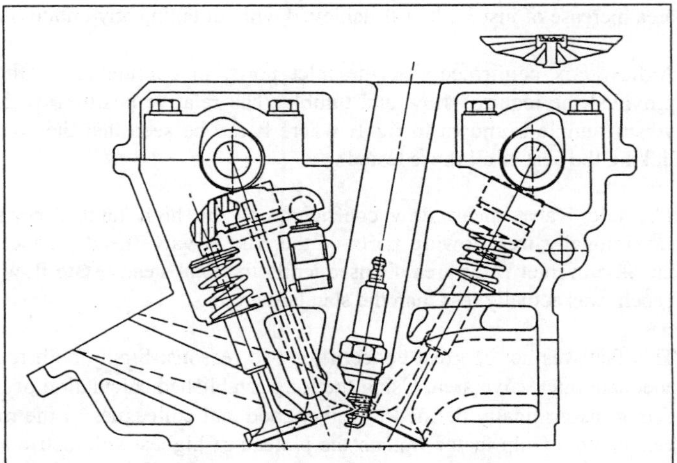

Figure 2. Valvegear Layout of 5v head

Of great interest was how much more inlet valve area could be achieved with the five valve layout. The total area of the valves is limited by the outside diameters of the valve seat inserts and the acceptable minimum distance between them. The minimum radial thickness of the seat inserts and the bridge width between any two is not a direct proportion of cylinder nor valve size thus as the bore diameter increases the inlet valve area as a percentage of bore area also increases. A graph of inlet valve area against bore size, using typical figures, is shown in fig 3.

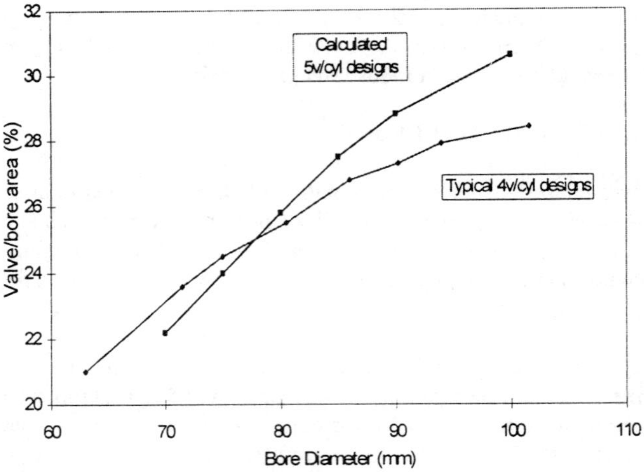

Figure 3 Typical Inlet Valve/Bore Area comparison

Here it would seem that the 5 valve head does not offer any valve seat area advantage over 4 valves on bores less than 75 - 80mm. This is using robust Ford Corporate design rules for the radial thickness of the valve seat inserts and the valve bridges.

Other designs of five valve head have shown a greater improvement in valve area but this has been partially achieved by the use of more delicate engineering which could have equally well been applied to 4v heads.

The engine has a bore of 82.4mm and the standard head provides a large valve area for a four valve layout. With such a good baseline it is difficult to make improvements but finally an inlet valve seat area increase of just 3.5% was achieved without taking any undue risks on the valve bridges.

A flow box confirmed that the inlet ports, in conjunction with the combustion chamber shape, provided the required flow and tumble. The relative position of this flow box on the chart of flow versus tumble is shown in fig 1. where it can be seen that the combined performance is on the line linking the best of all those tested.

The inlet valve mean flow coefficient of the high tumble ports was inevitably lower than the exceptionally free flowing ports of the base 4 valve head, in fact by 7 % which is more than the increase in inlet valve area. Consequently the total steady state flow capability, as measured on a flow bench, was actually less than the standard engine.

This fact was not of great concern for two reasons. Firstly both four and five valve versions offered adequate inlet valve area. Using the long established calculation of Inlet Valve Mach Number or Gulp Factor, as originally defined by Livengood and colleagues in the mid 1940's, it showed that on both versions the Gulp factor was on the plateau of highest volumetric efficiency. Secondly it was known that 5 valve heads generally produce improvements in breathing capability due to the rapid opening and closing of valve curtain area which enhances the dynamic effects but which is not shown on the steady state flow bench.

4. ENGINE TEST BED DEVELOPMENT

4.1 Baseline data from 4 v engine

The baseline tests on the standard 4 valve per cylinder engine showed how very effective the IMRC is in modifying the charge motion with its consequential effect on burn rate. In fact the low load combustion was such that it set a very difficult target to equal.

4.2 Performance development

The 5 valve per cylinder version suffered very few of the typical teething problems during initial break-in and soon produced the target power from the very first build configuration.

The primary runners (or ram pipes) were changed slightly to optimise the breathing and produce a small increase in performance at lower engine speeds. Subsequently more work was undertaken on the induction system to modify the internal volumes of the twin plenum chambers and to provide an interconnection between them. This was aimed at smoothing the mid speed torque curve by moving the commonly experienced adverse resonance trough between the two beneficial peaks.

Some of the configurations tested have produced a BMEP as high as 13.2 bar but generally these high figures have been associated with unacceptably peaky torque curves. A typical example of some more useful curves measured to date is shown in Figure 7. Here it can be seen that the 5 valve version produces around 10-12% more torque than the standard engine over the important mid speed range and a maximum power increase of approximately 8%.

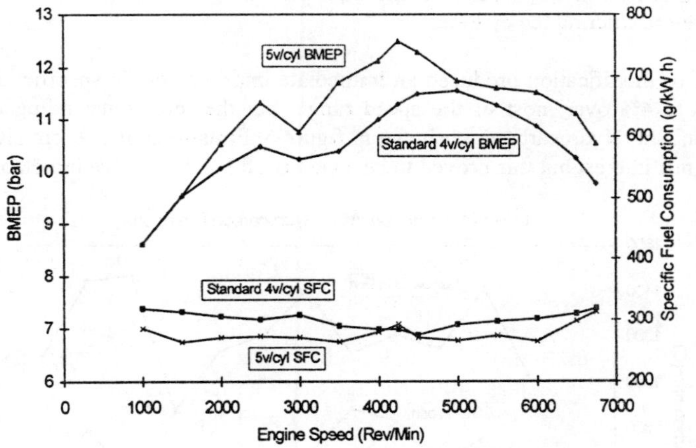

Figure 4. Maximum Performance and SFC

Interestingly the specific fuel consumption of the 5 valve version at WOT is approximately 10% better than the baseline 4v when each was set to optimised fuelling and spark timing. The test results indicate that combustion chamber activity affects the degree of enrichment beyond stoichiometric AFR at which maximum performance is attained. Intense activity which promotes mixing of the fuel and air, as in this tumbling five valve configuration, minimises the power losses resulting from running at stoichiometric AFR rather than leanest for best torque (LBT) fuelling. This is of great interest for the forthcoming emission regulations where the three way catalytic converter will need to be active right up to full load, yet manufacturers will wish to be able to quote the highest maximum power figures.

4.3 Rates of pressure rise

The downside of a high activity chamber running at WOT is that the fast combustion typically causes a harshness which is difficult to totally insulate from the driver. This undesirable characteristic may prove to be a real problem with this, or indeed any type of engine, which cannot modulate the amount of activity within the chamber. However very interestingly the subjective harshness of this particular Vee6 engine was not as great as expected in an in line 4 cylinder unit with similar rates of pressure rise, an observation which may indicate some areas worthy of future investigation.

4.4 Fuel injector evaluation

4.4.1 Injector effects at wide open throttle

Initial break-in and power testing was carried out with a single, fairly narrow, spray aimed at the central inlet valve. With this arrangement the power output was a little disappointing at normal Air / Fuel ratios and it was subsequently shown that we had unwittingly made a stratified charge engine in which not all of the air was being combined with fuel.

To establish a best case baseline for steady state conditions some wider angle spray injectors were installed mid way along the inlet ram pipes to give more time for the fuel and air to mix thoroughly before entering the cylinder.

This modification produced an immediate improvement in specific air consumption of approximately 3 to 4% over most of the speed range, but the most outstanding change was in the emissions of unburnt Hydrocarbons as shown in figure 5. Emissions of unburnt HC at full load are not normally of great interest but this proved to be a very useful and sensitive indicator of air/fuel mixing.

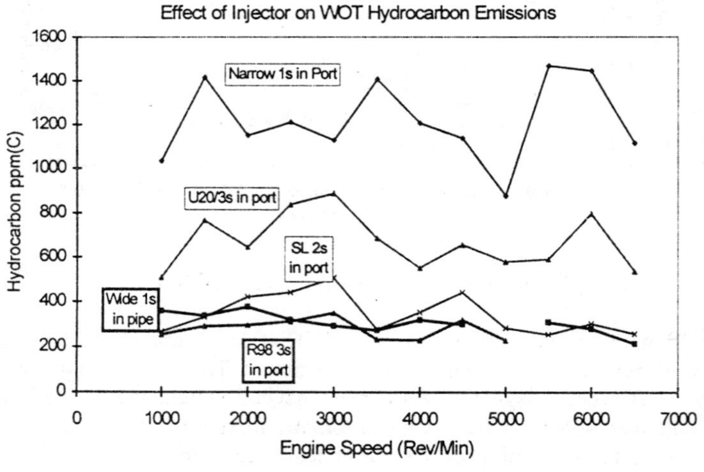

Figure 5 WOT HC Emissions with different injectors

Unusually high HC emissions were recorded with the single narrow beam injector in the inlet port. The wider cone injector installed back up the ram pipe reduced this by 75% i.e. to one quarter of the original, in fact significantly lower than the baseline 4v HC emissions which are influenced by the necessity of aiming the single injector at just one of the two inlet valves of each cylinder.

A current production twin spray unit with claimed good atomisation lowered the HC towards the in-pipe results but was not considered optimum for transient conditions in a vehicle application.

The first attempt at making a tri-spray injector produced three separate sprays but with some imbalance of the individual beam flows and with rather poor atomisation. On the power curve this produced HC emissions around half that of the original single narrow spray cone injectors but still twice that of the in-pipe installation. The second version tri-spray injector had a much better atomisation and beam to beam fuel distribution than the first. The HC emissions at wide open throttle were generally slightly lower than the in-pipe injector.

This proved that it is feasible to design and make an injector that will spray evenly over the back of all three inlet valves and enable the tumbling multivalve engine to achieve good fuel mixing at wide open throttle. In doing so the HC emissions are considerably lower than can be achieved when inlet manifold runner control or valve deactivation necessitates injection down only one inlet port of a pair.

4.4.2 Injector effects at part load

A similar comparison of injector types was made at 1500 rev/min at a particular low load site commonly known as the Ford World Wide Mapping Point (WWMP). The results may be seen in Figure 6 which shows the results of EGR loops carried out at this condition.

The single narrow cone spray in the inlet port was the worst of all for HC emissions with most other injector configurations giving similar results to each other at around 85% of the single beam HC emissions.

Up to an EGR rate of 15% the baseline 4 valve engine had slightly lower HC emissions than any of the 5 valve configurations tested. It should be remembered that at this load condition only one of the two inlet ports is effective and thus the fuel and air are entering the cylinder together.

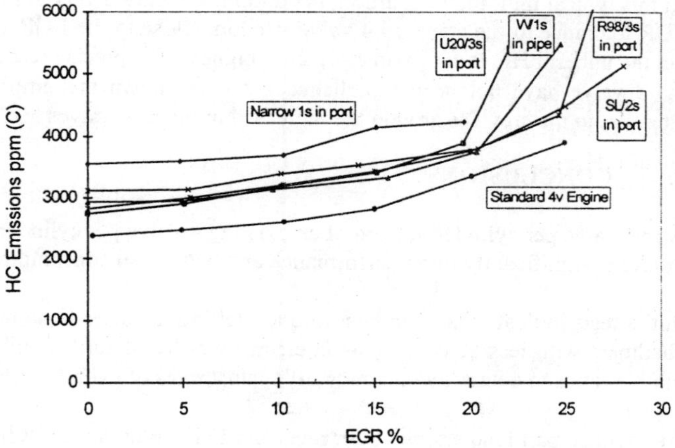

Figure 6. HC Emissions during light load EGR loop

4.5 Part load burn rate and EGR tolerance

At the Ford WWMP the burn rates achieved by the tumbling 5v engine matched or exceeded those of the base 4v with IMRC activated to produce axial swirl. Figures 7 and 8 show further results from the EGR loops run at this condition. Figure 7 shows that the tumbling 5 valve version closely matched the 0-10% burn angles of the base 4v engine. The 10-90% burn has a different response rate to EGR concentration but was slightly faster up to 20% EGR. Figure 8 shows the comparative specific emissions of unburnt HC and NOx for the same tests. The tumbling 5 valve produces slightly more HC but lower NOx.

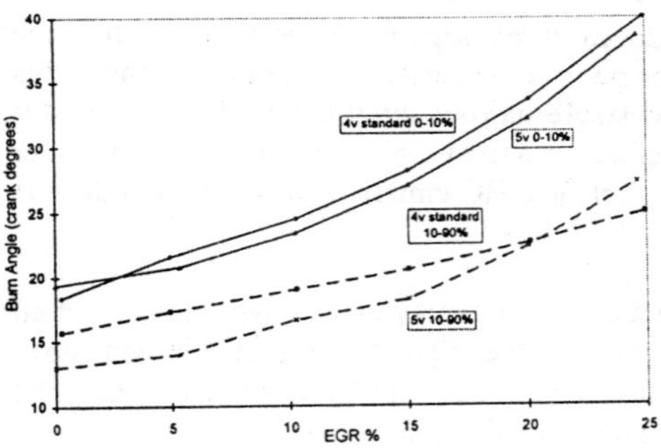

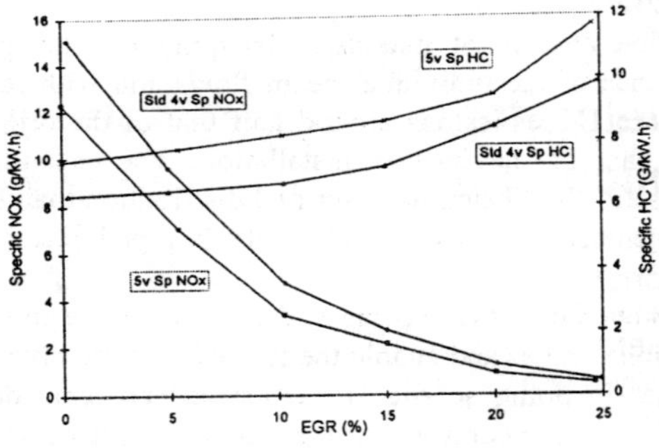

Figure 7. Burn rate at low load point.

Figure 8. HC and NOx at low load

Although not shown here both the specific fuel consumption and the stability, as measured by the Coefficient of Variation of IMEP, were almost identical up to 20% EGR flow.

At this typical light load condition, the tumbling 5v has a very similar burn rate, fuel consumption and EGR tolerance to the standard 4 valve version. Close to the EGR limit the NOx emissions are similar but the unburnt HC is approximately 20% higher. The precise reasons for the inferior HC emissions of the 5 valve have not been established but it is known that improvements can be made by careful attention to the crevices around the increased number of valves and optimisation of the burn rate.

5. CONCLUSIONS

A five valve per cylinder version of an existing 4 valve per cylinder Vee6 engine has been made which produces significantly more performance and better fuel consumption.

Initial tests indicate that lean best torque fuelling is closer to stoichiometric AFR than in combustion chambers with less activity. This is clearly worthy of further investigation including testing various tumble ratios to determine the trade off against rates of cylinder pressure rise and their effect on NVH.

At a typical part load site the burn rates and EGR tolerance are very similar to the 4v baseline engine. Emissions of unburnt HC from this first development engine are some 20% worse although the NOx emissions are generally slightly better.

The combination of the volumetric efficiency advantages of a 5 valve per cylinder layout with vigorous tumbling charge motion has resulted in an engine with the potential for matching or bettering the base engine without the cost and potential problems of the IMRC and its control systems.

ACKNOWLEDGEMENT

The author is grateful to the Advanced Powertrain Department of Ford Motor Company, Dearborn for providing financial and technical support for this project.

This paper is largely based upon SAE Paper 950815 © 1995 Society of Automotive Engineers. Inc. from which certain sections are reprinted with permission.

Development of the powertrain calibration process – application to engine idle speed control

R F DOREY BSc, PhD, CEng, MIMechE and **S DELASALLE** BSc, PhD, MIEE
Ford Motor Company Limited, UK
A T SHENTON BSc, PhD, CEng, MIMechE and **Z SHAFIEI** MS, PhD
University of Liverpool, UK

SYNOPSIS

There is considerable potential to improve the powertrain calibration process and reduce development time through the application of PC based control system design and analysis tools. This paper describes the features and functionality of a PC based calibration system that has been developed to tune a feedback controller in-vehicle from powertrain test data collected on-line. The system offers a generic capability for controller design and analysis using software that has been interfaced to the input/ output system of the powertrain electronic control unit. The paper describes the application of the system through the tuning of an engine idle speed controller.

NOTATION

Ne	engine speed in rev/min
prbs	pseudo random binary sequence
prts	pseudo random ternary sequence
vi	mark to space ratio of input signal to ISC valve, dimensionless
z	z-transform argument

1 INTRODUCTION

In recent years there has been an enormous change in the capability of hardware and software tools to monitor the operation of powertrain control strategies and their effect upon the performance of the powertrain and vehicle. Early monitoring and development systems for in-vehicle and dyanamometer work enabled little more than keypad input and the led display of calibration parameters and calculation variables, and many data acquisition tasks relied upon specially developed instrumentation and data loggers. This situation changed with the widespread use of the PC, particularly: its availability in laptop and notebook forms with powerful processing and large disk storage capabilities, and with the amount of good applications software available for data acquisition, signal processing and simulation (1). Although the commercial PC hardware and software packages are not necessarily compatible with automotive manufacturers' equipment they have certainly provided the catalyst for manufacturers to develop their own PC based systems and interfaces. This is the case with Ford Motor Company, who has developed a PC product to calibrate the powertrain electronic control unit and monitor the operation of the powertrain and vehicle.

The basic requirement of modern powertrain control is to regulate the amount of fuel injected into the engine and to set its spark ignition timing in order to minimise vehicle fuel consumption subject to the legislation governing exhaust emissions; secondary control objectives are concerned with the vehicle performance and driveability during steady-state and transient operation. To meet all the control objectives additional engine subsystems are regulated associated with the throttle, the braking system, exhaust gas recirculation, evaporative emissions, and secondary airflow into the exhaust (2, 3).

To minimise cost, and restrict the number of sensors and actuators significant portions of powertrain control strategies have been devoted to model the physical behaviour of the powertrain and infer its operational states, and component and system performances have been represented by look-up tables and rule based relationships. Over the years the complexity of powertrain strategies has increased, with features added to provide refinement and fix particular problems, with the consequence that strategies have become time consuming and difficult to calibrate.

The strategy associated with the control of engine idle speed is complex, and is a good example to choose to demonstrate what can be achieved by using a control theory based approach to help to simplify the tuning process and optimise the calibration of the strategy. In the following a PC tool is described that provides specially targeted system identification and controller tuning capabilities that are intended to optimally tune the engine idle speed control system.

2 THE ENGINE IDLE PROBLEM

When load is applied or removed from an engine at idle its speed dips or flares and this changes the structural borne and airborne vibration levels in a vehicle. Idle quality is related to the sensory perception of vibration through the steering column, seats and fascia, and to the presence and intensities of particular sound frequencies. Control of the idle speed to minimise the speed excursions caused by changes in load is of benefit in helping to reduce the excitation of unwanted resonances. Poor idle speed control has a detrimental effect upon fuel consumption and emissions, and can introduce instabilities that result in surges of engine speed when loads are applied or removed {and which can be made considerably worse by driver interaction} (4).

The idle speed control problem is essentially one of disturbance rejection. Significant loads are applied by the mechanical and electrical loads associated with power steering, air conditioning, heated screens, fans, windows and lights that unless compensated can stall the engine. The principal control elements of the idle speed system are the idle speed control (ISC) valve, and the spark ignition timing unit. Of these the valve has the major effect; although the dynamics of the air loop take a little longer to act than spark, air carries much more authority. To help offset the disturbance caused by large loads the strategy may employ some load sensing in order to provide a feedforward control action in anticipation of large loading effects. This helps to reduce the requirements of the feedback action in reacting to and correcting speed changes, but carries a cost penalty.

3 PC BASED CONTROL SYSTEM DESIGN

The key features of the Ford PC based calibration and monitoring system are its shared memory interface with the powertrain control unit, and its data acquisition software. Together these allow it to monitor, display and change parameter values associated with the operation of the strategy and the powertrain. There is the means to disable parts of the strategy and link in user written pieces of code so that particular tests can be performed on the powertrain. In the work reported here a suite of software routines was written that built upon these features to provide system excitation, idenitification, controller tuning and implementation capabilities. These were applied to optimally tune the PID feedback controller in the section of strategy concerned with the air loop of the engine idle speed control system, Fig 1.

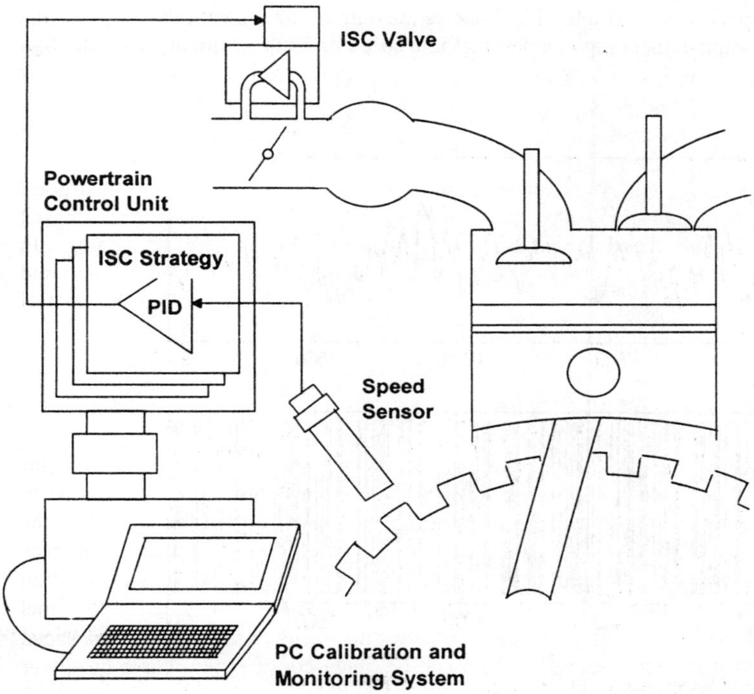

Fig 1 Idle speed control development system

3.1 System Excitation

The basis for the controller design and tuning work is a linear model relating changes in engine speed to changes in the control valve input. This is determined with the system operating in open loop at a particular steady state condition. However, it is not an easy matter to determine the model, the engine's performance is non-linear, time varying, noisy and subject to cyclic and random variations (4). There are definite issues associated with the selection of the excitation signal for subsequent system identification, and care is needed if consistent and reliable results are to be obtained. During the preliminary stages of the calibration process it is useful to excite the system with different signals in order to develop an understanding of the performance. To assist this the software supports: step, square, prbs, prts (5), and random input signals. Of these step and square are largely investigatory and helpful visually; whilst the prbs, prts and random signals are dynamically rich and able to excite the system over a broad frequency band.

3.2 System Identification

For the purposes of explaining the system identification procedure it can be assumed that a small amplitude prbs signal is applied to the input of the ISC valve and that the system response is measured from the output of the speed sensor. The excitation signal is applied at every 180 degrees of crankshaft movement, synchronous with each combustion event; it is derived from a software implementation of a shift register (6), whose number of bits and duration are set by the user based upon some simple knowledge of the system response (obtained from step tests for example). Fig 2 shows the output of a typical prbs test, obtained from a 1.6L Zetec engine operating at a nominal idle speed of 880 rev/min and a steady light load.

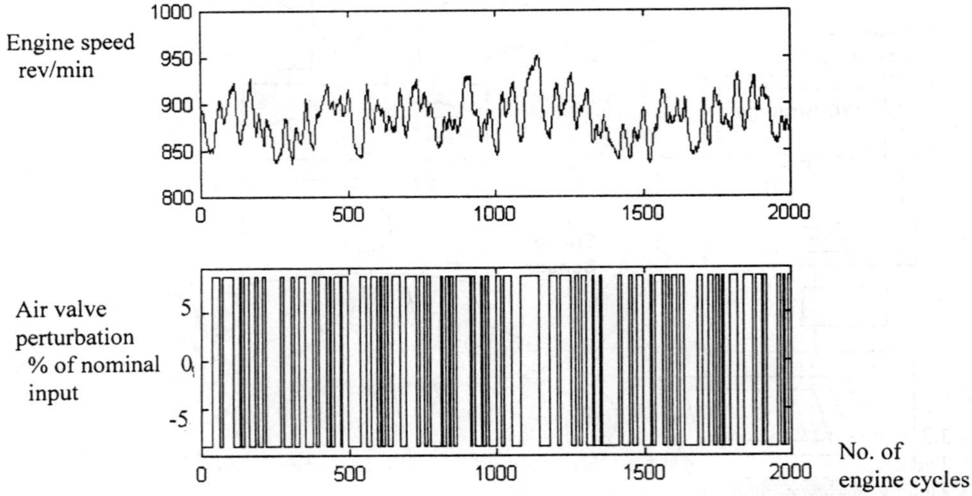

Fig 2 Engine Speed Response with PRBS Excitation Applied to the Input of the ISC Valve

The software operates on this data to determine a model to relate changes in speed to changes in ISC valve setting. The system identification process applies a least squares polynomial fit moving window algorithm (7) to prefilter the output speed and then a recursive Kalman identifier to operate on the input-output sequence (8). These procedures were developed for their ease of implementation and their ability to provide results comparable to the ARMAX model identification procedure in Matlab™ (9). The software provides the user with advice about the model delay and order to start the identification, and it employs Akaike's Final Prediction Error criterion (10) to provide a measure of the goodness of fit of the model.

For the idle speed control problem it is known that a second order model with time delay is able to represent the dynamic response of the air loop. This simplifies the amount of investigation required and it is relatively straightforward to optimise the model. For the data shown in Fig 2 the following discrete model was found:

$$\frac{N_e}{v_i} = \frac{58}{z^4(z^2 - 1.759z + 0.773)}$$

The predicted output of the model is compared with the measured speed signal over the first 1000 engine cycles in Fig 3. This shows a very good level of agreement.

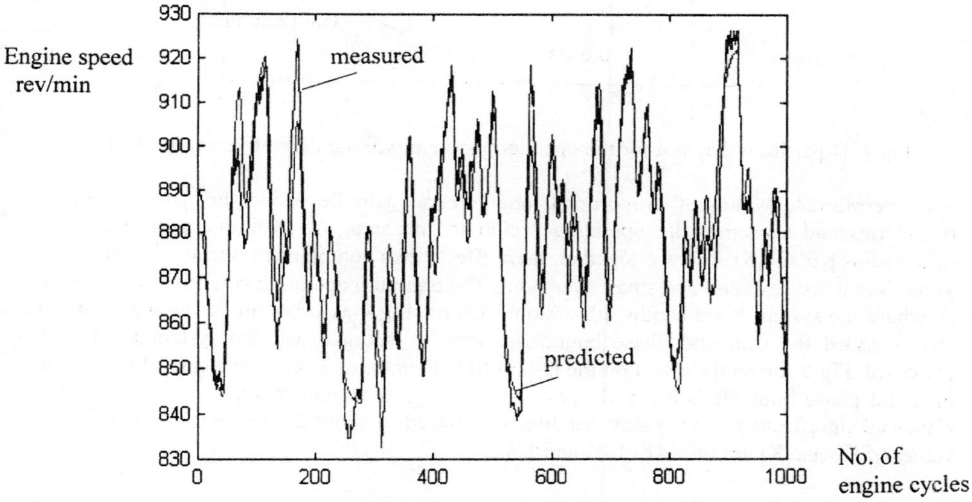

Fig 3 Comparison of predicted and measured engine speed

3.3 Controller Tuning

The software allows the gains of the Idle Speed PID controller to be tuned using either a Ziegler Nichols approach (11) or the open-loop D-partition method (12, 13). The Ziegler Nichols approach is included because it is commonly understood and provides the user with a point of comparison, but it does not lead to a properly tuned controller. For the idle speed problem this is mainly due to the performance requirement which emphasises the disturbance rejection characteristics of the controller and the need to avoid oscillatory behaviour and overshoot, and the significant time delay present within the system.

The D-partition method is based upon frequency response specifications and satisfying gain and phase margin criteria. It lends itself to simple graphical interpretation and allows the user to see the effect of individual controller gains on stability margins. Representative gain and phase margins for the idle speed system are 10 dB and 75 degrees. Assuming these values and the model developed in the previous section, Fig 4 shows the D-partition diagram which expresses the relationship between the values of proportional and integral gain to meet the specified gain and phase margins.

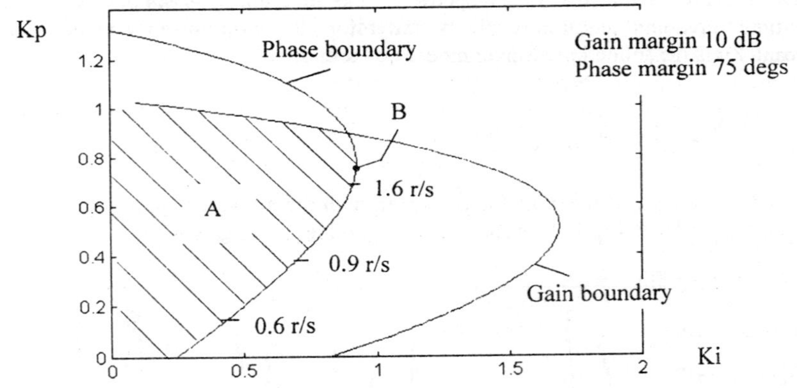

Fig 4 D-partition diagram for the open loop system with no derivative control action

Permissible values of proportional and integral gain lie inside the gain and phase boundaries and constrain the operating region to the area, A. Desirable performance is achieved as Kp and Ki increase together, their effects then complement one another and tend to produce a fast response and small overshoot. The optimum operating condition lies at point B, where the system has a bandwidth of some 1.6 r/s. Fig 4 assumes no derivative action, if this is added the gain and phase boundaries can be enlarged and the system bandwidth increased. Fig 5 shows the effect on the D-partition diagram of a derivative gain of 0.165; the gain and phase boundaries are both moved to the right (the gain boundary in particular is displaced significantly), the system bandwidth is raised to about 2.1 r/s, and the permissible values of Kp and Ki are raised to 1.05 and 1.55.

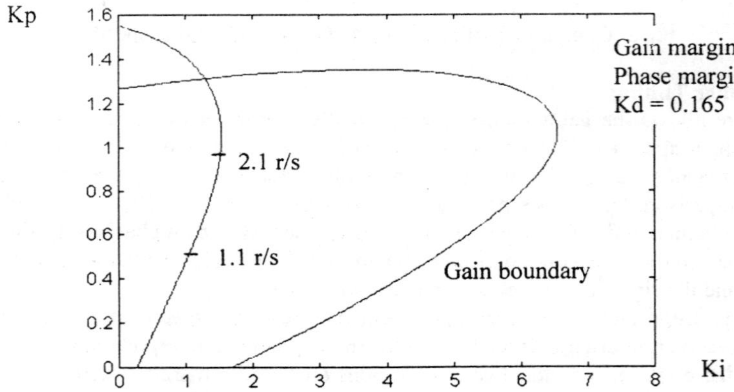

Fig 5 D-partition diagram for the open loop system with derivative control action

3.4 Controller Implementation

The engine has essentially non-linear characteristics and performance for which a simple linear time invariant controller is insufficient. The engine's characteristics change significantly with load and speed; over a range of idle conditions it has been shown (14) that the natural frequency and damping ratio of the engine's idle speed control system change by more than a factor of 2. This variation in performance can be taken into account, however, with a scheduled form of control, and in the powertrain control unit the nominal controller gains found above are scheduled with operating condition in order to increase the performance whilst keeping the high stability margins.

Fig 6 shows the result of implementing the optimised controller in the powertrain control unit. Two electrical loads comprising the interior fan and the heated rear window are simultaneously switched on and then off; the engine speed dips and flares in response to the size of the load changes and the action of the controller.

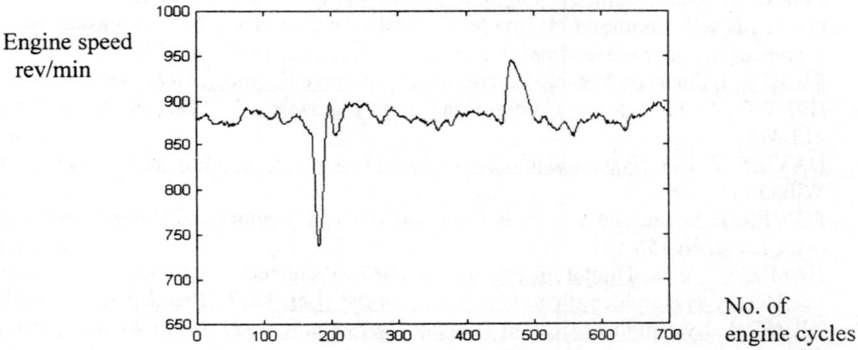

Fig 6 The effect of load on the engine with the optimised controller

The benefit obtained from the optimised controller may be seen by contrasting these results with those shown in Fig 7, which were obtained using current calibration practices to tune the controller gains. The optimised controller reduces the magnitude and duration of the speed excursions by up to 50%, and produces greater consistency of operation.

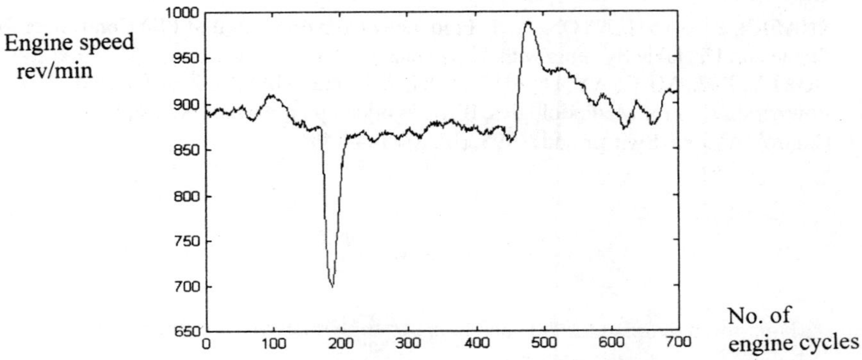

Fig 7 The effect of load on the engine with the conventionally tuned controller

CONCLUSIONS

Very promising PC based control system tuning tools have been developed for a powertrain control unit. Preliminary findings suggest they can be applied to achieve significant performance benefits; in particular to reduce the magnitude and duration of speed excursions arising from the disturbances made to an idle speed control system. These benefits accrue from a more highly tuned controller. The tools contribute to a more consistent and robust powertrain calibration process.

REFERENCES

(1) DOREY, R.E. A Computer Aided Approach to the Development of Powertrain Control Systems, Engine Design International '95, 1995, to be published.

(2) POWELL, J.D. Engine Modelling for Control - An Overview, IFAC Workshop on Advances in Automotive Control, Ascona, Switzerland, 1995, (IFAC).

(3) RICHES, I. Focus on Vehicle Emissions, Automotive Engineer, 1993, Vol 18, 48-61.

(4) HEYWOOD, J.B. Internal Combustion Engine Fundamentals, 1988, (McGraw-Hill) 413-427.

(5) DAVIES, W.D.T. System Identification for Self-Adaptive Control, 1970, (John Wiley) 111-119.

(6) POWER, H.M. and SIMPSON, R.J. Introduction to Dynamics and Control, 1978, (McGraw-Hill) 156-157.

(7) HAMMING, R.W. Digital Filters, 3rd Ed. 1989, (Prentice-Hall) 42-50.

(8) LJUNG, L. System Identification - Theory for the User, 1987, (Prentice-Hall) 86-88.

(9) LJUNG, L. System identification Toolbox User's Guide, 1991, (MathWorks) 2,10-2,11.

(10) LJUNG, L. System Identification - Theory for the User, 1987, (Prentice-Hall) 420-421.

(11) POWER, H.M. and SIMPSON, R.J. Introduction to Dynamics and Control, 1978, (McGraw-Hill) 224-225.

(12) SHENTON, A. T. and SHAFIEI, Z. Relative Stability for Control Systems with Adjustable Parameters, Journal of Guidance, Control, and Dynamics, 1994, Vol 17, 304-310.

(13) SHAFIEI, Z. and SHENTON, A. T. Frequency Domain Design of PID Controllers for Stable and Unstable Systems with Time Delay, to be published.

(14) DOREY, R.E., MACLAY, D., SHENTON, A.T. and SHAFIEI, Z. Advanced Powertrain Control Methodologies, IFAC Workshop on Advances in Automotive Control, Ascona, Switzerland, 1995, (IFAC) 144-149.

C498/22/258/95

A concept for a lightweight engine cylinder block

P HALL, D H TIDMARSH BSc, PhD, CEng, FIMechE, MBIM, MIMgt, and **G ATKINSON**
University of Central England, UK

1 SYNOPSIS

This paper outlines concept work undertaken in developing a lightweight engine by replacing the block with a skeletal structure, reducing mass, warm-up-times, energy wastage, setting-up and production costs.

Water-cooled engines are traditionally built around a block, almost invariably cast from iron or aluminium. Setting up a block-line for a new engine involves capital investment of millions of pounds. There is heavy environmental cost and the bulk contributes to fuel inefficiency. Porosity means that the scrap rate of castings can be high and is usually detected only after adding value.

2 INTRODUCTION

The cylinder block has been at the heart of the engine for about a century now, and has been virtually indispensable for most of that time. Several attempts have been made to replace it with a welded or moulded facsimile. The Crosley-Taylor engine (1) in 1946 consisted of a number of pressings brazed together to produce a virtual clone of a cast block. The plastic engine (2) developed by Polimotor Research, New Jersey, in 1982 was merely a cast block reproduced with much of the metal replaced by plastic.

It may be that these engines were destined to fail from the outset because their designers did not go far enough back in the design cycle. The block has been rooted in the design of the engine from the outset, when many of the functions it performed could not be performed by other means. As technology has advanced over the years, many of those functions have become redundant, and it is arguable that the technology now exists to find alternative methods of performing all of its prior functions.

3 ANALYSING THE BLOCK'S FUNCTIONS

3.1 Redundant Functions

As the engine has evolved over the past century much of the mechanical control gear has been replaced by electronics, using wires to transmit signals instead of levers. The fuel pump, carburation, ignition and spark distribution can all been replaced by electronics, either as a full engine management system or as discrete components, attached to the cylinder head. Valve gear is now exclusively housed in the cylinder head, as are most cam-shafts. There are no longer any push-rods passing through the block, as over-head cam-shafts have become the norm, with belt or chain drive from the crank-shaft.

Most modern engines employ electric cooling fans, and lubricant pumps are readily mounted on the nose of the crank-shaft.

3.2 Other Identifiable Functions

If all of the above, currently accepted practices in various engines, are employed, then the only remaining functions that the block performs are as follows:

i) A support for the engine mountings and other peripheral devices.

ii) A top bracket for the crank bearings.

iii) A means of keeping the crank fixed in relation to the head and liners.

iv) A structure that provides rigidity and strength to the whole engine assembly.

v) A locator for the cylinder liners.

vi) A water jacket.

4 MINIMISING THE BLOCK'S FUNCTIONS

Having identified the functions that still exist, it is necessary to find alternative ways of fulfilling these functions before the block can be eliminated. Taking each of the items i) to vi) in section 3.2 above:

It is a simple matter to eliminate i) [*A support for the engine mountings and other peripheral devices*] - there are plenty of other places to locate equipment, including the engine mountings.

ii) [*A top bracket for the crank bearings*] can be eliminated if a separate bearing ladder is utilised, replacing both upper and lower sections in modern engines, similar to the arrangement shown in **Figure 1**.

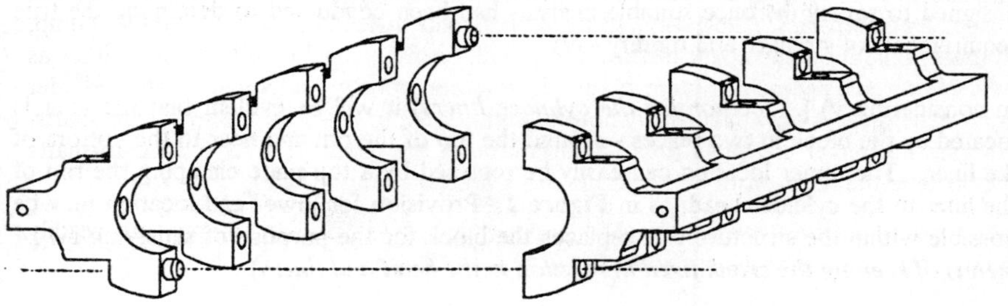

Figure 1 - Crank-bearing Ladder

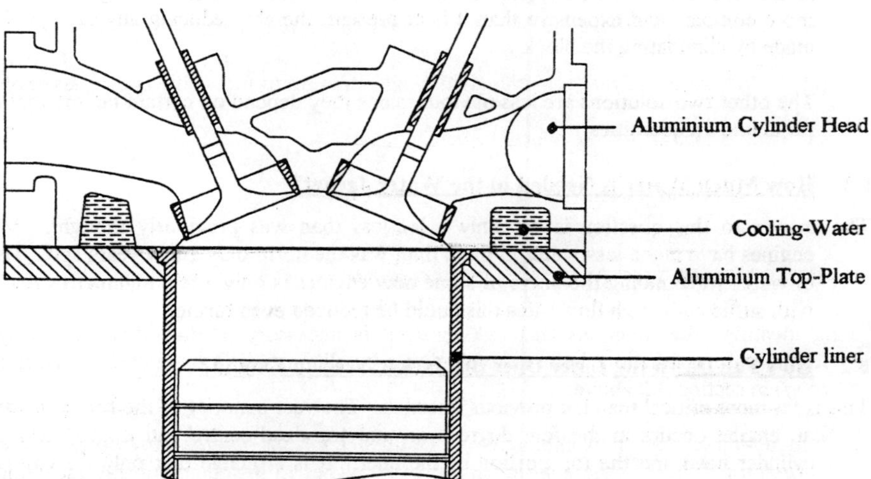

Figrue 2 - Retaining the Cylinder Liner

As far as iii) [*A means of keeping the crank fixed in relation to the head and liners*] and iv) [*A structure that provides rigidity and strength to the whole engine assembly*] are concerned, it is highly likely that the strength and rigidity provided by a cast block are way in excess of requirements. A much lighter and less complex structure could be designed to satisfy iii) once suitable analysis had been conducted to determine the true requirements of strength and rigidity - iv).

In considering v) [*A locator for the cylinder liners*] it will be evident that the liner is located by the block in two places - around the top of the rim and near to the bottom of the liner. The upper location can easily be replaced by a top plate clamping the rim of the liner to the cylinder head, as in **Figure 2**. Provision for lower end location may be possible within the structure that replaces the block for the purpose of satisfying iii) [*A means of keeping the crank fixed in relation to the head and liners*].

The sole remaining function of the block is vi) [*A water jacket*

5 ELIMINATION OF THE WATER JACKET

Three possible solutions are evident when it comes to water cooling of the liners. One solution is to incorporate a water jacket into the liner itself, fed from the top plate which retains the liner. The main drawback with this is that the liner casting becomes much more complex and expensive than it is at present, thereby reducing any savings that are made by eliminating the block.

The other two solutions are less obvious, since they depend on certain factors that are so far unknown quantities.

5.1 How Much Water is Needed in the Water Jacket?

The answer to this question is certainly a lot less than was previously thought. Modern engines have much less water in them than was the norm only a few years ago. The film of water surrounding the liners of some new engines is only a few millimetres thick, and with sufficiently high flow rates this could be reduced even further.

5.2 How Far Down the Liner Does the Jacket Need to Extend?

This is far more critical than the previous question. The vast majority of the heat generated by an engine occurs in the few degrees around top-dead-centre and is absorbed by the cylinder head and the top portion of the liner. It is arguable that only the top two or three centimetres of the liner is really in need of water cooling, and so the water jacket could be incorporated into the top plate retaining the liner, similar to that in Figure 3.

It is possible that the temperatures below this point are such that oil cooling would be sufficient, even if this meant using much higher volumes of oil than is current practice.

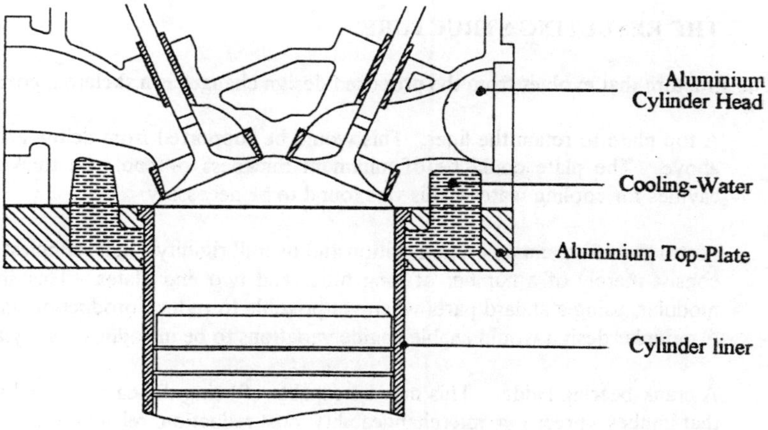

Figure 3 - Incorporating Water-Cooling Into the Top-Plate

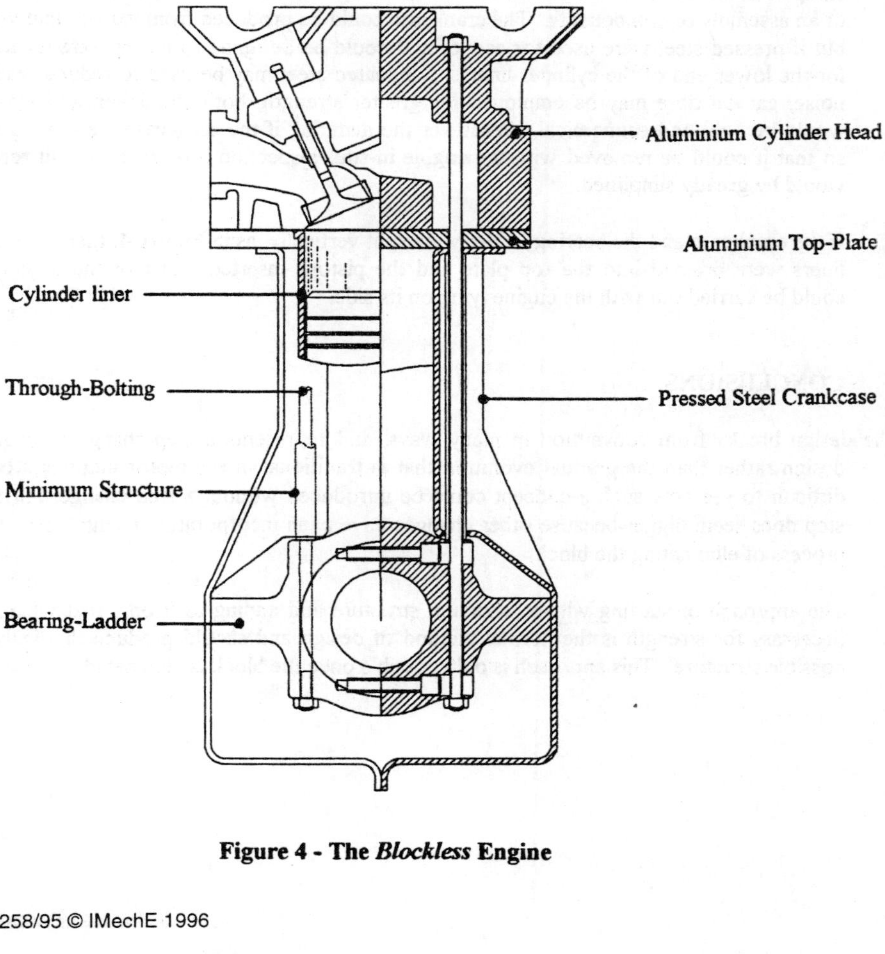

Figure 4 - The *Blockless* Engine

6 THE RESULTING STRUCTURE

The structure that evolves from the proposed design changes is a skeleton, consisting of:

A top plate to retain the liner. This would be supported from below and clamped from above. The plate could be of minimum thickness or could be thickened to enclose cavities for cooling water if this was found to be necessary.

A structure to maintain crank position and overall rigidity. In its simplest form this could consist merely of a number of long tubes and two end plates. This should ideally be modular, using standard parts wherever possible to reduce production and tooling costs. A modular design would enable engine variations to be introduced easily and cheaply.

A crank-bearing ladder. This may be capable of being die-cast, with all the benefits that that implies - precision, interchangeability, cost reduction, reliability, etc.

Through-bolting to hold the whole assembly together.

A crankcase that encloses everything below the cylinder head. This could be a simple sump or could contribute to the structure of the engine, and could be a single component or an assembly of components. The crankcase could be produced from various materials but if pressed steel were used for economy it could be designed to incorporate location for the lower end of the cylinder liners. Laminated steel may be used to reduce engine noise, carbon fibre may be employed for greater strength, both aluminium and plastic would be light and economical. Whatever the material, if the crankcase were designed so that it could be removed with the engine in-situ, inspection, maintenance and repair would be greatly simplified.

If the crankcase and the bearing ladder were split vertically, as in **Figure 4**, then once the liners were pressed into the top plate and the pistons inserted, most of the assembly could be carried out with the engine lying on its side.

7 CONCLUSIONS

The design breaks from convention in many ways, and represents a step-change in engine design rather than the gradual evolution that is traditional in the motor industry. It is difficult to see how such a concept could be introduced without a step-change, but the step does seem bigger because other changes have been incorporated, facilitated by the process of eliminating the block.

The approach of starting with a minimum structure and adding to it only that which is necessary for strength is the normal method of design and should produce the lightest possible structure. This approach is only possible once the block is eliminated.

8 REFERENCES

(1) KLOTSCH, P. Copper Brazed Crosley Motor, Society of Automotive Engineers, Detroit, 1947.

(2) McCANN, M. K. Engine Materials: Lightening the Load Underhood, Automotive Industries, 1982, January, pages 49 - 52.

Driveline Developments

C498/24/115/95

GI tripode joints – calculation of durability

J-P LEY
GKN Glaenzer Spicer SA, France

SYNOPSIS

Durability of Constant Velocity Joints is one among the several main factors influencing the selection of an adequate size of joint to be mounted on a car. In this paper, we shall focus on CV plunging joints of tripode type (GI Joints),and analyze the durability of this type of joint. A methodology will be described, which is now implemented in a software package, and which allows modelling of the relevant phenomenon, and prediction of the durability of GI Joints, when service conditions are defined.

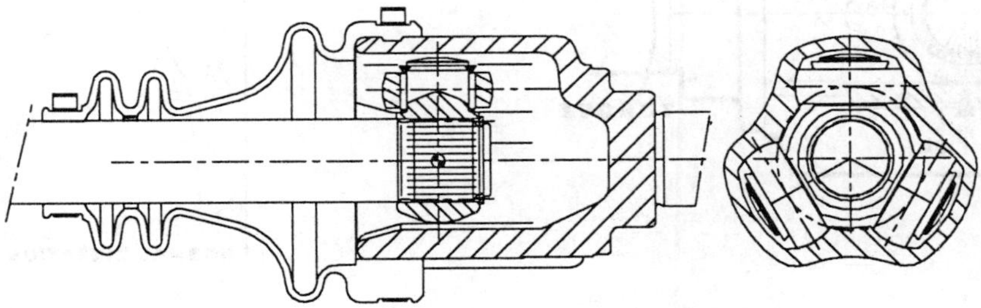

Figure 1 GI JOINT

1. INTRODUCTION

The GI joint is now the most common type of plunging Constant Velocity Joint (CVJ) which is used on the gearbox side of halfshafts, on front wheel drive cars, to transmit movement and torque from the gearbox to the wheel.

Durability of this mechanism is a key factor (among several others) involved in the general improvement of the reliability of cars.

2. UNDERSTANDING THE KEY PHENOMENON

Experience, especially from numerous endurance tests, has shown that the durability of GI joints is mainly driven by the **rolling contact fatigue** phenomenon occuring on the working surfaces of the pegs of the tripode.

The rolling contact fatigue phenomenon, consists of cycles of subsurface shear–stress, leading to subsurface cracks, and, later, to surface spalling.

i. The **load** is the local consequence of the torque transmitted through the joint, and it is distributed on rolling elements according to a law specific to the tripode type of joint (figure 2).

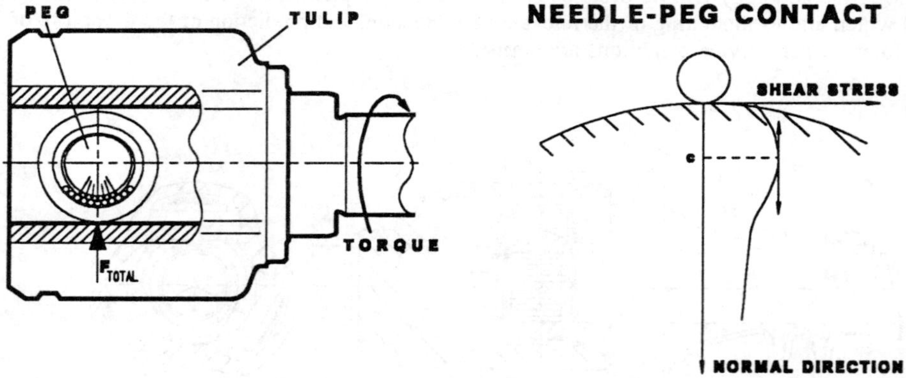

Figure 2 GI JOINT INTERNAL FORCES

ii. The **cycles** come from the alternate rotating movement of the set of needles around the peg, when the tripode joint operates with a bending angle. The kinematic movement of the roller, relatively to the peg, is a complex combination of a plunging movement parallel to the peg's axis, and an alternate rotation of the roller around the peg's axis (see figure 3). This last movement generates an alternate movement of the set of needles around the peg.

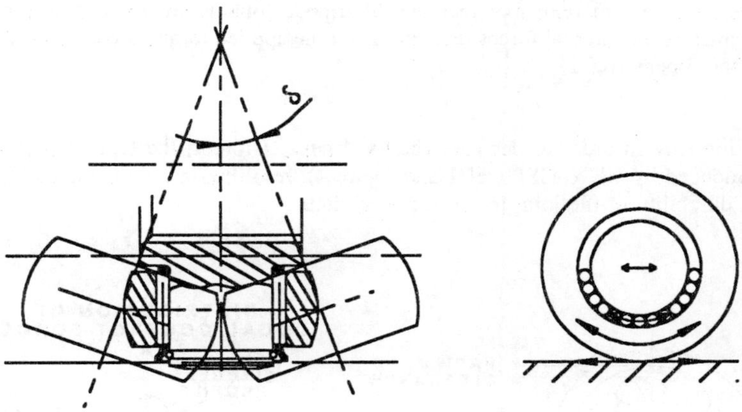

Figure 3 NUMBER OF LOAD CYCLES

3. QUANTIFYING THE PHENOMENON

3.1 Number of load cycles

The number of load cycles is related to the number of revolutions, by the following formula:

number of load cycles = number of revolutions * sin (δ) * K

where δ = bending angle of the GI joint
K = analytical factor depending on the dimensions of rolling elements

Note This formula is a continuous function of the bending angle. At first sight, this may seem contradictory due to the fact that, within a short period of time and with a small bending angle, some areas of the surface of the peg is run over by needles, and some are not (there is no overlap of the movement of two consecutive needles). In fact, the actual movement of the needles, during operation of a normal GI joint, is the result of the superposition of this "short term" movement with a "long term" movement which is a slow rotation of the set of needles around the peg. On top of that, there are also random rotations, due to the variations of working conditions.

3.2 Level of the load

As for a needle bearing, the local load experienced by the material of the peg's surface is the **Hertz pressure**. This pressure is the result of the **very specific equilibrium of forces** that takes place in the tripode joint when a torque is applied.

-> it is important to remenber that the GI tripode joint is an isostatic system, which means that the internal forces resulting from the applied torque are equally distributed on the 3 pegs (ref 1).

-> additionally in order to decrease the local contact forces, the Optimum Profile was introduced by GKN-GSP (ref 1 and figure 4), resulting in a tremendous increase of the durability of the joint in service conditions.

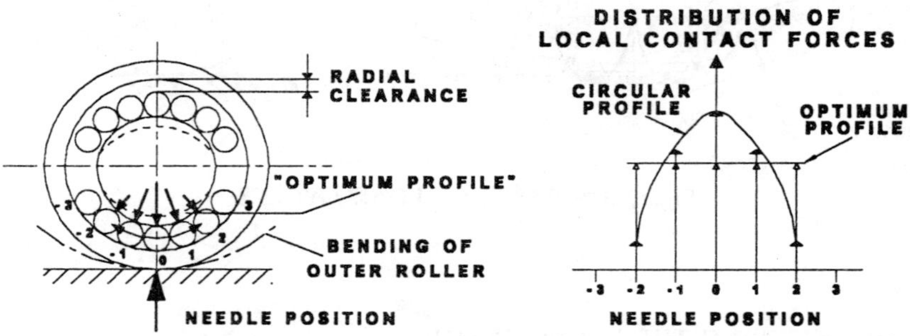

Figure 4 CONTACT FORCES AND OPTIMUM PROFILE

GKN-GSP has developed a model, and a corresponding computer program, to simulate and **calculate the load distribution within the peg-needles-roller assembly**. This model takes into account:

- The geometry of the joint
- The Optimum Profile definition
- The applied torque

This model also takes into account the local contact deformations, the radial clearance in the assembly, and the fact that the roller cannot be considered as a "thin curved beam". The corresponding computer program was validated with analytical tools such as FEA.

In addition to the durability application, this core program can be used in different ways as a design tool :

- For a given applied torque, it can determine the number of loaded needles, the load values and the bending stress in the roller.
- For a given typical endurance torque value, it can determine the corresponding definition of the optimum profile, that could, then, be tuned to the application.

3.3 Definition of the "Cumulative Damage"

The theoretical concept of cumulative damage gives the ability to quantify the fatigue phenomenon of a material submitted to a high number of load cycles.

Considering the obvious analogy with a needle bearing, we use a similar law of cumulative damage, as defined by :

$$\text{cumulative damage} = \sum_{STEPS} \underbrace{\left(\text{hertzian pressure}\right)^{10/3}}_{\text{Load}} \times \underbrace{\left(\text{NUMBER OF REVOLUTIONS}\right) \times \sin(\delta) \times K}_{\text{number of cycles}}$$

This damage is a "local criteria". It means that it characterizes what happens locally in the material, and that it is, consequently, independent from the joint size.

4. CALIBRATION OF THE THEORY

The calibration of the above theory is done using the following process :

1) A typical endurance test is selected, which fulfills 2 conditions :

   * being representative of typical actual service conditions (torque, angle, speed) of the joint on the car.

   * having a corresponding statistically significant database of test results.

   For this test, a Weibul analysis can be done, giving the relationship between the number of test hours, and the failure rate.

2) The above theory allows us to relate the number of test hours and the corresponding theoretical cumulative damage.

At this stage, through these 2 steps, we have got the **fundamental relationship** between the **theoretical cumulative damage** associated with the service load conditions, and the corresponding **actual failure rate** (figure 5).

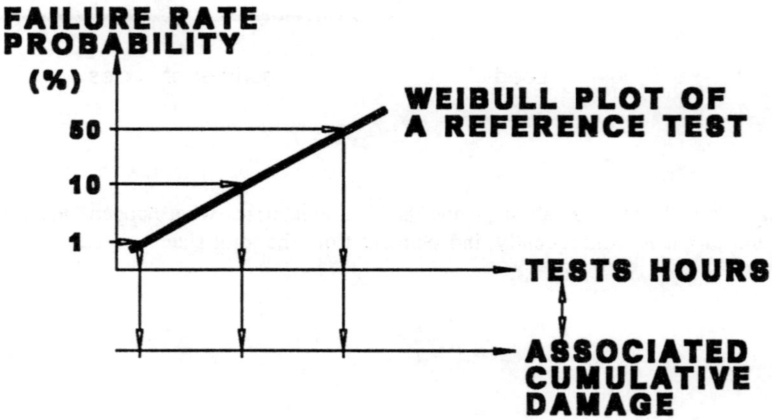

Figure 5 CALIBRATION OF THE THEORY

5. HOW TO PREDICT GI JOINT DURABILITY

Making the hypothesis that this fundamental relationship remains valid for other service load conditions, it is very easy to use this theory in order to predict GI joint durability.

When given service load conditions, the following steps must be followed (figure 6).

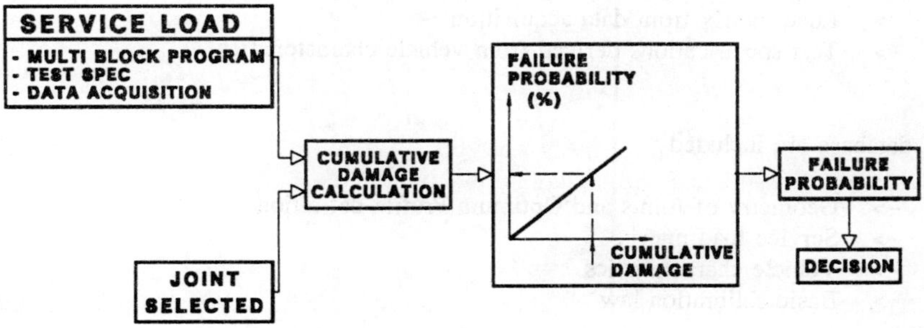

Figure 6 PREDICTION OF GI JOINT DURABILITY

- Transform the input data, if necessary, into the right format, which gives the number of revolutions for each running condition (torque and bending angle),
- Select a GI joint size.
- Calculate the theoretical cumulative damage associated with the input service load,
- Derive the predicted failure rate, from both the cumulative damage and the fundamental law established before.
- Make the decision to confirm or change the selected size of joint.

6. SOFTWARE IMPLEMENTATION

A software package has been developed within GKN, based on a modular approach, in order to perform all the above mentioned calculations easily and efficiently.

Service loads can be input in the most common existing formats :

-> Multi Element Block Program
-> Load matrix from data acquisition
-> Test specifications derived from vehicle characteristics

Several database are included :

-> Geometry of Joints and Optimum Profile definition
-> Service load matrix
-> Vehicle characteristics
-> Basic calibration law

7. CONCLUSION

GKN owns an operating sophisticated tool to predict the durability of GI joints. This software tool is used both in Research and Developement, for the evaluation of design improvements, and for Application Engineering, to help in selecting the joint size, according to the customers' requirements in durability.

J.P. LEY

REFERENCE

(1) General theory and experiment of homokinétic joints
M. ORAIN, 1977

C498/24/119/95

Efficiency of constant velocity joints for automotive application

L WENNING
GKN Automotive AG, Germany

SYNOPSIS

The losses of modern CV-joints are lower than 1 % when operated at normal articulation angles. This energy loss causes heating of the joints which together with additional heat sources in the vehicle can lead to a reduction in the durability of the joints. The parameters influencing joint efficiency are investigated and the potential for improvement shown.

1 INTRODUCTION

Modern CV-joints already have efficiencies of more than 99% when these are used in cars at articulation angles smaller than 10 degrees. Although the losses are smaller than 1% the power dissipation causes a heating of the joints. In front wheel driven cars the joints may be additionally heated by the catalytic converter, the gearbox and the brakes. Furthermore, reduced air cooling as a result of encapsulated vehicle floorpans reduces the thermal emissions and increases the joint temperatures. These effects cause high operating temperatures which tend to shorten the lifetime of the joints.

The target of the efficiency investigations is to reduce the power losses and as a result of this the heating of the joints in order to improve their life time. Therefore the influence of different designs, modified joint parameters and of new low friction greases on the efficiency of fixed and plunging joints was investigated.
For the efficiency investigations a new test-rig was built to enable measurements with the highest accuracy.

When the investigations were started, the joints were found to have several parameters which have an influence on their efficiency. The efficiency of new joints changes during running-in. Therefore before starting further investigations the running-in procedure had to be defined exactly. Additionally, for comparison of joints of different size and of different joint families, the joint specific running-in and test torque had to be defined. Also an exact definition for the direction of the torque input and the direction of rotation (the "operating quadrant") had to be made. Temperature influences the joint efficiencies in different ways for joints of different families. The investigation of different greases in the same joint requires a new running-in after the grease change to eliminate the influence of the previous grease. Furthermore an influence of the speed on the joint efficiency was expected. Only by regarding all these aspects for every single joint, are results with the necessary accuracy achievable.

To date a large number of fixed and plunging joints have been investigated for comparison purposes. More detailed investigations have been carried out with VL 93 ball plunging joints and with the fixed ball joints RF 91, RF 91c and AC 91.

2 Explanation of CV-Joints

2.1 CV-Joints in cars

Every car has a minimum of four CV-joints.

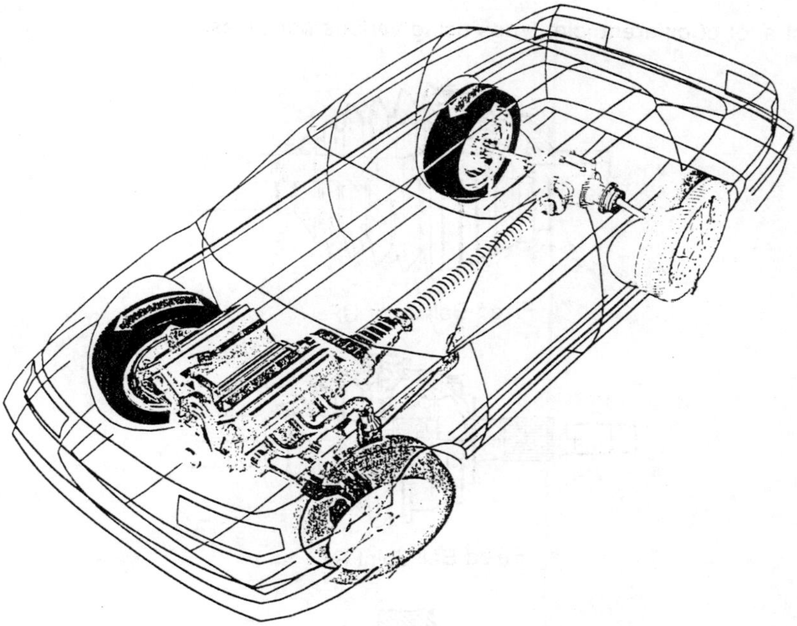

Fig 1 CV-joints in a car

In front wheel driven cars (FWD) there are four CV-joints. Two of them are fixed joints mounted at the wheels. These joints can be articulated up to 45° (e.g. for parking). The other two joints are plunging joints mounted at the gearbox. Plunging joints enable an variation of shaft length by movement of the inner components in axial direction relative to the outer race. A disadvantage of these joints is the maximal articulation angles of 22°.

Rear wheel driven cars (RWD) can have the same joints as FWD cars. This means every sideshaft consists of a fixed and a plunging joint. Besides this sideshafts with two plunging joints, self centering are possible. Two plunging joints can be used because a RWD car requires smaller maximal articulation angles than a FWD car. Additionally a RWD car may have two CV-joints in the propshaft. At least one of these has to be a plunging joint to enable an variation of the propshaft length.

Four-wheel driven cars can have up to 14 CVJ´s depending on the construction of the gearbox and the propshafts. This can mean a high expenditure in weight and cost.

2.2 CV-Joint types

There exist a lot of different joint types using various principles.

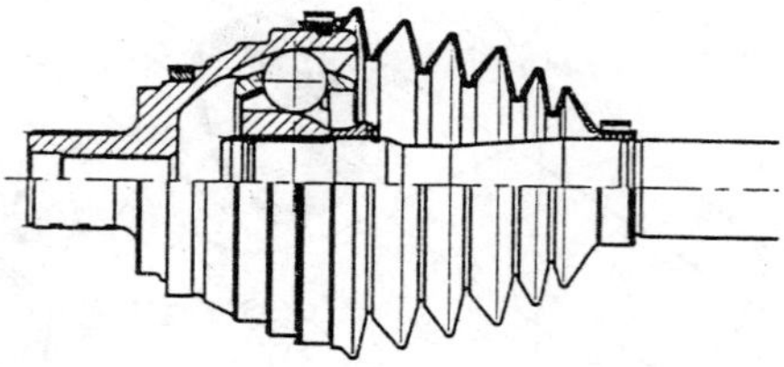

Fixed Ball Joint UF

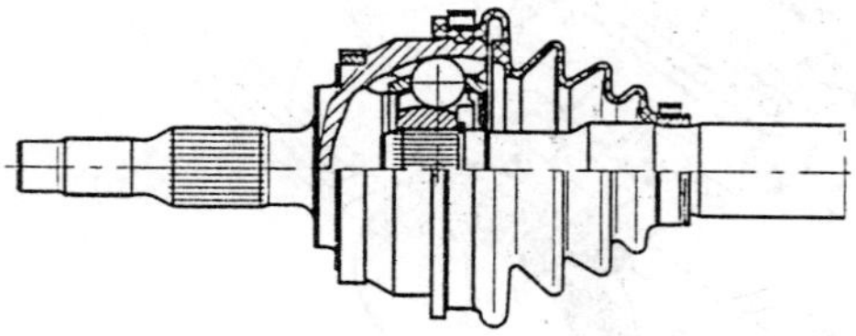

Fixed Ball Joint RF

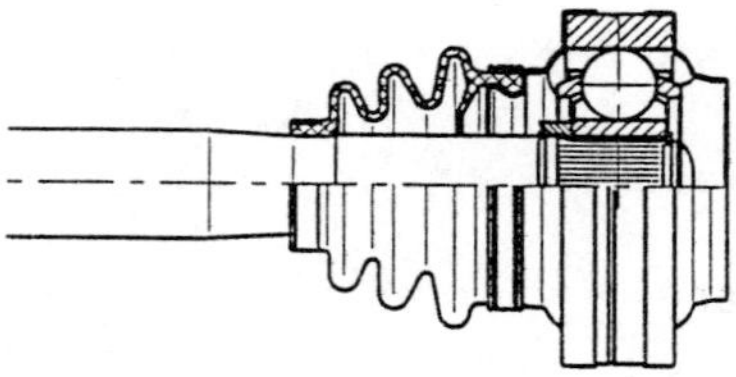

Plunging Ball Joint VL

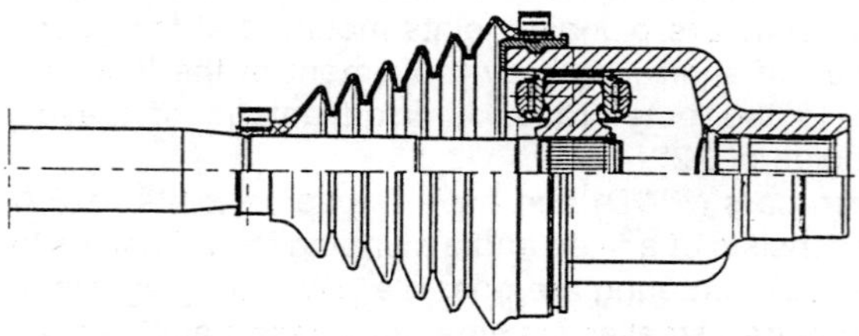

Plunging Tripod Joint AAR

Fig 2 A selection of different CV-joints

Firstly it has to be distinguished between fixed and plunging joints. Fixed joints don`t allow any axial movement between their inner and outer races. A variation in shaft length is not possible. Their advantage are their large articulation angles, up to 45°.

These joints have to be used in FWD-cars at the wheel side to enable large steering angles for parking. Fixed joints are available having two different principles, ball joints and tripods. The ball joints are identified by six balls and a cage, the tripod joints by a tripod with three trunnions.

Plunging joints enable an variable shaft length by moving the inner race relatively to the outer race. This is required accomodate shaft length variation caused ba suspension movement. The plunging joints also have to be separated into ball joints and tripodes.

3 Test rig and measurement procedure

3.1 Test rig

For the joint efficiency investigations a new test rig was built to enable measurements with the highest accuracy.

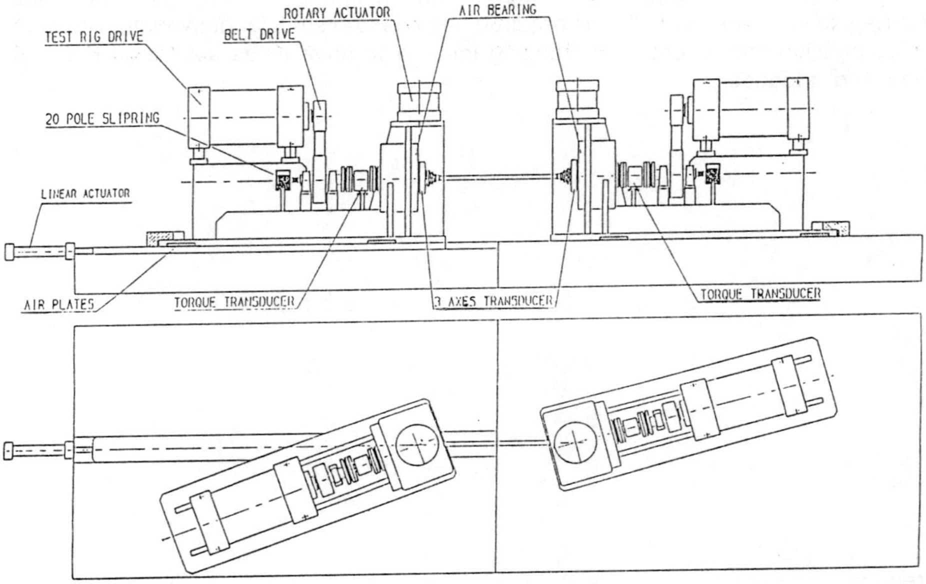

Fig 3 CVJ-Efficiency test rig

The test rig consists of two speed controlled electric motors which are connected to a driveshaft with two poly V belts. Depending on the choosen transmission ratio the rig enables rotational speeds from 50 - 2000 rpm and torques up to 500 Nm. The measured shaft is positioned between two air bearings to have the lowest mechanical losses possible. A complete sideshaft with a fixed and a plunging joint can be tested. At the left and at the right end of the test shaft there are two torque transducers to measure the torque input and output on each side of the sideshaft. A further torque transducer in the test shaft is used for the control of the motors. To test joints at any articulation angle one half of the test rig can be rotated up to 50°. The electronic control system enables the joints to be tested in each rotational and torque direction ("quadrants") at any speed, torque and articulation angle.

3.2 Torque In / Torque Out method

For testing the joints the "Torque in-Torque out" method is used.

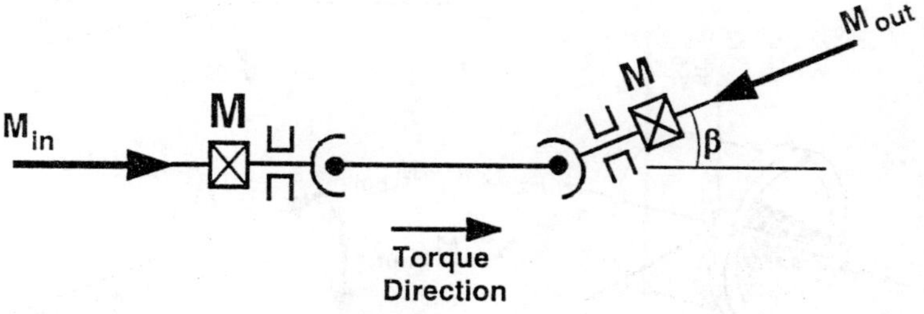

$$\text{Efficiency } \eta = \frac{Mout}{Min}$$

Fig 4 "Torque in - Torque out" principle for efficiency measurement

After mounting a sideshaft, consisting of a fixed and a plunging joint, the required torque and revolution speed is adjusted in the desired quadrant. While the joint is ran at an articulation angle of zero degrees the two torque transducers for the input and output torque are zeroed. This specifies that both joints of the mounted sideshaft have no torque losses when running at zero degrees. Furthermore the losses in the air bearings are compensated.

After articulating one half of the rig a torque difference between the torque input and torque output side occurs. The differential signal of the input and output signals shows oscillations. Therefore this differential signal is stored for 15 seconds and an average value is calculated. Using this evaluation method the results can be obtained with a high accuracy of ± 0,1 % of the transmitted torque. This accuracy enables us to find the smallest differences between joints or to indicate influences of changed joint or test parameters.

3.3 Secondary Couple Method

A second method to measure joint efficiencies is the "Secondary Couple Method".

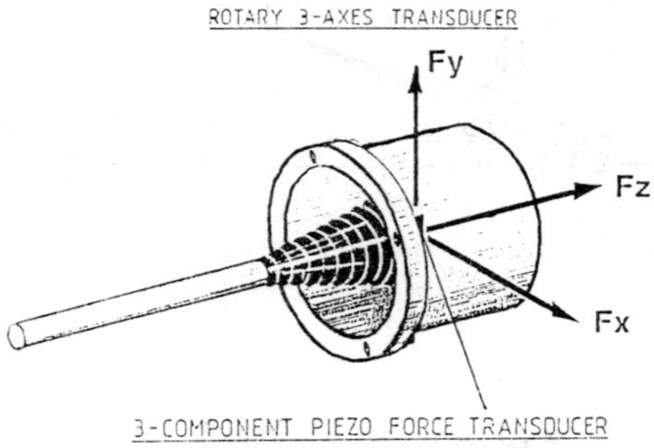

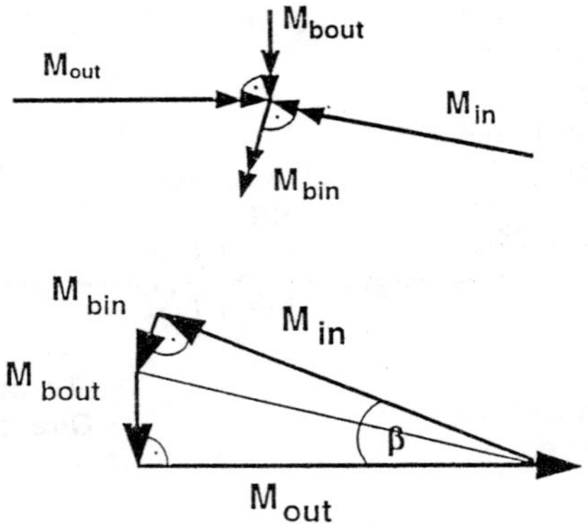

$$\eta = \cos\beta + \frac{M_{bin}}{M_{in}} \cdot \sin\beta$$

Fig 5 Secondary Couple Method

This method uses a load cell with four piezo elements which is attached to one of the air bearings of the test rig. This load cell is used to measure the forces in all three axes and the moments about all three axes. From these the efficiency of the joints can be calculated.

The advantage of this method is the higher accuracy for small joint articulation angles in comparison to the torque-in torque-out method. The disadvantage is the complicated measuring device which has many technical problems especially when used at higher revolution speeds.

4 Efficiency influencing parameters

When the efficiency investigations were started a lot of unexpected effects were found which have an influence on the joint efficiencies measured. These effects have to be taken account of exactly during the tests to get results with a high accuracy.

The parameters influencing joint efficiency are e.g. the running-in. The efficiency of new joints changes during the running-in. Therefore before starting investigations the running-in procedure had to be defined exactly. Furthermore a test torque making joints of different size and families comparable had to be defined. Changing the running-in torque and the test torque causes a different joint efficiency. The rotational speed during the tests was espected to have an influence, the joint temperature had to be held constant during the tests and had to be the same for all tested joints. The quadrant the joints are tested in, the grease and the variance of several joints even of the same joint type and the same batch had to be regarded. Only when all these parameters are known and taken account of exactly during the measurements results with a sufficient accuracy for joint comparisons can be achieved.

4.1 Running-in behaviour

When the efficiency of a new joint is investigated the efficiency improves continuously because of the running-in effect on the joint.

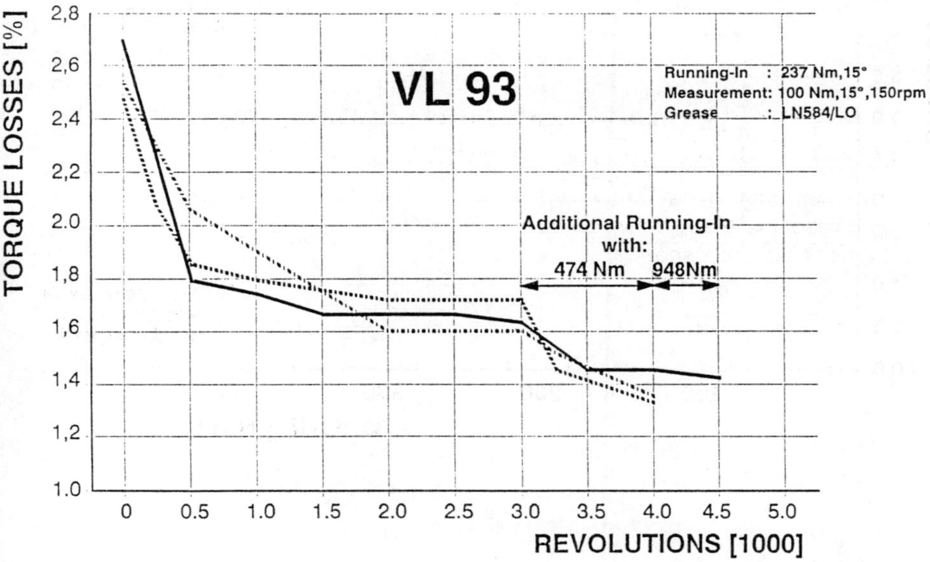

Fig 6 Torque losses during running-in

This behaviour of all joints requires the definition of a running-in torque and a running-in time. The running-in torque was applied until the efficiency of the joints remained constant within the measurement accuracy. The running-in a torque was set at 11,84% of the torque classification of each joint. This percentage was used because 5 x GT-Factor of an UF 107 joint is 11,84% of it´s torque classification (3300 Nm). The running-in torque depends on the joint size because the running-in torque has an influence on the joint efficiency when it is tested with a test torque lower than the running-in torque. Fig. 6 shows that a second running-in with the torque doubled increases the efficiency, when the efficiency is measured at the same test torque.

4.2 Test torque

One aim of the efficiency improvement investigations is to compare joints of different size, e.g. a 1300 size with a 3300 size, and to compare joints of different families, e.g. a GI with a RF joint. A parameter which would enable such a comparison, which was already available and was suitable for all joints seemed to be their torque classification. Fig. 7 shows the influence of the applied test torque on the percentage torque losses.

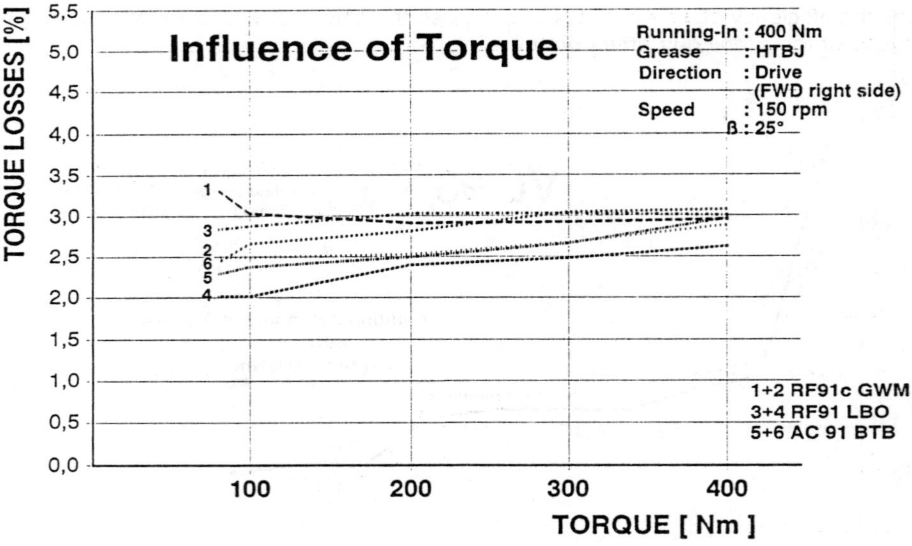

Fig 7 Influence of test torque on percentage losses

When the torque applied is increased a decrease of the percentage losses was expected. But in contrast to the expected decrease of the percentage losses each joint showed a different influence from a changed test torque. Some joints showed no measurable influence when torques between 80 - 400 Nm were applied, other joints showed an increase of the percentage torque losses as a result of an increased test torque. Therefore a definition of the test torque for all joints was required to make joints of different size and of different types comparable. To allow high speeds at high articulation angles 5 % of the joint torque classification was used for the test torque for comparison measurements.

4.3 Torque and Rotational direction ("Quadrant")

Before starting the joint efficiency investigations it had to be clarified which rotational and torque directions were to be used ie. which "quadrant" the joints should be tested in. There are different definitions existing each saying the same as shown in Fig. 8.

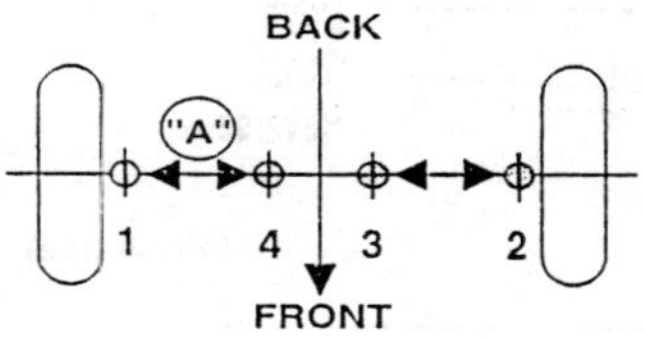

TOP VIEW

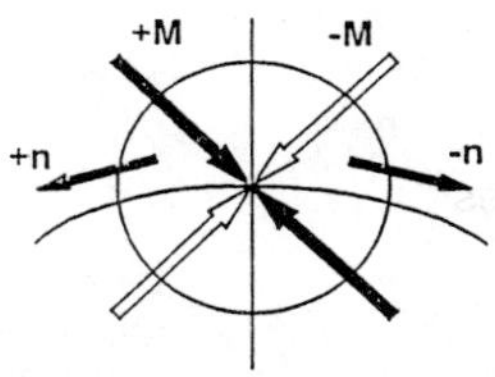

VIEW FROM "A"

Fixed joint (1)			Plunging joint (4)
drive	+n	+M	coast reverse
reverse	-n	-M	coast
coast	+n	-M	reverse
coast reverse	-n	+M	drive

Fig 8 Definition of joint "quadrants"

For the efficiency tests it was decided to test the joints in the same way as they are used in a FWD car on the right side. This means when looking from the direction of the shaft to the joint, plunging joints mounted at the gear box have to be driven clockwise with the torque input on the outer race. Fixed joints mounted at the wheel have to be rotated anticlockwise with the torque input on the shaft. For both joint types this is the "Drive" condition.

One AC 91 joint was tested in all of it`s 4 quadrants. Before testing the running-in was done in the "Drive" and "Reverse" position, ie. the torque input was on the shaft in both cases but the load was on both sides of the track. Fig. 9 shows an advantage for the torque input on the shaft for both cases, and the obtained differences depending on the tested quadrant. When testing joints of different joint types and even several joints of the same type no common tendency could be found.

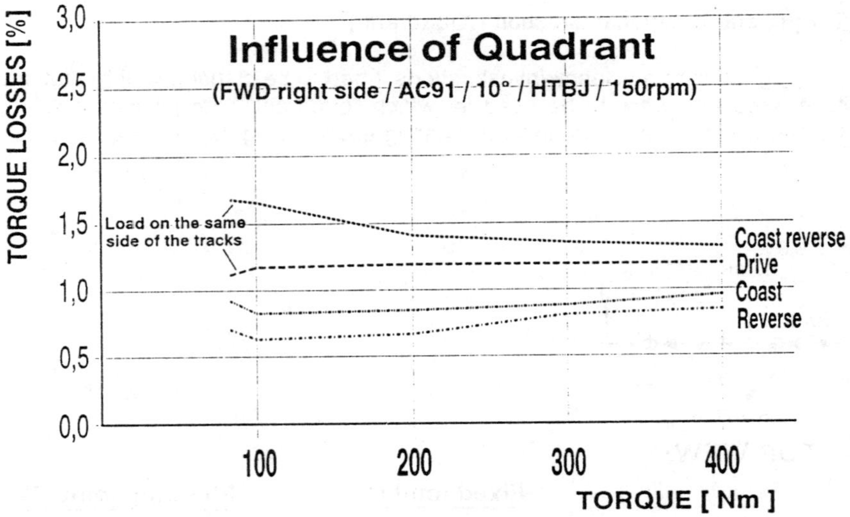

Fig 9 Influence of "Quadrant" on torque losses

4.4 Rotational speed

The test rig available was built to enable the testing of the joints with rotational speeds between 20 - 2000 rpm. It was expected that there would be a dependence of the joint efficiencies on the speed because of variance in the frictional behaviour of the individual joint components. A lot of joints of nearly all joint families were tested up to 1500 rpm and articulation angles up to 10°. When testing at these conditions no joint showed a dependency on the speed. Fig. 10 shows as an example the results of an UF 107. All joints tested showed comparable results.

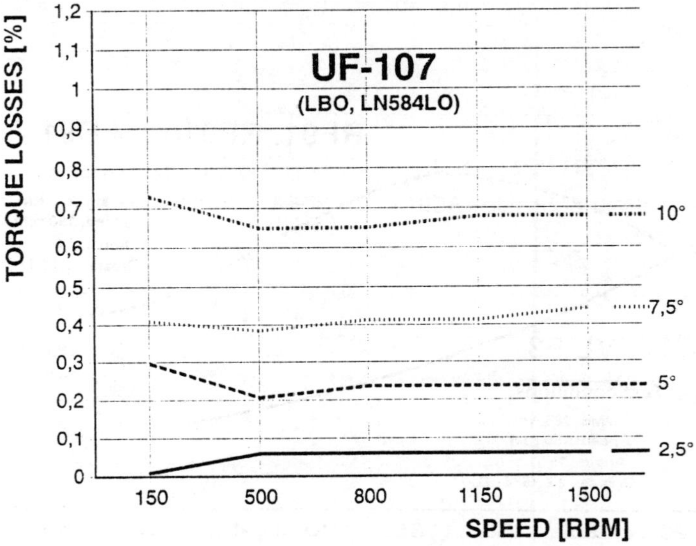

Fig 10 Influence of rotational speed on efficiency

4.5 Joint surface temperature

It was expected that the joint temperature would have an influence on the efficiency of the joints. To get reproducible efficiency results during the tests the joint temperature has to be indicated. Furthermore it is necessary to define how and where the joint temperature is taken. During all measurements the joint surface temperatures were measured with a pyrometer.

During testing the joints were additionally air cooled. All joint efficiencies quoted are for results obtained for joint surface temperatures between 20 - 25°C so that the influence of the temperature can be neglected. It has to be mentioned that there is a temperature difference between the joint surfaces and their inner components.

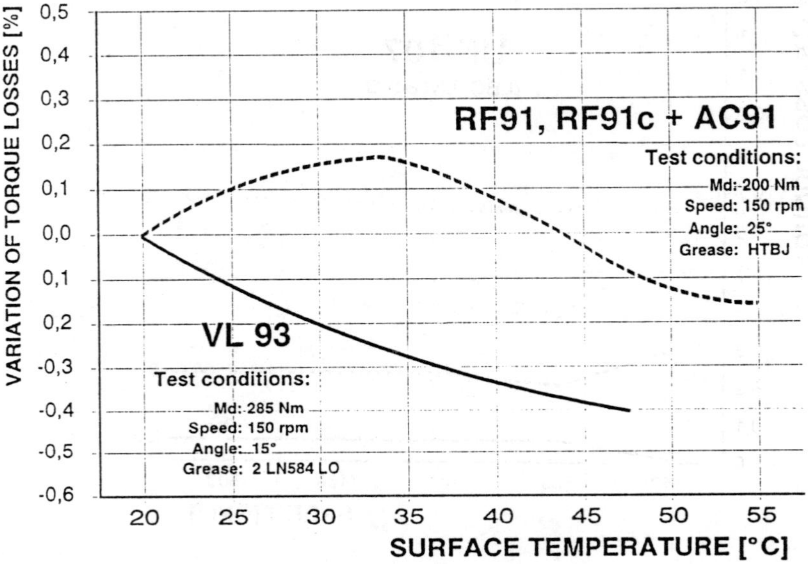

Fig 11 Influence of surface temperature on joint efficiency

The results showed that using the grease LN584 LO in a VL joint effected decreasing torque losses while the joint surface temperature increased. The investigation of the influence of the joint surface temperature on the efficiencies of RF and AC joints gave rise to a totally different behaviour in comparison to the VL joints. As the joint surface temperatures started increasing the torque losses increased by about 0,15 - 0,2 % up to temperatures of about 35°. Running the joints at higher temperatures caused a decrease of the torque losses. At the surface temperature of 55°C, the maximum obtainable with the test conditions applied, the efficiency was about 0,15% lower than at room temperature. This demonstrates the significance of always using the same temperature for comparison measurements. The reason for the torque losses increase at increasing surface temperatures can not be explained definitely. It may be that the additives of the grease are responsible for this behaviour.

4.5 Grease variation

The influence of the grease on the efficiency of a joint is of great interest. Because of the variance in joints of the same batch, different greases have to be tested in a single joint. If several joints of the same type or even of the same batch were used, it may happen that the influence of a different grease is smaller than the influence of the variance in the joints.

A VL 107 joint has been tested with several different greases applying the procedure described above. It could be confirmed that the low friction grease LFJ-HT causes a definite reduction of the torque losses in comparison to the other greases.

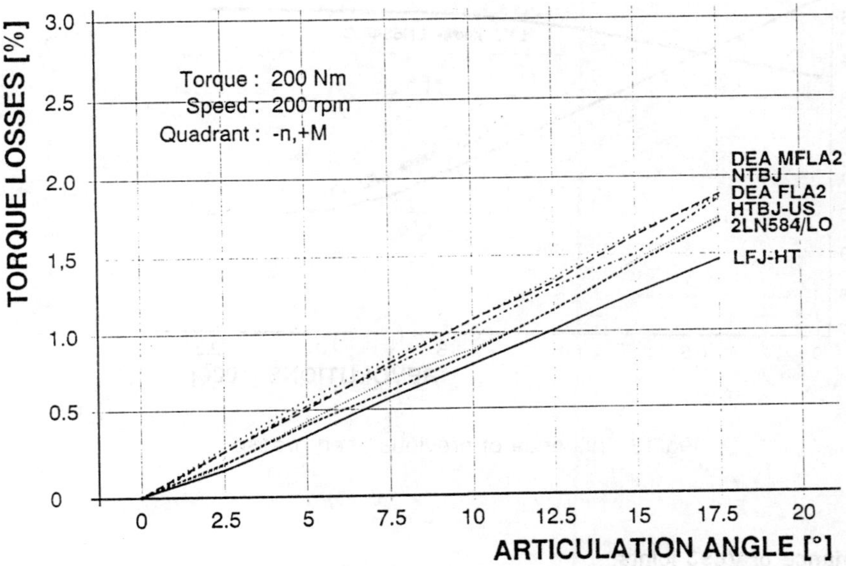

Fig 12 Torque losses of a VL 107 joint for different greases

When the grease is changed the joints have to be cleaned carefully, which is done in a washing machine at 70°C and by hand. Although after washing the joint components look absolutely clean after filling the joint with a different grease the efficiency test showed an influence of the previous grease used.

Especially when changing the grease from a standard type to the new low friction grease LFJ it takes some time before the efficiency results become constant once more. This can be explained by a chemical reaction of individual grease additives with the steel surface so that a layer is built. This layer can not be removed by cleaning such as washing with a grease solvent fluid or by hand cleaning with a rag, but only by a strong frictional load. The friction between the joint components has to remove the layer to enable reactions of the additives in the new grease with the steel. Especially for the LFJ grease this reaction causes lower friction and as a result of this a higher efficiency of the joint.

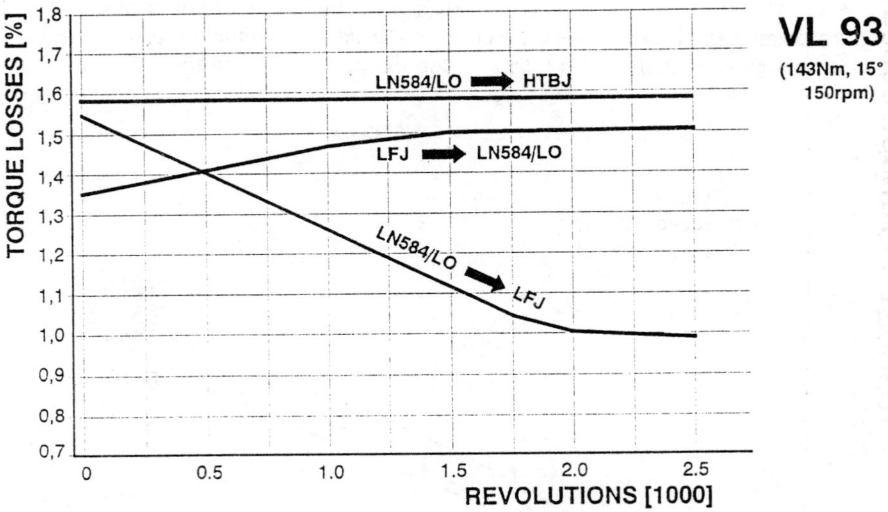

Fig 13 Influence of previous used grease

4.7 Variance of VL93 joints

If the efficiency of different joint types is to be compared it is not sufficient to test only one joint of each type. Sometimes the difference is obtained only by averaging the results of testing several joints. This has also to be taken account of when it is intended to investigate modifications of the design or surface treatment. Only statistically verified results - minimum 3 joints of every type - can be taken for comparisons. After testing five VL 93 joints from the same production batch after running-in, the joints showed a efficiency variance of about 25 %. This result shows that it is absolutely necessary to use the same joint to investigate the influence of for example different greases. Otherwise if two joints were to be be used it could be that the variance between the joints is bigger than the influence of the changed joint parameter.

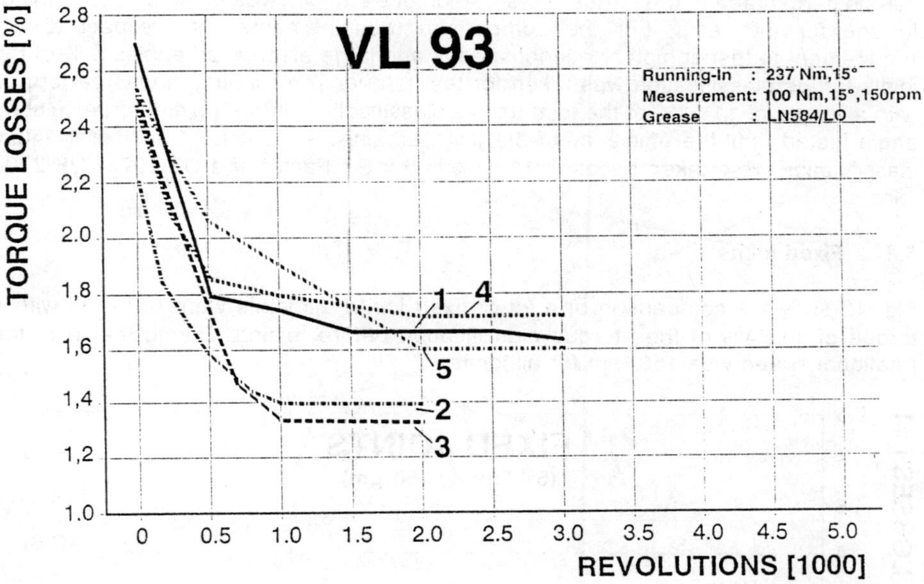

Fig 14 Efficiency variance of five VL 93 joints from the same production batch

5 Joint comparison

For comparison and for creating a data base a lot of fixed and plunging joints were tested. The most suitable and understandable procedure appeared to be to take a certain percentage of the torque classification of each joint was taken for the running-in and for the tests. For the comparison measurements with respect to the requirement to test at high rotational speeds and large articulation angles 5 % of the joints torque classification was taken for test torque. The running in was performed with a torque of 11,84% of the joint torque classification at the maximum articulation angle tested until the efficiency of the joint remained constant. 11.84% of the joint classification were taken because 11,84% is 5 x GT Factor of a UF 107 = UF 3300 joint.

5.1 Fixed joints

Fig. 10 shows a comparison of a lot of fixed joints. All joints were run in in with a torque of 11.84% of their torque classification before testing. During the tests the rotational speed was 150 rpm for all joints.

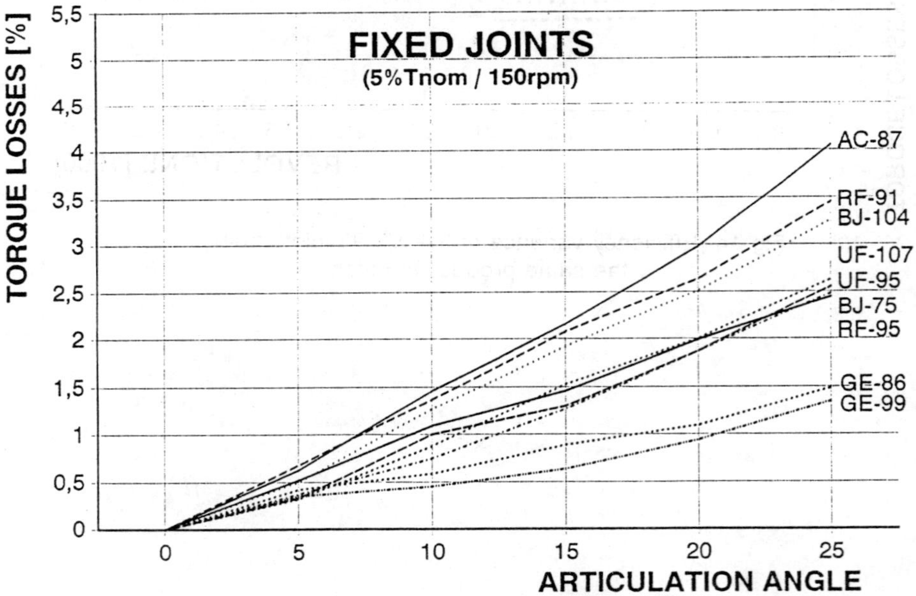

Fig 15 Comparison of fixed joints

The GE 99 showed the best results especially at bigger articulation angles. An AC 87 joint showed the worst result. It is necessary to consider that for understanding the results joints of the same batch can have a variance in the efficiency of 25%.

That means that an AC 87 is not necessarily worse than a RF 91 and that AC 87 or RF 91 joints of other batches may be nearly as good as UF joints. For comparison two NTN joints were tested, a BJ-75 and a BJ- 104.

5.2 Plunging joints

Fig. 11 shows a comparison of plunging joints. All joints were run in at 11,84% and tested at a torque of 5% of their torque classification. The rotational speed for all tests was 150 rpm. The tripod joints GI and the standard AAR showed comparable efficiencies up to articulation angles of 10°. Between 10 - 15° the AAR showed a definite advantage. The ball joints showed worse results and specially for bigger angles the DO has high torque losses. The torque losses of the tripod joints are significantly lower in comparison with plunging ball joints. The losses of the tripods are so small up to articulation angles of 10° that a variance between different joints is not measurable. The ball joints showed a variance of 25% similar to the variance of the fixed joints.

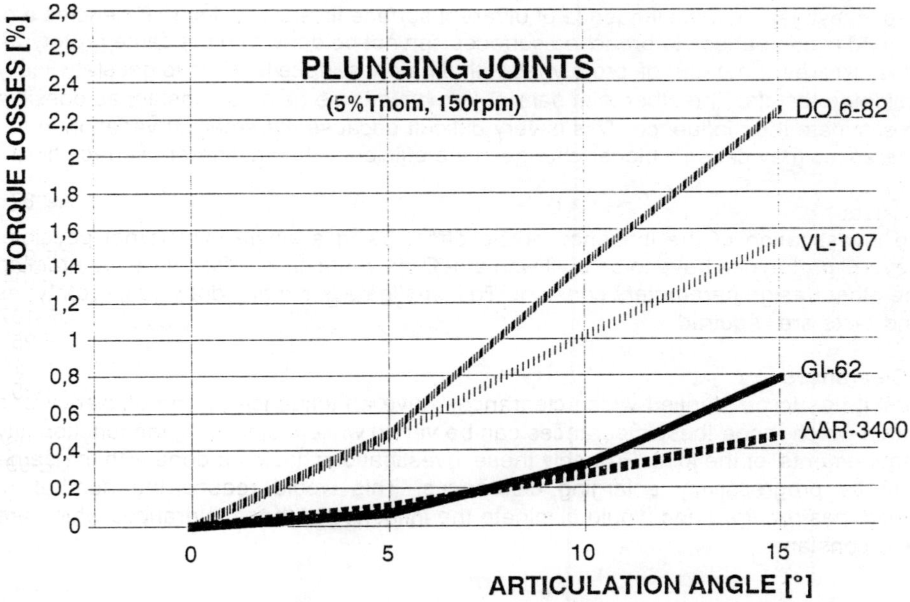

Fig 16 Comparison of plunging joints

6 Outlook

To date a lot of basic work has been done and much knowledge about joint efficiency behaviour has been collected. The complexity of all the joint efficiency influencing conditions has been observed as a result of doing these investigations. These conditions are investigated extensively with VL ,RF and AC joints. In addition to this the efficiency of a joint is a result of the many requirements for function and performance. An optimisation of a joint efficiency can not be performed without taking account of all the other joint specific requirements.

At the current stage of the investigations several features which improve the joint efficiencies are known e. g. the use of new low friction greases or assembly tolerances.

There are several additional aspects which haven´t been assessed up until now. Due to their potential for efficiency optimisation these are intended to be examined in the future. These outstanding items are:

- Surface finish

The investigation of the influence of different surface finishes on joint efficiencies e.g. ground in comparison to broached surfaces can not be done with the same prototype. This requires a number of prototypes with different surface finishes to get statistically validated results. The other joint parameters have to be held as constant as possible to eliminate their influence. This is very difficult because the smallest variation in the tolerances may cause a bigger change in the efficiency than a different surface finish.

- Offset

An investigation of the influence of the offset using a single joint is not possible. Several prototypes have to be built with a defined variation in offset while maintaining the other design parameters constant. To investigate this many drawings, prototypes and tests are required.

- Clearances

First it has to be clarified which clearances have an influence on the efficiency and within which range these clearances can be varied while maintaining the functionality requirements of the joint. Probably these investigations could be done with the same joint by progressively enlarging clearances. This would reduce the number of prototypes required and would eliminate the influence of those tolerances which are held constant.

C498/24/127/95

Factors influencing the performance of rubber CVJ boot materials

I S HOPE BSc, CEng, MIM, MWES
GKN Technology Limited, UK

SYNOPSIS

Constant velocity joint boots operate in one of the most demanding environments of any rubber component on a vehicle. High strains, grease, ozone, and extremes of temperature, cause deterioration of the material during service. An overview is given on the composition and manufacture of polychloroprene rubber boots and how individual constituents influence properties and environmental resistance. The performance of boots in service has been evaluated to quantify the changes in material properties and composition with mileage/age. Examples of typical boot failures and factors which influence service life are discussed.

1 INTRODUCTION

Constant Velocity Joints (CVJs) are used to transmit torque from the gearbox to the wheels whilst accommodating steering and suspension movements. Two types of joint are generally used on front wheel drive applications, see Figure 1.

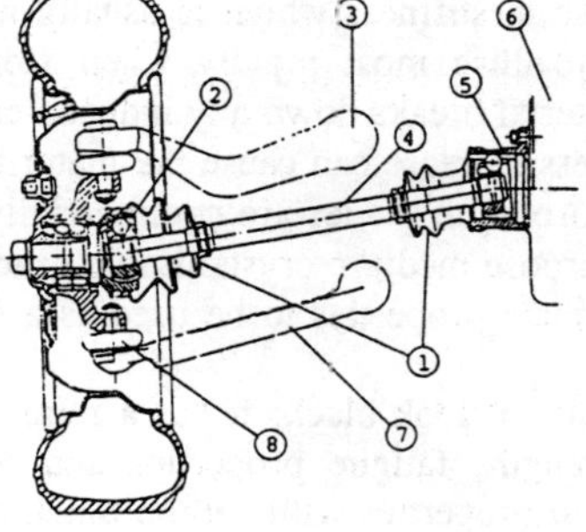

Fig 1 Location of constant velocity joints used on front wheel drive vehicles

The plunge joint, which accommodates changes in shaft length with suspension movement, is located on the gearbox side of the assembly. They are typically designed to allow up to 55 mm plunge movement and up to 25° articulation. The fixed outboard joint located by the wheel has to accommodate much larger angles of up to 47°. Examination of boots used on vehicles shows that the operating angle, and hençe strain applied to the boot, has the most significant influence on service life.

The function of the boot is to retain the joint lubricant and prevent the penetration of dirt and water into the joint. If the boot fails, the joint deteriorates very quickly. Consequently, the boot must be designed to outlast the joint. The environment in which the boot operates is the one of the most demanding for any rubber component in the vehicle and includes:

- complex fatigue loads
- wide operating temperature range, e.g. -40°C to +120°C.
- attack from ozone, heat and grease.
- impacts due to road debris.

Polychloroprene rubber is still the principle material used for CVJ boots. Thermoplastic elastomers have replaced polychloroprene for some boot applications due to superior performance but the boot is more expensive and has higher plunge resistance. Silicone and ethylene acrylic rubbers are used to provide superior high temperature performance. To ensure good performance in service, the specification of the material and product performance is critical. This paper addresses how the material constituents, manufacturing process and service environment influence the performance of polychloroprene CVJ boots.

2 COMPOSITION OF POLYCHLOROPRENE BOOT MATERIALS

Polychloroprene rubber has been used for CVJ boots for over 30 years; one of the first applications was the 'Mini'. Worldwide, there are over 100 different commercial grades of polychloroprene rubber; each have specific benefits with respect to ease of processing, mechanical performance or low temperature performance. In consequence, a blend of grades is often used in order to achieve the correct balance of properties.

The chemical structure of the polychloroprene repeat unit is:

$$(-CH_2 - CH = CCl - CH_2 -)$$

Polychloroprenes are one the few rubbers, including natural rubber, which crystallise under certain conditions. This results in an increase in stiffness which is usually measured as an increase in hardness. Polychloroprenes crystallise most rapidly when stored at -10°C; increasing the temperature or flexing the material breaks down any induced crystallinity. At very low temperatures, however, this hardness increase can cause the material to behave in a brittle manner. Low crystallisable polychloroprene grades are commercially available but can be 40% more expensive than general purpose medium crystallisable grades.

The rubber alone does not have adequate properties to be used as a CVJ boot so it is modified with various additives:

- Carbon black is added not just to make it look black; it has a reinforcing role and significantly increases hardness, strength, fatigue properties and tear resistance. Figure 2 shows the typical variation in properties with carbon black content (1).
- Plasticisers are added to help achieve the correct hardness value and also to improve processability and low temperature performance.
- Protective agents are added to improve resistance to heat ageing (antioxidant) and ozone (antiozonant).

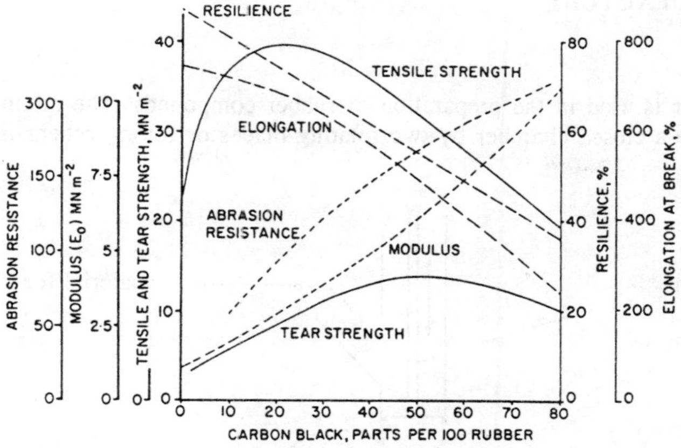

Fig 2 Variation in properties with carbon black content for natural rubber (1)

A typical CVJ boot composition is given in Table 1 (2). It can be seen that the rubber content is usually only about 50% of the material composition.

Table 1 Typical polychloroprene CVJ boot composition (2)

Constituent Material	Quantity, %
Polychloroprene rubber	47
Carbon black	33
Plasticiser	11.6
Antioxidants	1.9
Antiozonants	2
Cure agents	4.3
Processing aids	0.2

There are many different grades of carbon black, each classified in terms of particle size and structure. The individual carbon black particles do not exist discretely but combine to form chains or clusters, like a 'bunch of grapes'. The size of the carbon black cluster defines the level of structure. The carbon black content, particle size and structure all have a significant effect on processing and material properties.

The quantity/grades of plasticiser and protective agents have an important role in determining the life of a CVJ boot in service. The protective system works by producing a thin sacrificial surface layer. During processing the chemical agents are soluble in the polymer matrix; after vulcanization, their solubility is much reduced and they migrate to the free surface. Here the protective agents form a thin barrier layer which prevent attack by reacting with ozone and/or oxidative species. If this layer is broken, fresh protective agents migrate to the surface. Eventually, this reservoir of protective agents will be depleted, after which time the rubber undergoes degradation.

3 BOOT MANUFACTURE

3.1 Mixing

An internal mixer is used in the preparation of rubber compounds. The mixing process is carried out inside a closed chamber by two rotating blades or rotors, schematically shown in Figure 3.

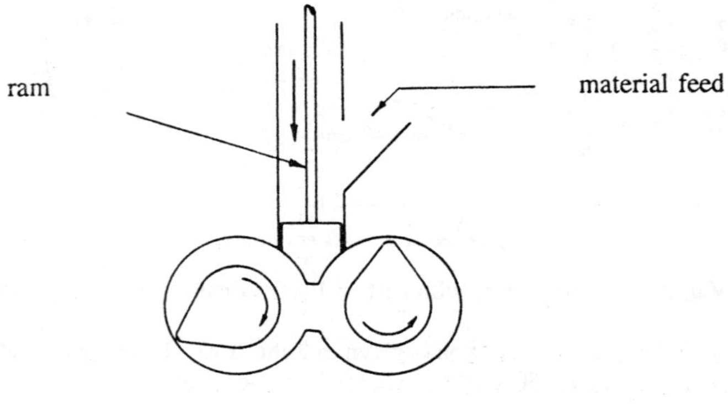

Fig 3 Schematic diagram of an internal mixer.

The sequence used to mix the rubber depends on the compounder. Either the polymer is added first and masticated for a short period (2-3 minutes) before the carbon black and plasticisers are added, or the carbon black and plasticisers are mixed first for a short period before adding the rubber; referred to as 'upside down' mixing. Once the rubber, carbon black and plasticisers are added, the ram is gradually lowered into the mixer. Sufficient dispersion occurs after 4 to 8 minutes, depending on the mixer and batch weight. Towards the end of the cycle, the 'smalls' are added which include the protective system and cure agents. These ingredients are mixed for a further period to ensure good dispersion. As the temperature of the compound begins to rise rapidly, the compound is quickly discharged and further processed into strip or pellets for moulding. The mixing process should be controlled by temperature, time and power consumption to ensure consistency between batches.

Good dispersion of the constituents materials within the rubber is critical for consistent material properties. A simple method to check the dispersion of the constituents is by optical dispersion measurements (in accordance with ASTM D2263).

3.2 Moulding

The majority of rubber CVJ boots are manufactured by injection moulding, but compression moulding is used by some suppliers. All CVJ boot injection moulding machines work on a reciprocating screw/ram principle with either vertical or horizontal clamping. The injection process is schematically shown in Figure 4.

The rubber stock in strip form is pulled into the machine by the reciprocating screw. During this stage the rubber is heated and plasticized. The rubber is fed into a space in front of the screw and simultaneously pushes the screw back until a preset volume is achieved. The screw now acts as ram and injects the rubber through a nozzle into the runner system and tool cavities. Precision steel tooling is typically used for CVJ boots with split female halves and a male core. An alternative type of injection moulding machine uses a separate screw and injection ram arranged in a V-head configuration. The moulded rubber compound is vulcanized under temperature and pressure for a specified time. During vulcanization, the next charge of material is being prepared. The boots are removed from the male cores using compressed air and the moulding cycle is repeated.

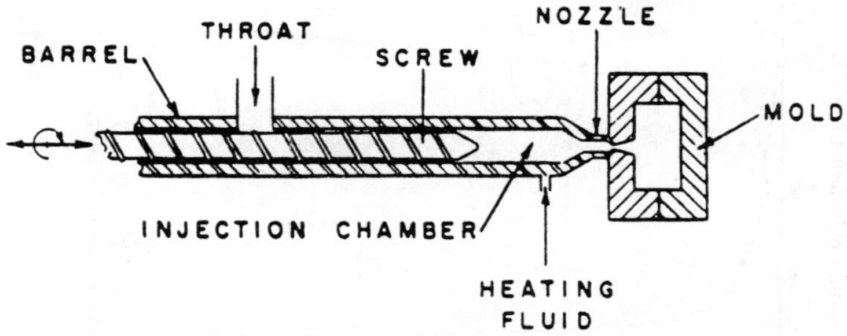

Fig 4 Reciprocating screw machine for injection moulding

4 INFLUENCE OF THE SERVICE ENVIRONMENT

4.1 Changes to Composition

The interaction of the rubber and the grease has a significant effect on the composition of the rubber. The solubility of the plasticiser and protective agents are similar to that of the base oils used in the grease. During service, the plasticiser and protective agents become leached from the rubber and the grease base oil is absorbed. The diffusion rate of the plasticiser and the grease base oil has been determined for some grease/rubber combinations. The rate of plasticiser loss can be a factor of 2x greater than the uptake of the base oil. The diffusion rate depends on the concentration gradient, temperature and the solubility of the material in the rubber. Although, the base oil acts as a plasticiser and lowers hardness, it does not impart improved low temperature performance. The amount of plasticiser present in the material is not easily determined but is best indicated by the glass transition temperature, Tg i.e the temperature at which the material changes from a rubbery to brittle material. The value depends, however, on the analysis technique used.

The composition of boots taken from vehicles has been evaluated and the level of protective agents has been found to reduce considerably with service life, stabilising to about one third of the original level after 40,000 km, as shown in Figure 5. The variation in glass transition temperature with mileage gives further indication of the fast diffusion rate, since there is a significant increase in Tg after 14,500 km which will result in a reduction in low temperature performance.

4.2 Changes to Material Properties

The fatigue performance, tensile properties and hardness have been determined for samples cut from boots with various mileage and age. The ageing process has been found to be surface dominated, bulk properties such as tensile strength and fatigue performance show no strong trend, only hardness significantly varies with mileage and age. The internal and external micro-hardness has been measured to show the influence of heat ageing and grease during service, see Figure 6. The initial hardness specification for boot examined is 63±3 IRHD. The external surface hardness increases due to heat ageing, the internal surface hardness does not increase so much due to the proximity of the grease and the diffusion of the base oil into the boot.

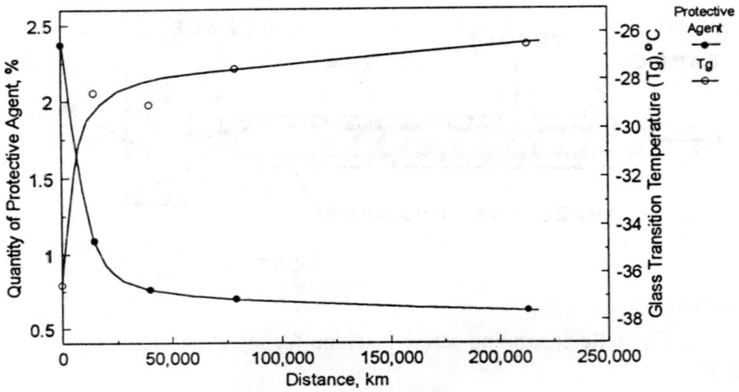

Fig 5 Variation in the level of protective agents and glass transition temperature with service life

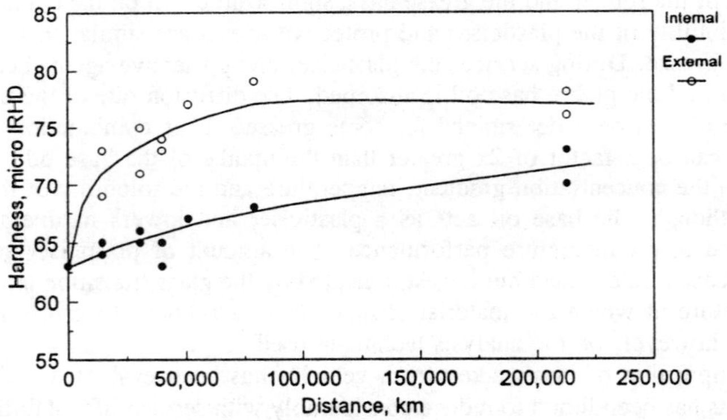

Fig 6 Variation in the internal and external surface hardness with service life

5 BOOT FAILURE

The age/mileage when a boot fails depends on the material/grease used, the boot design and vehicle application conditions i.e. installation angles, speed, packaging constraints and proximity to heat sources. Boots fail due one of three reasons:
- environmental fatigue
- collapse due to negative pressure
- accidental damage.

Boots do not generally fail by wear because they spend only limited time at high angles during service.

5.1 Environmental Fatigue

Environmental fatigue is a general term used to describe a material undergoing fatigue loads whilst being affected by the service environment i.e. heat ageing, cold temperature, ozone attack and the interaction of the grease. Environmental fatigue cracks occur in regions of high strain, typically in the convolution valleys. Figure 7 shows a boot which has failed due to environmental fatigue cracking. The crack is usually crazed in appearance which is thought to be caused by imperfections in the surface barrier layer of protective agents. Once the crack is formed on the surface it grows through the thickness of the boot ultimately causing a split and grease loss. Improved resistance to environmental fatigue is obtained by using a compound with a high rubber content and the use of protective agents which are not readily extracted by the grease and provide continuous barrier layers which do not crack under fatigue loads.

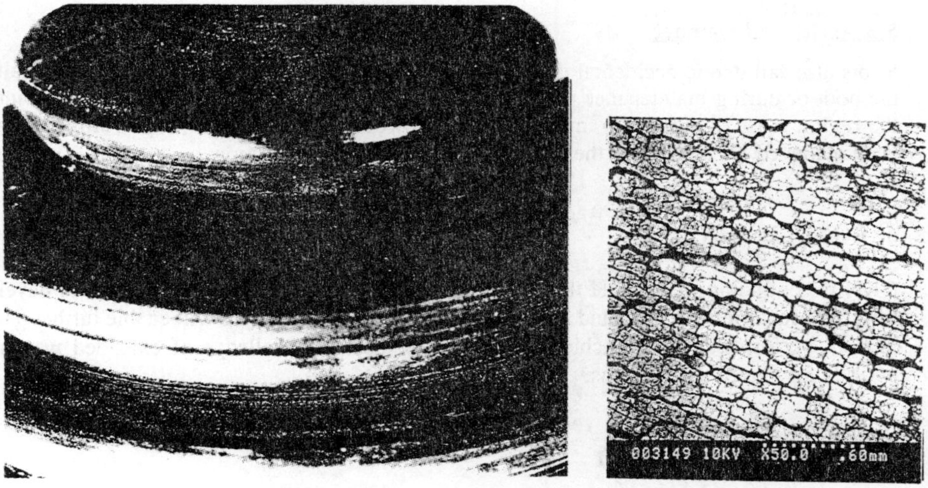

Fig 7 Environmental fatigue cracking showing detail of the cracks

5.2 Negative Pressure

Negative pressure effects are principally thought to be caused by the change of air permeability in polychloroprene with temperature. The volume of air sealed inside the boot expands when the joint temperature increases; air passes through the boot wall to equilibrate internal and external pressures. When the vehicle stops and the joint cools, the volume of air inside the boot decreases, however, the air permeability of the rubber has also reduced resulting in a partial vacuum being created. Initially this causes the boot to be sucked in towards the joint but can ultimately cause a convolute to collapse, see Figure 8. This mechanism is complex and is also related to the type/quantity of grease (volatiles) and the heat ageing resistance of the rubber. Improved resistance to negative pressure is usually achieved by increasing the convolute wall thickness and hence overall boot stiffness. The limitation with this method is that the strains in the convolute valley are also increased which will reduce boot durability.

Fig 8 Collapse due to negative pressure

5.3 Accidental Damage

Boots also fail due to accidental damage either as a result of road debris being flung against the boot or during maintenance operations. The area where the boot locates onto the joint is the region most vulnerable to impact damage, however, improvements in performance have been achieved by increasing the thickness of the boot in this region.

6 CONCLUSION

GKN is continuing to evaluate the performance of CVJ boots in service in order to develop improved materials, designs and more meaningful test methods. This will enable further gains in boot performance to be achieved in order to meet the challenge of extended warranty periods for future applications.

7 ACKNOWLEDGEMENT

The author gratefully acknowledges the permission of GKN Automotive Driveline Division to publish this paper and wishes to thank the colleagues who contributed to the activities reported.

8 REFERENCES

(1) FREAKLEY, P.K., PAYNE, A.R., Theory and Practice of Engineering with Rubber, 1978, Applied Science Publishers Ltd., Page 20.

(2) BRYDSON, J.A. Rubbery Materials and Their Compounds, 1988, Elsevier Applied Science, Page 434.

C498/24/128/95

Development of advanced constant velocity joint lubricants

G FISH BSc, PhD, DIC, ARCS, CCHEM, MRSC, MSTLE
GKN Technology Limited, UK

SYNOPSIS

Constant velocity joint (CVJ) lubricants are outlined from their introduction in the Austin Mini through to today's more advanced products, which are designed to meet increased customer needs, environmental legislation and the desire for competitive advantage. In particular, developments at GKN Technology have led to the introduction of three new greases: the first two enhance the life and performance of CVJs used in Motorsport applications; the third is a grease for use in high angle high speed propshaft CVJs in all wheel drive vehicles.

1 INTRODUCTION

Constant velocity joints (CVJ's) were invented by William Whitney in 1908 (1). However it was not until 1958 that CVJ's were accepted into their first main market in the Austin Mini. GKN is the world's largest manufacturer of CVJ's, the majority of which are used in front wheel drive passenger cars. Early CVJ lubricants included lithium greases containing molybdenum disulphide (MoS_2), calcium greases and gear oils. Comparatively little change in composition took place until the late 1980's, when higher installed angles and higher temperatures in applications placed new demands on lubrication.

In this paper, the lubrication conditions of CVJ's, grease technology and grease development will be discussed, highlighting in particular, recent developments that have given GKN market advantage.

2 LUBRICATION REQUIREMENTS OF CVJ'S.

A lubricant is designed to do four things:

1. cool
2. separate the surfaces
3. reduce friction
4. remove debris

However, a grease, being a semi-solid, does not cool very well and does not remove debris.

The early suppliers who used gear oils found that the joints survived satisfactorily, but the key problem was the then short life of the boot. The failure of the joint occurred very soon after the failure of the boot. The solution was to use grease, which is essentially a liquid lubricant which has been solidified to stop it escaping from the point of application.

The mechanism of grease lubrication is not fully understood. In some applications e.g. in the bearing industry it is believed that the grease acts as a sponge. When the contact approaches, the oil is squeezed out of the matrix of the grease, and is responsible for production of the lubricant film, which separates the surfaces. In many other applications it is believed that the grease as a whole passes into the contact. It has been shown (2,3) that soap particles from the grease lay down on the surface of the contact and provide assistance is building films. Figure 1 illustrates the thicker lubricant film formed compared to the base under steady-state uni-directional rolling.

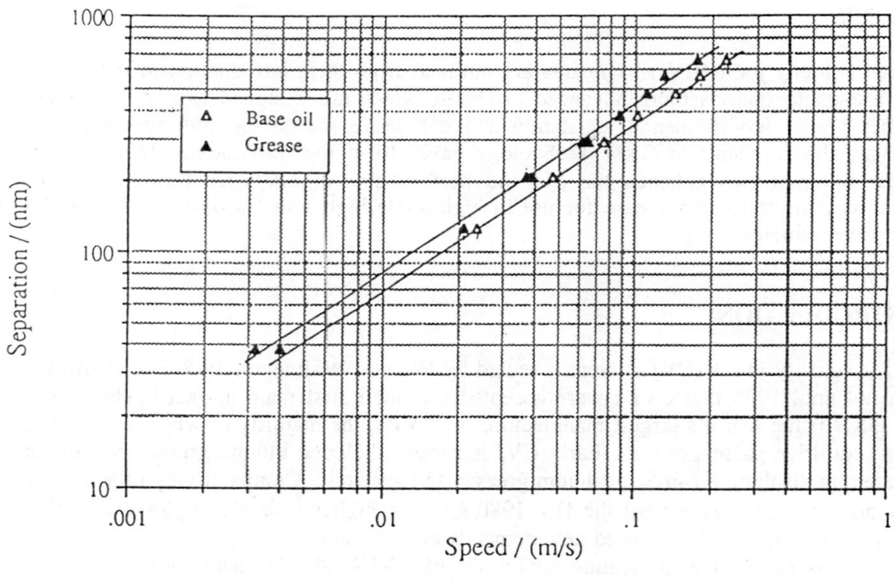

Fig 1 Speed/film thickness effects of grease and its base oil (22.9°C, 0.46 GPa) (2)

Research work in collaboration with Imperial College, London, has been funded by GKN to establish the detailed mechanism of lubrication inside a CVJ under the reciprocating rolling and sliding conditions inherent in their functioning.

It has been widely accepted that there are 10 key elements for grease lubrication. SAE J310 gives a list of those requirements (4). For ease of explanation, this SAE list has been modified to encompass the general requirements of CVJ's (table 1).

Table 1. General requirements for CVJ joint greases

1	Provide adequate lubrication to reduce friction and to prevent harmful wear.
2	Prevent the premature initiation of rolling contact fatigue
3	Protect against corrosion
4	Resist objectionable change in structure or consistency with mechanical working
5	Not stiffen excessively to cause undue resistance to motion in cold weather
6	Have suitable physical characteristics for the method of application
7	Be compatible with elastomer seals and other materials of construction in the lubricated component
8	Tolerate some degree of contamination, such as moisture, without loss of significant characteristics
9	Have suitable oxidation and thermal stability for the intended application
10	Not allow oil to separate from the matrix of the grease either in storage or at elevated operating temperatures.

Most CVJ's have specific additional requirements. For example, in ball joints (BJ's), the contacts between the balls and the tracks produce contact stresses between 1 and 6 GPa. The ball velocities are low, resulting in poor elastohydrodynamic film formation. The initial roughness is such that the specific film thickness is 1 to 2 orders of magnitude smaller than in bearings (5). The sliding between the cage and cage sphere of the inner and outer raceways results in poor lubricant film retention; MoS_2 and surface active polymers have been added to ball joint greases since 1960 to combat this. However the recent price increase in the cost of the MoS_2 has made it necessary to develop MoS_2-free greases for BJ applications.

In tripod joints(TJ's) the contact conditions are nominally much less arduous than in ball type CVJ's. The highly conformal contacts between the roller and track give rise to much lower stresses and the tendency of the rollers is to roll with only a minor component of sliding. The optimised trunnion profile also reduces the severity of the needle, trunnion and roller inner surface contacts. Modified bearing type greases, such as lithium, anhydrous and complex calcium extreme pressure (EP) greases are satisfactory for most applications.

A major problem for in-board CVJ's in particular, is the close proximity of exhaust pipes and catalytic convertors, which combined with under bonnet shielding, produces high local temperatures. These resulting high temperatures force GKN to supply joints with silicone rubber boots and high temperature grease in such applications.

In summary GKN, categorises greases into seven different classes (table 2).

Table 2 Classes of CVJ Greases

CVJ Greases	Temperature	Joints	Notes / Additives
Ball Joint (BJ)	-40 to +120°C	all normal BJs	EP + solids
Tripode Joint (TJ)	-40 to +120°C	all normal TJs	EP
High Temperature Tripode Joint	-40 to +150°C	TJs	Close to Catalytic Convertor, Exhaust Pipes
High Temperature Ball Joint	-40 to +150°C	HT BJs	Downsized Joints EP + solids
Low Friction(LF)	-40 to +120°C	high angle TJs	EP + LF
High Temperature Low Friction	-40 to +150°C	high angle TJs	All Elastomer Compatible EP + LF
High Efficiency	-40 to +150°C	High Speed	High Installed Angle BJs

3 GREASE TECHNOLOGY

A grease actually consists of an oil or synthetic base fluid, a thickener and additives to enhance performance. In order to develop new greases to meet the needs of new joints, and more severe applications, it is necessary to understand the influence of each of the constituents, and their synergistic effects.

3.1 Base Oils

The main influences of the base oils are on the prevention of wear and rolling contact fatigue and on low temperature performance. Selection of a base oil is usually a compromise. Mineral oils offer typically an operating range for the grease of -40°C to +120°C. For wider temperature ranges mineral oil/synthetic fluid mixtures can be used and fully synthetic fluids are adopted for extreme conditions. Today, most base oils used for normal temperature CVJ greases are paraffinic/naphthenic mixtures. This allows the required thickener fibre structure to be formed during manufacture, makes them compatible with the elastomers used and prevents the grease from being too hard at low temperature.

In order to choose a base oil, the requirements of the application and key base oil parameters would be examined:- viscosity at 40°C and 100°C; viscosity index; flash point; pour point. In addition, base oils should be compatible with the boot elastomers such that there is a low volume change and only a small change in hardness after immersion at elevated temperature.

3.2 Thickeners

The main grease thickeners are soaps (metal salts of fatty acids), clays and polymers. The thickener forms a lattice structure and its choice can depend on many things including the required shear stability and oil retention. Availability of base oils has an influence on thickener formation in manufacturing. For lithium soap greases it is fairly easy to select the type of fatty acid to be used in making the soap. Traditional anhydrous calcium greases use only one type of fatty acid. Complex soap-thickened greases (using mixtures of short and long chain organic acids) are difficult to make and good production controls are needed to ensure that satisfactory grease is manufactured.

If a higher temperature grease is required, then a higher temperature thickener is essential. It can be selected from the list of lithium complex, polyurea or clay. Clay greases are not widely used in CVJ's. Polyurea thickened greases are very common for higher performance greases, but they are difficult to make and the raw materials used, diisocyanates and amines are hazardous. Much work has gone into the study of polyurea and lithium complex thickeners. At this stage neither grease type is preferred and is best selected on the basis of the application, the cost and the base oils available.

3.3 Additives

At GKN Technology there is an extensive fundamental study taking place to understand additives in detail and to determine their optimum concentration levels. This will allow grease formulation to meet the particular performance criteria. Additives are basically chemical and physical in nature.

Chemical additives are those which undergo a chemical reaction when they perform their required function. Standard CVJ greases contain anti-oxidant, anti-wear, extreme pressure, and corrosion inhibitor additives, while friction modifiers are a further requirement of low friction greases. An important effect to take into account is that some additives show synergy while others can interfere with beneficial effects.

Physical additives consist of solids lubricants (e.g. PTFE, graphite, MoS_2) and polymers which function as tackiness agents, pour point depressants, and viscosity index improvers. All types of these compounds are used in CVJ greases today, their use arising either from the need to modify the base oils, to act as boundary additives, or to help the chemical additives function.

4 DEVELOPMENT

The development pressures on CVJ lubricants come from 3 main sources: The first is to satisfy the advanced needs of customers, the second is Environmental legislation and the third is to gain a competitive advantage in the market place.

Increased reliability, improved refinement and lower cost are the main pressures from customers on CVJs. Grease design plays a significant role in meeting these needs.

For Example, GKN has developed a new generation of halfshafts which are smaller, lighter and lower cost. However, by transmitting the same amount of power through a smaller component, the conditions inside the joint are much more severe and the lower grease volume is worked much harder. GKN has developed new greases which give the customer the performance they demand.

For many years environmentally undesirable additives were used in many automotive components, including greases. Lead additives give excellent performance in the prevention of wear, seizure and failure. GKN has taken a decision to not use lead in any new grease and is actively pursuing an environmentally proactive policy on all materials. Biodegradability is probably the most difficult area for the near future but such products may be available early in the next decade.

Before any grease is made, the target application is analyzed and a series of targets set. Tests are then defined to check out these requirements. A grease to meet these targets is then manufactured in the laboratory and tested.

The first tests carried out are using simple tests such as an oscillating reciprocating friction and wear testers (fig 2).

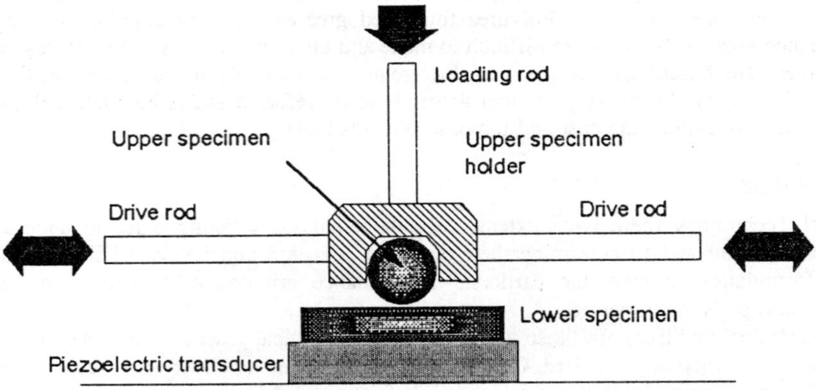

Fig 2 Oscillating friction and wear tester geometry

A series of loads and speeds are set accordingly to the application requirements. If little wear and a good stable friction trace (fig 3) is obtained the grease is then put forward for the next round of testing.

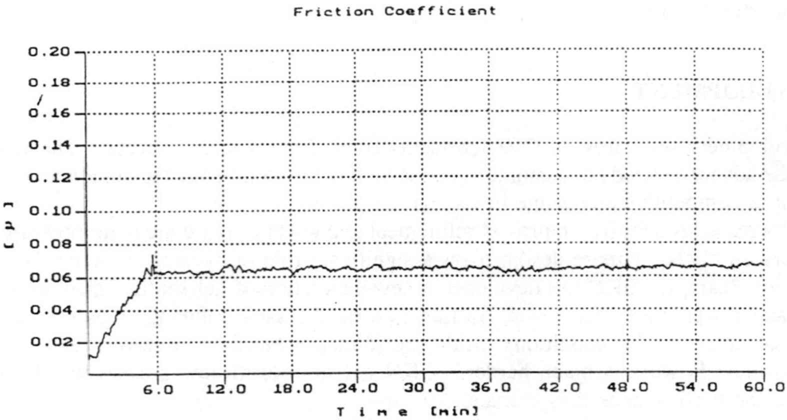

Fig 3 Grease with stable friction characteristics

The next stage after carrying out screening tests on the laboratory made sample is to have a small commercial batch made. At this stage, production quality assurance requirements can be identified and documented. This ensures that the significant characteristics of the grease are controlled. The screening tests are then repeated to ensure that the results obtained were valid.

Trial greases are subjected to a range of application tests. For an application requiring a high level of refinement, axial force testing, plunge resistance and in-car NVH ratings are carried out. If the grease is for a high efficiency application, a heat generation test would be carried out.

Elastomer compatibility tests will then be carried out with the fully formulated grease. If this is satisfactory then the grease will be tested for durability in high torque and high speed tests. If the joint rating (extent of wear and RCF) is satisfactory the grease is typically tested under a 400 hour multi-element block programme. After each series of tests the results are reviewed and compared to current greases in similar applications. The final stage of the development is to have a full production scale sample manufactured, with all the necessary quality steps and controls.

5 CASE STUDIES

The survival of a CVJ joint is dependent on the grease type and delivery to the working surfaces of components. GKN is actively developing its own range of greases, which produces differentiation and a competitive advantage in the market place.

5.1 Motorsport Development

For many years GKN has supplied special joints and driveshafts for the racing car and motorsport Industries. In 1991, an approach was made to GKN Technology to develop a new grease, targeted particularly at Formula 1 racing. Discussions were held with customers as to what was required. The specification was for a grease to survive at temperatures of up to 200°C, at speeds up to 2800 rpm, with a load of 1050 Nm, and a contact stress of 4 GPa.

At the time of writing the specification, tripod CV Joints were widely used in Formula 1, but plunging ball joints (VL/Löbro) and fixed ball joints (AC and UF) were used in other formulae. This necessitated two greases, one for use in ball joints and one without solids for tripods.

Many other requirements: high efficiency; low friction; thermal stability; oxidation resistance; low wear; good rolling contact fatigue resistance; elastomer compatibility; corrosion resistance; and no chemical additive which would have required special labelling.

A medium viscosity poly-alpha olefin with an upper operating temperature of >200°C was selected as the base oil. A stearate/azelate lithium complex thickener with a dropping point above 300°C was also chosen.

Mixtures and combinations of additives were tried and eventually two formulations were arrived at which gave both low friction and low wear. These combinations are now the subject of an international patent application. The laboratory-made combinations were manufactured in larger 50kg batches, tested in a CVJ rig and in cars at a major circuit. The new greases lasted significantly longer than the previous product and met with customer satisfaction. They also performed well in other smaller racing formulae cars.

Following this commercial batches of both Tripod and Ball joint greases were manufactured and have found success in many diverse motorsport applications.

5.2 Propshaft Grease

In the USA, there has been a sharp rise in the level of refinement expected by owners of vans and light trucks. These vehicles were traditionally rear or all wheel drive and had Hookes joints on the main propshaft, which gave rise to NVH problems. By changing the transmission layout, the refinement could be improved, but CV Joints were needed on the shaft to allow it to operate at an installed angle (figure 4) of up to 10°.

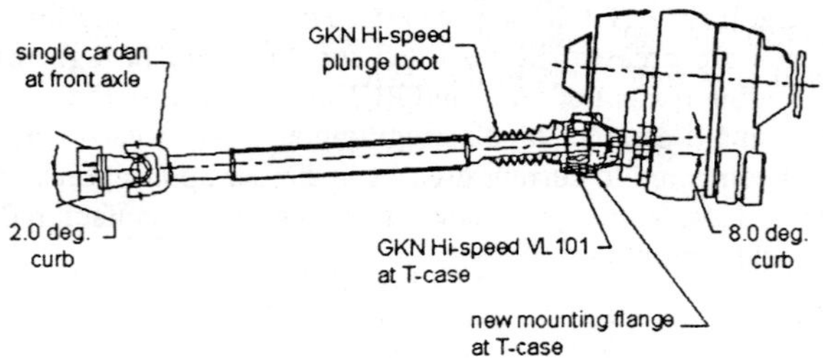

Fig 4 High speed high angle propshaft CVJ application

However as the installed angle of the joint increases the efficiency drops and a higher efficiency grease is needed. An operating range of -40°C to +150°C was demanded. In addition the grease needed to be: soft but shear stable; contain no solid additives which might be adversely affected by centrifugal force; have all components listed with no substance included which would cause the grease to be labelled or result in hazardous waste; thermally and oxidatively stable; high dropping point; low oil separation on storage and at high temperature; low volatility up to 150°C; compatibility with thermoplastic elastomers; able to absorb small amounts of water and still give protection against corrosion; not be corrosive or promote wear in the metal components.

GKN Technology was then approached to find a solution to the problem. A competitive supplier grease contained solid additives which at 4800 rpm centrifuged out to create a thick insulating layer on the surface of the outer raceway preventing lubrication. This suggested that liquid additives were preferable, including those which could help bind the grease to the surfaces and maintain good lubrication under high centrifugal forces. Cost ruled out a fully synthetic base oil, similar to that used in the motorsport grease, but many of the lessons learned in that development were applied to the propshaft grease. The resulting grease was very sticky and when tested under severe heat generation tests performed satisfactorily, and at 30% of the cost of the fully synthetic grease. Figures 5 & 6 shows the relative performance of the two GKN greases with the references greases.

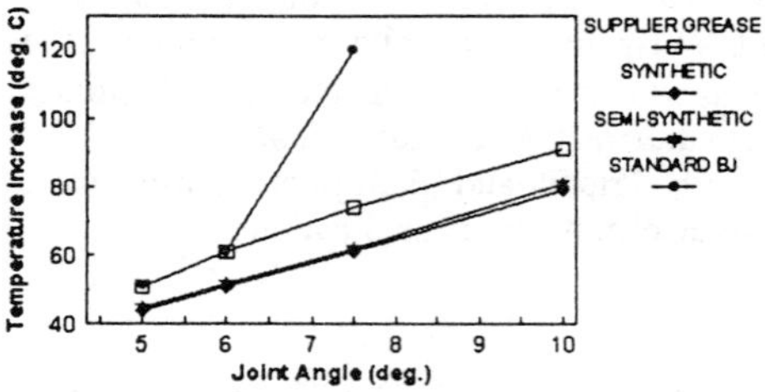

Fig 5 Heat generation in constant velocity joint (high speed, low torque)

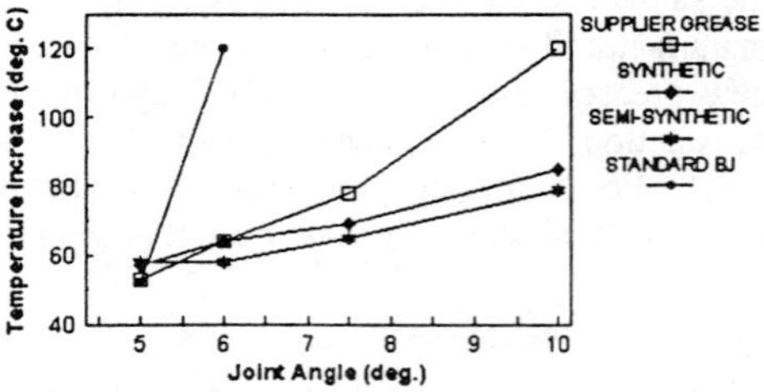

Fig 6 Heat generation in constant velocity joint (med speed, high torque)

The key requirement under the heat generation test is that the outer CVJ temperature should not exceed 100°C. This is because the maximum ambient temperature is 50°C and the boot melts above 150°C total temperature. The grease has now been made in commercial quantities and is in field trials in the USA.

6 CONCLUSION

In the five years since the grease activity was set up at GKN Technology significant progress has been made on CVJ lubricants. Recent advanced applications demonstrate that GKN now has the technology and capability to develop greases to meet customer expectations, satisfy legislation and our own environmental policies.

7 ACKNOWLEDGEMENT

The author gratefully acknowledges the permission of GKN Automotive Driveline Division to publish this paper, and wishes to thank the colleagues who contributed to the activities reported.

8 REFERENCES

(1) SCHMELZ, F., COUNT SEHERR-THOSS, H.-CH., AUCKTOR, E. Universal Joints and Driveshafts, 1992 (Springer Verlag), pp29-37
(2) JOHNSTON, G.J. Private Communication, 1992.
(3) CANN, P.M.E. Private Communication, 1995.
(4) SAE J310. Automotive Lubricating Greases, Society of Automotive Engineers, 1979.
(5) Life Adjustment Factors for Ball and Roller Bearings, an Engineering Guide Sponsored by the Rolling Elements Committee, the Lubrication Division of the American Society of Mechanical Engineers, 1971.

Transmissions

C498/30/116/95

Torque sensing and speed sensing limited slip differentials in front wheel drive vehicles

H HUCHTKÖTTER and **H TAUREG**
GKN Automotive AG, Germany
W ROHREGGER
GKN Viscodrive GmbH, Germany

1 SYNOPSIS

Limited Slip Differentials (LSD) are known mainly as driveline components in rear wheel drive and four wheel drive vehicles. Developments in recent years, however, point toward the probability that these devices will become major players in mainstream front wheel drive applications. Production applications in European and Japanese front wheel drive vehicles have demonstrated that speed sensing Limited Slip Differentials (i.e. Viscous Coupling) provide substantial improvements not only in traction on slippery surfaces but also in handling and stability even under normal driving conditions. Torque sensing Limited Slip Differentials have just entered the market in a few applications.

This paper presents a series of proving ground tests which investigate the effects of Limited Slip Differentials in front wheel drive vehicles on traction and handling. These tests were performed with a GKN Viscodrive Viscous Coupling as an example for a speed sensing type of LSD and with a torque sensing type called POWR-TRAK which is also produced by GKN Viscodrive. Testing demonstrates traction improvements but also the influences on torque steering. Factors affecting this torque steering in front wheel drive vehicles are described. Key vehicle design parameters are identified which greatly influence the compatibility of Limited Slip Differentials in front wheel drive vehicles.

Cornering tests show the influence of the Viscous Coupling and the POWR-TRAK on the self steering behaviour of a front wheel drive vehicle. Further testing demonstrates that Limited Slip Differentials can improve stability under acceleration and throttle-off manoeuvres during driving in a bend.

2 THE VISCOUS COUPLING

The Viscous Coupling is a well known component in drivetrains [1] [2] [3]. In this paper only a short summary of its basic function and principle shall be given.

The Viscous Coupling operates according to the principle of fluid friction and the slip limiting effect is thus dependent on speed difference. As shown in Figure 1 the Viscous Coupling has slip controlling properties in contrast to torque sensing systems.

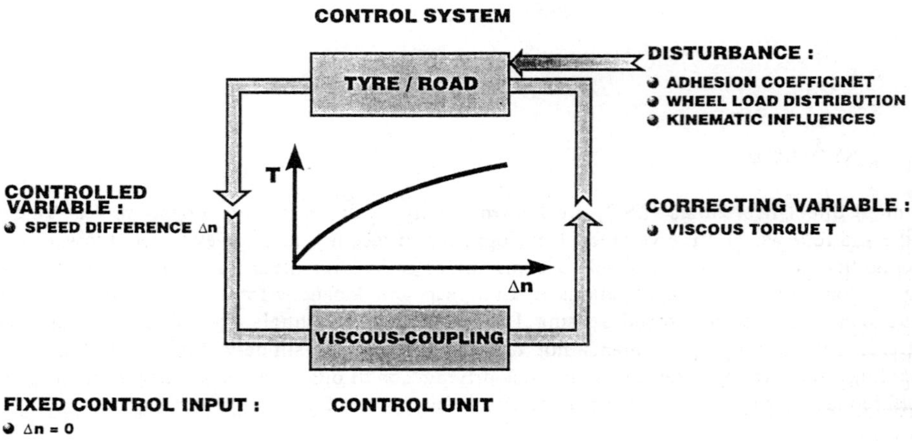

Figure 1: The Viscous Coupling as a Speed Sensing Control Unit

This means that the drive torque which is transmitted to the front wheels is automatically controlled in the sense of an optimized torque distribution.

Front wheel drive vehicles are often limited in space for the differential. So internal Viscous Couplings are limited in their maximum torque capacity or design changes to the gearbox are necessary. An external Viscous Coupling which is shown in Figure 2 can often overcome this problem.

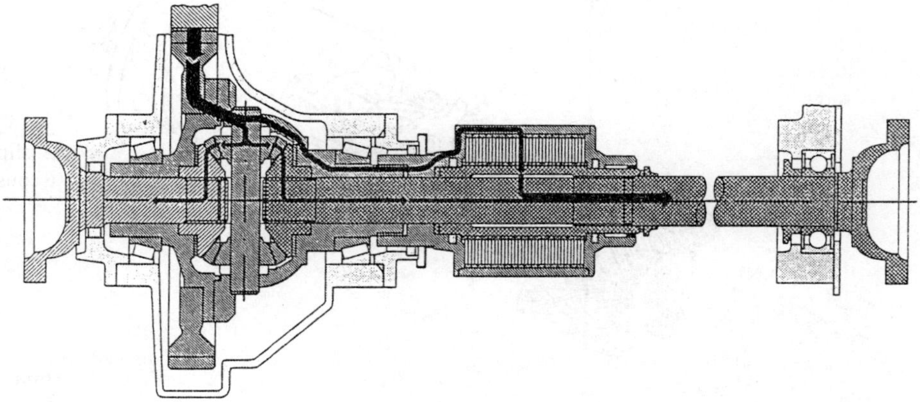

Figure 2: Cross-Section of an External Viscous Coupling in a Front Wheel Drive Vehicle (arrows show examples of power flow through the system)

This layout has some significant advantages over the internal solution. First, as mentioned before, there is usually enough space available in the area of the intermediate shaft to provide the required torque capacity. Further, only minimal modification to the differential carrier and transmission case is required. Inhouse production of differentials is thus only slightly affected. Introduction as an option can be made easily, especially when the shaft and the Viscous Coupling is supplied as a complete unit. Finally, the intermediate shaft makes it possible to provide for sideshafts of equal length which is important to reduce torque steer (shown later in section 4).

3 THE POWR-TRAK DIFFERENTIAL

The POWR-TRAK is a torque sensing limited slip device which was described in several publications [4], [5], [6]. So this paper gives only short information about its design and function.

The POWR-TRAK is a helical gear type LSD which locks the differential as a function of the drive torque. The design of it is shown in Figure 3.

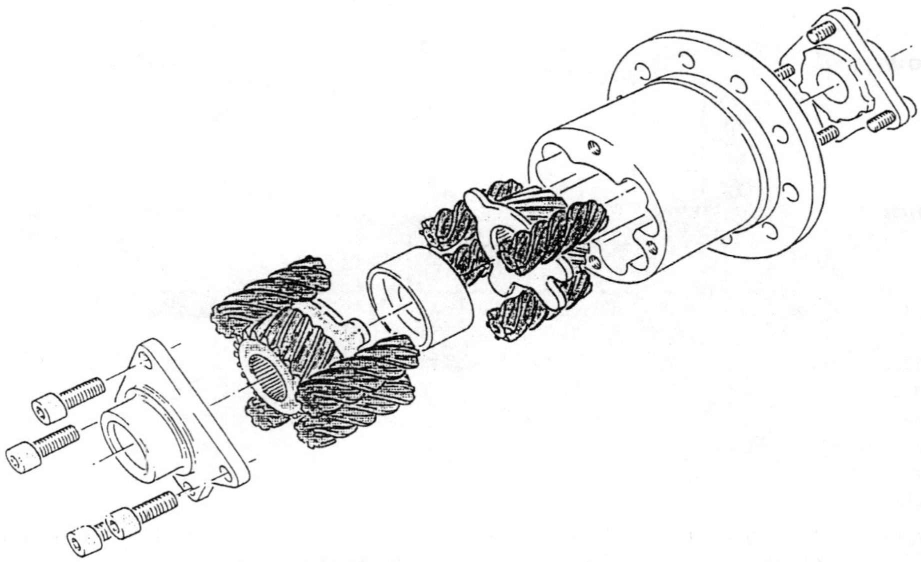

Figure 3: POWR-TRAK Helical Gear Limited Slip Differential

The drive torque generates a mechanical friction torque in the frictional surfaces of the POWR-TRAK which prevents the wheels from spinning until a certain torque ratio is reached.

The POWR-TRAK gives excellent possibilities to adjust for the Torque Bias ($T_b = T_{out,hi}/T_{out,lo}$) within a large range ($T_b$ = 1.35 to 3.5). In contrast to worm gear type LSD also relatively low locking ratios can be achieved with the POWR-TRAK which is an advantage especially in front wheel drive vehicles. The production cost is low for high volume production. Further the backlash can be adjusted to the car manufacturers requirements (1 degree to 3 degrees between the output gears in new condition). An adequate service life with almost unchanged properties (backlash, torque bias) was proven after a short running-in phase. Finally, no negative impact on the NVH characteristics of the vehicle was observed.

3 TRACTION EFFECTS

An open differential provides nearly equal tractive forces to both driving wheels because it works as a torque balancing device. It allows each wheel to rotate at different speeds during cornering without torsional wind-up in the drivetrain. These characteristics, however, can be disadvantageous when adhesion variations between the left and right sides of the road surface (split-μ) limits the torque transmitted for both wheels to that which can be supported by the low-μ wheel.

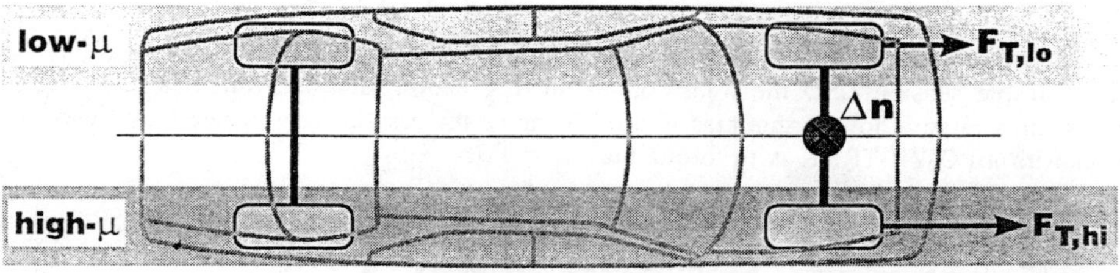

Figure 4: Unequal Tractive Forces on a Vehicle with Front LSD on a Split-μ Surface

When, for example, the maximum transmittable torque for one wheel is exceeded on a split-μ surface or during cornering with high lateral acceleration, a speed difference between the driving wheels occurs with an open differential. The resulting self-locking torque of the LSD avoids this speed difference (torque sensing) respectively resists any further increase in speed difference (speed sensing) and transmits the appropriate torque to the wheel with the better traction potential.

It can be seen from Figure 4 that the difference in the tractive forces results in a yawing moment which tries to turn the vehicle in to the low-μ side. To keep the vehicle in a straight line the driver has to compensate this with opposite steering input. Through the fluid-friction principle of the Viscous Coupling and the resulting soft transition from open to locking action, this is easily possible. With increasing steering angle the required steering torque increases accordingly. The appropriate results obtained from vehicle tests are shown in Figure 5.

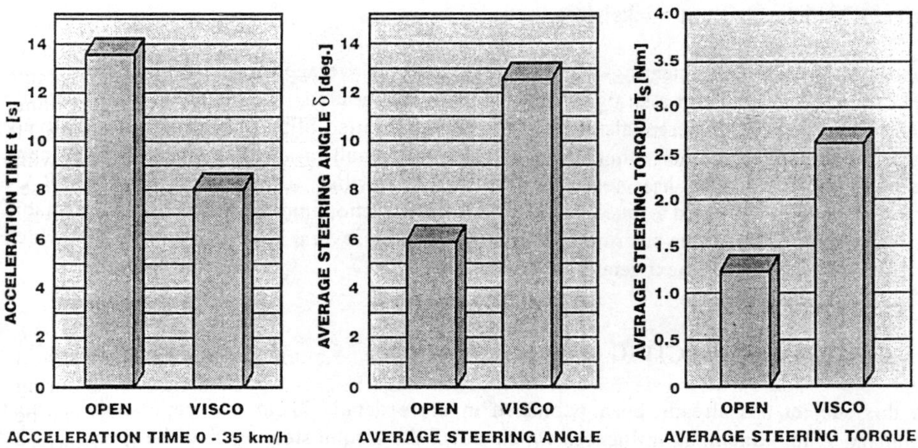

Figure 5: Traction and Steering-Influence Parameters while Accelerating on Split-μ (standing start, closed loop) with Viscous Coupling (Vehicle 1)

With a torque sensing LSD the transition to locking action is faster which is one of the reasons for a relative low torque bias. Figure 6 shows the results from acceleration tests on split-μ with a POWR-TRAK with torque bias T_b = 1.75.

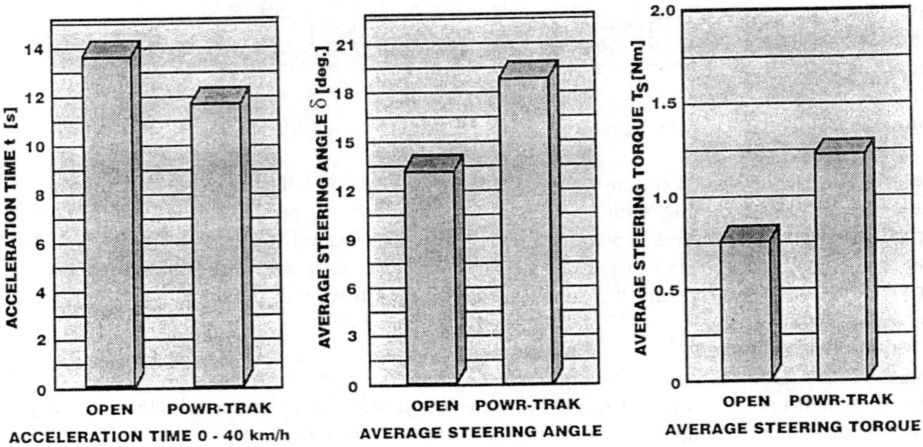

Figure 6: Traction and Steering-Influence Parameters while Accelerating on Split-μ (standing start, closed loop) with POWR-TRAK (Vehicle 2)

The differences between the results with open differential are given through the facts that different vehicles and test tracks were used.

Reported are the average steering-wheel torque T_S and the average corrective opposite steering input required to maintain a straight course during acceleration. The differences between the values with the open differentials and those with the Limited Slip Differentials are relatively large in comparison to each other. However, they are small in absolute terms. Subjectively, the steering influences are nearly unnoticeable. Further testing with POWR-TRAK differentials with increased torque bias showed further traction improvements but unacceptable torque steering. The torque steering is also influenced by several kinematic parameters which will be explained in the next section of this paper.

4 FACTORS AFFECTING TORQUE STEERING

As this subject has already been published in more detail [7], in this paper only a short summary of the kinematic influences which affect the torque steering shall be given.

The first and most important parameter is the toe-in elasticity of the front wheel suspension system. For differing tractive forces, which appear when accelerating on split-μ with Limited Slip Differentials (Figure 4), the toe-in response on the high-μ wheel is bigger than on the low-μ wheel. Unfortunately, this effect leads to an undesirable turn-in response to the low-μ side, i.e. the same yaw direction as caused by the difference in the tractive forces. Reduced toe-in elasticity is thus an essential requirement for the successful front-axle application of a Viscous Coupling and especially for torque sensing Limited Slip Differentials with a higher torque bias.

Second, the distance between the king-pin axis and wheel centre is influencing torque steering. This distance is named wheel disturbance lever arm. The tractive forces on the driven wheels result in steering torques at each wheel via the wheel disturbance lever arm. In the case of front wheel drive with open differentials the influence which can be felt by the driver on the steering wheel is only slightly noticeable, since the differences in tractive forces are small (torque balance).

For applications with Limited Slip Differentials, however, the influence is significant. Thus the wheel disturbance lever arm should be as small as possible.

Differing wheel loads also lead to an increase in torque steering so the difference should be as small as possible.

When torque is transmitted by an articulated CV-Joint, on the drive side and the driven side, differing secondary moments are produced that act around the king-pin axis. It is therefore apparent that not only differing driving torques but also differing articulations by various driveshaft lengths are also a factor. The conclusion from this is that for minimal steering influences it is important to have vertical bending angles of the sideshafts which are centered around zero (or negative, i.e. outboard joint lower than inboard joint) and sideshafts of equal length.

Of smaller influence is the distance between the wheel centre and the centre of the outboard CV-Joint (Joint Disturbance Lever Arm). For other reasons this value normally is small or zero so it is of minor influence on torque steering.

5 EFFECT ON CORNERING

Limited Slip Differentials also provide a self-locking torque when cornering, due to speed differences between the driving wheels. During steady state cornering, as shown in Figure 7, the slower inside wheel tends to be additionally driven through the LSD by the outside wheel.

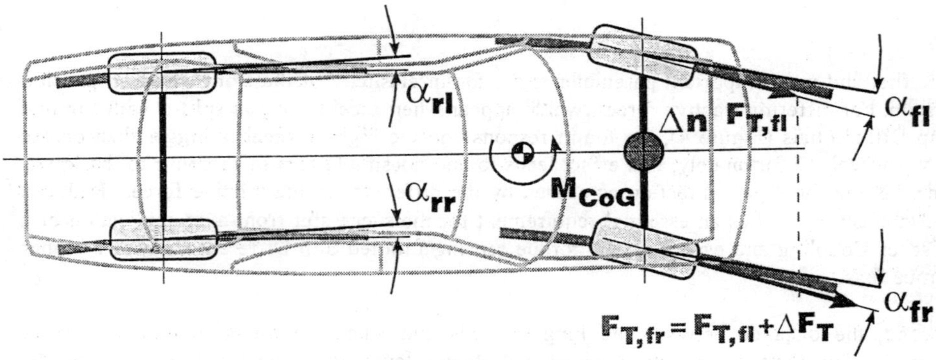

Figure 7: Tractive Forces for a Front Wheel Drive Vehicle with LSD during Steady State Cornering

The difference between the tractive forces ΔF_T results in a yaw moment M_{CoG} which has to be compensated by a higher lateral force, and hence a larger slip angle α_f at the front axle. Thus the influence of a LSD on self-steering tends towards an understeering characteristic. This behaviour is consistent with the handling bias of modern vehicles. Appropriate test results are shown in Figure 8.

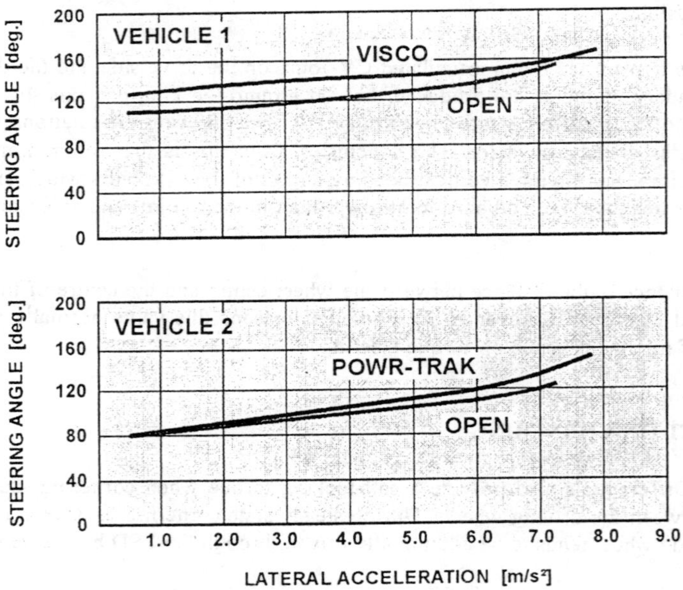

Figure 8: Self Steering Behaviour of Vehicles Fitted with Open Differentials in Comparison to Viscous Coupling (Veh. 1) and POWR-TRAK (Veh. 2) Vehicles

It is obvious with the viscous vehicle that the Viscous Coupling causes an increase in required steering angle especially at lower lateral accelerations. The differences become smaller with increasing speed due to the fact that the inner wheel gets into bigger wheel slip, which reduces the speed difference and thus the locking effect of the Viscous Coupling. The graph also shows that the car reaches higher lateral accelerations under stable conditions as the Viscous Coupling reduces the risk of spinning the inside wheel. This is the same with the POWR-TRAK which also offers the chance of reaching higher lateral accelerations under steady state conditions. The relatively small difference between the graph for the open differential and the POWR-TRAK for the concerning vehicle is due to the small torque bias of the LSD which was tested.

High powered front wheel drive vehicles with open differentials often spin their inside wheels when accelerating out of corners. In vehicles fitted with Limited Slip Differentials, this spinning is eliminated or minimized and the torque generated by the speed difference between the wheels provides additional tractive effort for the outside driven wheel, which is shown in Figure 9.

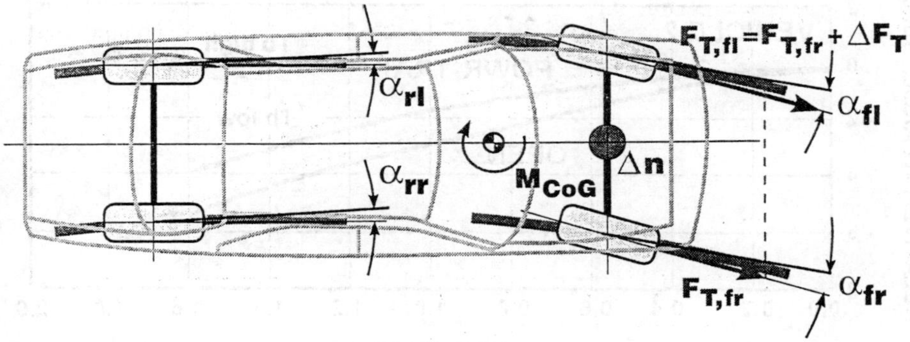

Figure 9: Tractive Forces for a Front Wheel Drive Vehicle with LSD during Acceleration in a Bend (open loop)

The acceleration capacity is thus improved, particularly when turning or accelerating out of a T-junction manoeuvre.

During acceleration in a bend, front wheel drive vehicles in general tend to understeer more than when running at a steady speed. The reason for this is the reduction of the potential to transmit lateral forces at the front tires due to weight transfer to the rear wheels and increased longitudinal forces at the driving wheels. In an open loop-control circle test this can be seen in the increase of the relative yaw angle (i.e. relative to the yaw angle which the vehicle would have maintained in case of continued steady state cornering).

Figure 10 shows the appropriate results of acceleration tests during steady state cornering.

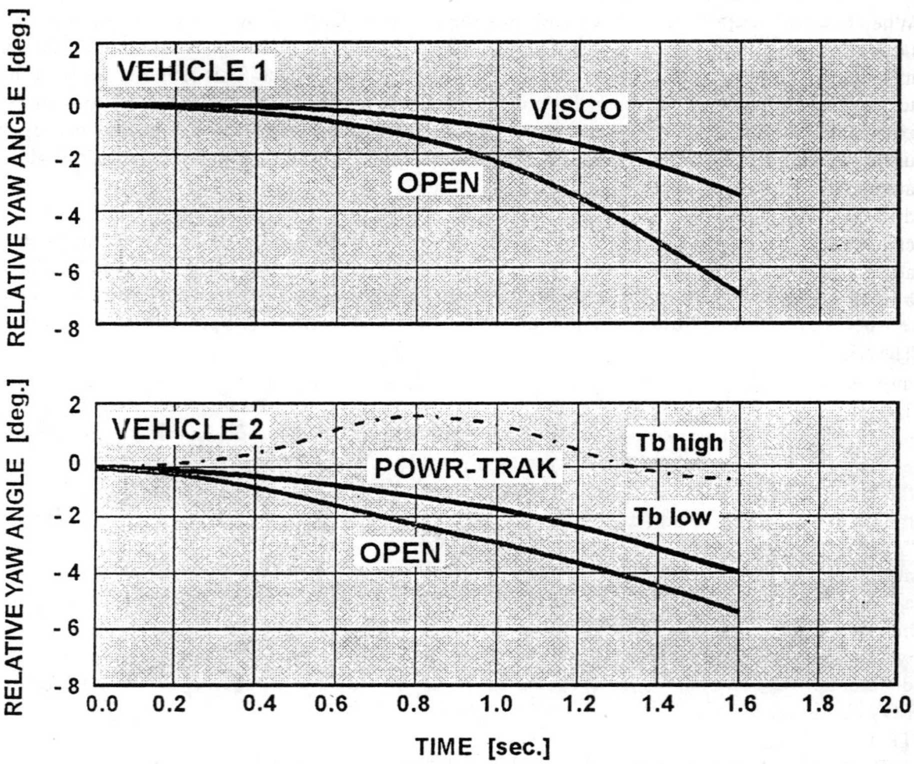

Figure 10: Acceleration Behaviour in a Bend of Front Wheel Drive Vehicles with Open Differentials in Comparison to Viscous Coupling (Veh.1) and POWR-TRAK (Veh. 2) Vehicles

It can be seen that the relative yaw angles of the open differential vehicles increases faster (in the direction of understeer) than the yaw angles for the LSD vehicles after starting to accelerate (time "0").

A car with Limited Slip front Differential thus has a more stable initial reaction under accelerating during cornering than a vehicle with an open differential, reducing its understeer. This is due to the imbalance in the front wheel tractive forces caused by the LSD as explained before. The tractive force difference ΔF_T (see Figure 9) causes a yaw moment M_{CoG} acting in direction of the turn, countering the understeer.

When the adhesion limits of the driving wheels are exceeded, a LSD vehicle understeers more noticeable than a comparable vehicle with open differential (here approximately 2 seconds after starting to accelerate). On very low friction surfaces, such as ice, stronger understeer is to be expected when accelerating in a curve with a LSD because the driving wheels - connected through the LSD - can be made to spin both at the same time more easily (power-understeering). This characteristic can, however, be easily controlled by the driver or by an automatic throttle modulation traction control system. Under these conditions a front wheel drive vehicle with a LSD is always much easier to control than a rear wheel drive car which can exhibit power-oversteering when accelerating during cornering. All things considered, the advantage through the stabilized behaviour of a LSD equipped vehicle during acceleration in a bend under normal conditions outweights the small disadvantage on slippery surfaces.

The acceleration of the Viscous Coupling equipped vehicle from steady state cornering was improved by about 15 % in comparison to the car with the open differential. The POWR-TRAK gave only very little improvements in acceleration which is due to the small engine power of that vehicle and the small torque bias of $T_b = 1.75$. Increasing the torque bias for better traction improvements can cause the following disadvantage: The difference between the tractive forces (ΔF_T in Figure 9) can become so big that the accordingly increasing yaw moment M_{CoG} over-compensates the normal understeer reaction (see Figure 10, graph for "$T_{b, hi}$"). This means that such a vehicle has an initial oversteering reaction followed by an abrupt change to heavy understeer which probably is hard to control by an average driver of a front wheel drive vehicle who is typically only accustomed to understeering self steering behaviour. The maximum speed differences caused by the spinning of the inside driven wheel which occurred during these tests was reduced from approximately 240 rpm with open differential to 100 rpm with the Viscous Coupling. The maximum speed difference in the POWR-TRAK vehicle was reduced from approximately 90 rpm to 70 rpm by the POWR-TRAK.

Throttle-off reactions during cornering, caused by releasing the accelerator suddenly, usually result in turning of the vehicle into the turn (throttle-off oversteering). High-powered models which can reach high lateral accelerations show the heaviest reactions. This throttle-off reaction has several causes such as kinematic influences, or as the vehicle attempting to travel on a smaller cornering radius with reducing speed. The essential reason, however, is the dynamic weight transfer from the rear to the front axle, which results in a momentary excess of lateral guidance at the front axle and a lack at the rear axle respectively, turning the vehicle towards a smaller circle. Because the rears are not transmitting driving torque, the influence on the rear axle in this case is greater than that of the front axle. The driving forces on the front wheels before throttle-off (see Figure 7) become overrunning or braking forces afterwards, which is illustrated for the LSD equipped vehicle in Figure 11.

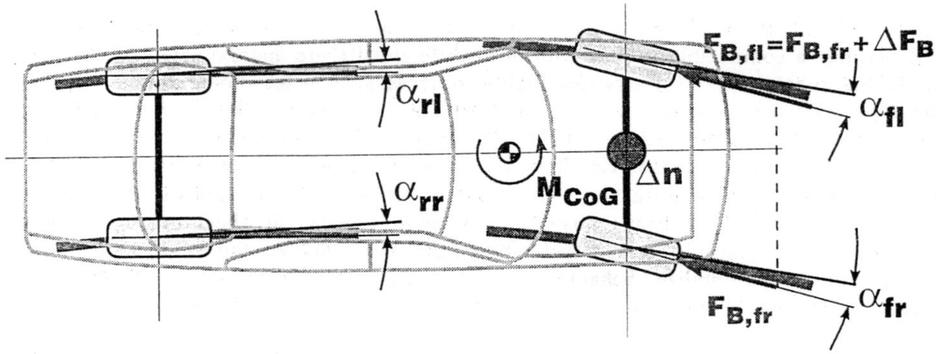

Figure 11: Braking Forces for a Front Wheel Drive Vehicle with LSD after Throttle-Off in a Bend (open loop)

As the inner wheel turns more slowly than the outer wheel, the LSD provides the outer wheel with the larger braking force F_B. The force difference ΔF_B between the front wheels applied around the centre of Gravity of the vehicle causes a yaw moment M_{CoG} that counteracts the normal turn-in reaction, i.e. the throttle-off oversteering is reduced.

The relative yaw angles after throttle-off (time "0") for the viscous vehicle (Veh. 1) in comparison to the same vehicle with open differential and for the POWR-TRAK car (Veh. 2) in comparison to the same vehicle with open differential are shown in Figure 12..

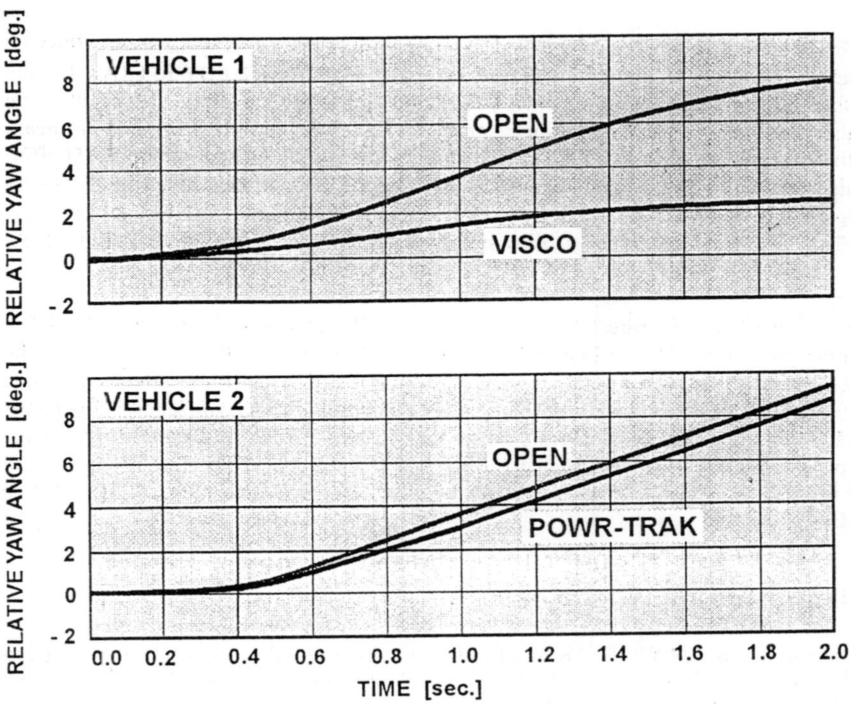

Figure 12: Throttle-Off Behaviour in a Bend of Front Wheel Drive Vehicles with Open Differentials in Comparison to Viscous Coupling (Veh. 1) and POWR-TRAK (Veh. 2) Vehicles

The relative yaw angles show a pronounced increase (in the direction of oversteer) after throttle-off (time = "0") for both vehicles provided with the open differentials. This increase of the yaw angle is significantly reduced in Vehicle 1 with a viscous LSD. Only slight stability improvements were found in the POWR-TRAK vehicle (Veh. 2). There are several reasons for the difference in the improvement potential of Viscous Coupling and POWR-TRAK. First, the torque bias of the POWR-TRAK was relatively small and with increased torque bias better results can be expected. The second reason is that the throttle-off effect is most pronounced in the area close to the maximum lateral acceleration. In this area the difference between the required steering angles for Vehicle 1 is small between the version with open differential and the version with the Viscous Coupling (see Figure 8). Thus the slip angles on the front wheels are also similar and the yaw moment provided by the unequal braking forces caused by the Viscous Coupling after throttle-off are fully used to counteract the oversteer reaction. In Vehicle 2 the required steering angle for the POWR-TRAK version is larger than for the version with the open differential and so are the slip angles on the front axle. Higher slip angles on the front wheels cause a heavier throttle-off reaction, that means the throttle-off oversteer with POWR-TRAK would be more pronounced without LSD effort after throttle-off than with the open differential. So the yaw moment caused by the difference in braking forces with the POWR-TRAK have to compensate a potentially heavier oversteer reaction. This results in the smaller stability improvements which were exhibited during these tests.

A normal driver of a front-wheel drive vehicle is usually only accustomed to neutral and understeering vehicle handling behaviour. The driver can then be surprised by sudden and forceful oversteering reaction after an abrupt release of the throttle, for example in a bend with an unexpected decreasing radius. This vehicle reaction is further worsened if the driver over-corrects for the situation. Accidents where cars leave the road to the inner side of the curve is proof of this occurrence. Hence a LSD and especially the Viscous Coupling LSD improves the throttle-off behaviour while remaining controllable, predictable, and safer for an average driver.

6 EFFECT ON BRAKING

A Viscous Coupling or POWR-TRAK in a front wheel drive vehicle without ABS (Anti-Lock Braking System) has only a very small influence on the braking behaviour. The influence can only be found on split-μ surfaces because on homogeneous-μ surfaces a LSD has no effect. Hence the front wheels are connected partially via the LSD, the brake pressure to lock the front wheel on the low-μ side is slightly higher than in a vehicle with an open differential. On the other side, the brake pressure to lock the front wheel on the high-μ side is slightly lower. These differences can be measured in an instrumented test-vehicle but are hardly noticeable in a subjective assessment. For the POWR-TRAK this is only affective if the vehicle is braked with a gear engaged. In neutral or with the clutch opened there is no overrunning torque transmitted to the differential input and so - as POWR-TRAK is a torque sensing device - no locking torque between the wheels can occur. The locking sequence of front and rear axle is not influenced by the LSD.

Most ABS offered today have individual control of each front wheel. Electronic ABS in front wheel drive vehicles must allow for the considerable differences in effective wheel inertia between braking with the clutch engaged and disengaged.

Partial coupling of the front wheels through a LSD should not therefore compromise the action of the ABS. The only theoretical exception to this occurs on a split-μ surface if a yaw moment build-up delay or Yaw Moment Reduction (YMR) is included in the ABS control algorithm. Figure 13 shows typical brake pressure sequences, with and without YMR.

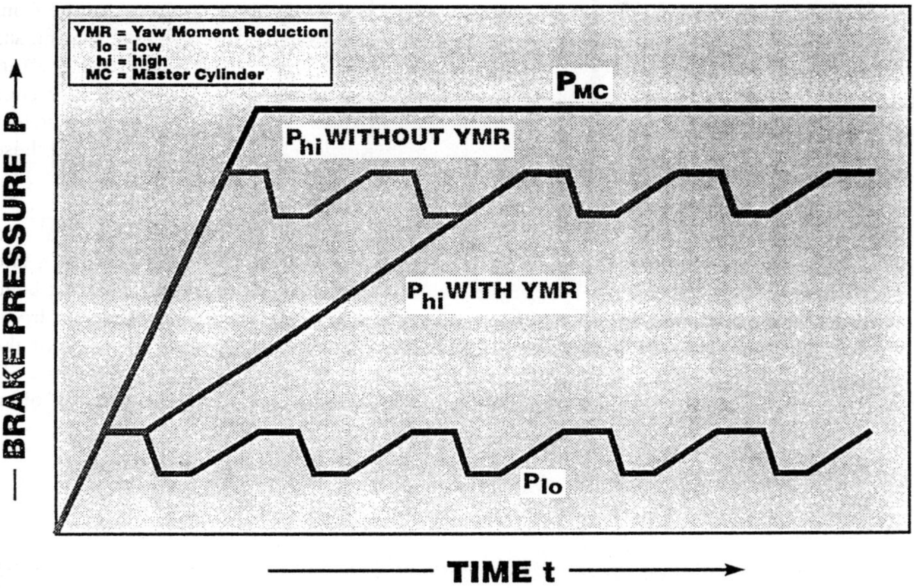

Figure 13: Brake Pressure Build-Up Characteristics for the Front Brakes of a Vehicle Braking on Split-μ with ABS

In vehicles with low yaw inertia and a short wheelbase, the yaw moment build-up can be delayed to allow an average driver enough reaction time by slowing the brake pressure build-up over the ABS for the high-μ front wheel. The wheel on the surface with the higher friction coefficient is therefore, particularly at the beginning of braking, under-braked and runs with less slip. The low-μ wheel, in contrast, can at the same time have a very high slip, which results in a speed difference across the differential. The resulting self-locking torque with LSD then appears as an extra braking force at the high-μ wheel which counteracts the YMR.

Although this might be considered as a negative effect and can easily be corrected when setting the YMR algorithm for a vehicle with a front LSD, vehicle tests have proved that the influence of a viscous LSD is so slight that no special development of new ABS / YMR algorithms are actually needed. Some typical averaged results are summarized in Figure 14.

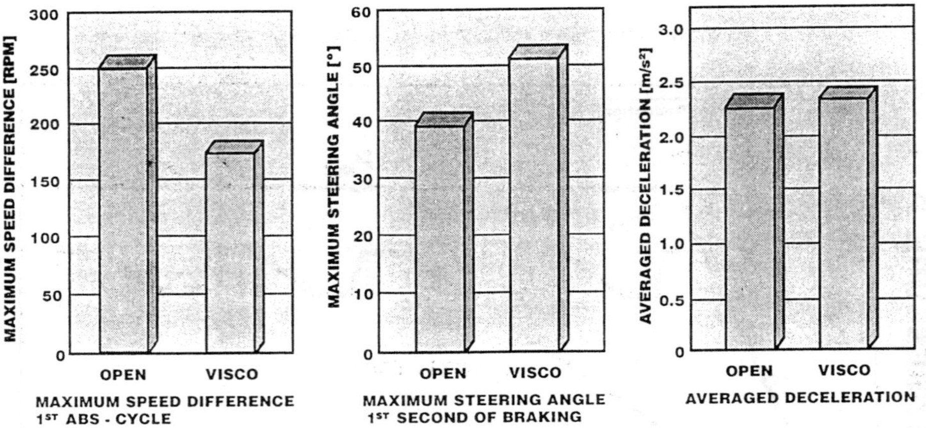

Figure 14: Results from ABS Braking Tests with YMR on Split-μ (v_0 = 80 km/h 3rd gear, closed loop)

Unfortunately no results from ABS braking tests with POWR-TRAK are available at the moment because the test vehicle was not equipped with ABS.

In Figure 14 on the left a comparison of the maximum speed difference which occurred in the first ABS control cycle - where YMR is fully affective - is shown. It is obvious that the Viscous Coupling is reducing this speed difference. As the Viscous Coupling counteracts the YMR, the required steering wheel angle to keep the vehicle in straight direction in the first second of braking increased from 39° to 51° (Figure 14, middle). Since most vehicle and ABS manufacturers consider 90° to be the critical limit, this can be tolerated. Finally as the self-locking torque produced by the Viscous Coupling causes in increase in high-μ wheel braking force, a slightly improved vehicle deceleration was maintained (Figure 14, right).

7 SUMMARY

In conclusion, it can be established that the application of a Viscous Coupling in a front-axle differential as well as of a POWR-TRAK does not only improve traction, as it is generally expected from a Limited Slip Differential. These devices also positively influence the complete vehicle handling and stability, with only slight, but acceptable influence on torque-steer.

To reduce unwanted torque-steer effects a basic set of design rules have been established:

- Toe-in elasticity of the front axle suspension system must be as small as possible.
- Distance between king-pin axis and wheel centre has to be as small as possible.
- Vertical bending angle-range should be entered around zero (or be negative).
- Vertical bending angles should be the same for both sides.
- Sideshafts should be of equal length.
- Wheel loads should be equal.
- Distance between king-pin axis and centre of outboard CV-Joint should be zero.

Braking with and without ABS is only negligibly influenced by a LSD. Traction is improved by LSD.

The self steering behaviour of a front wheel drive vehicle is slightly influenced by a LSD in the direction of understeer while the cornering speed limit is increased. The improved reactions to throttle-off and acceleration during cornering make a vehicle with LSD in the front axle considerably safer. These stability improvements are more pronounced with a speed sensing LSD like the Viscous Coupling than with a torque sensing device like the POWR-TRAK. Therefore the Viscous Coupling is the more suitable product in mass production front wheel drive vehicles which are used by normal drivers. The POWR-TRAK - especially with higher Torque Bias values - should be seen as a device for sporting front wheel drive vehicles because of its potential to change the reaction when accelerating in a bend from understeer to oversteer which requires a more experienced driver.

8 BIBLIOGRAPHY

[1] H. Taureg: Application and Function of Viscous Couplings in Automotive Transmissions AutoTech 1989 - C399/5

[2] H. Taureg / G. Herrmann: Application of Viscous Couplings for Traction Control in Passenger Cars SAE 890524

[3] H. Taureg / J. Horst: Self Induced Torque Amplification in Viscous Couplings SAE 900557

[4] P. Amborn / W. Hildebrandt: POWR-TRAK - Das drehmomentfühlende, selbstsperrende Differential Allradantriebe, Vieweg, 1992. ISBN 3-528-065453-6

[5] W. Rohregger / P. Amborn / O.A. Kaufmann: Advanced Development of Self-Controlled Torque Sensitive Limited Slip Differential by Means of Helical Gears SAE 940733

[6] P. Amborn / W. Hildebrand: Advanced Development of Self-Controlled Torque Sensitive Locking Differentials by Means of Spur-Gears SIA Congress 1991, Straßbourg

[7] H. Huchtkötter / H. Taureg: The Effect of a Viscous Coupling Used as a Front Wheel Drive Limited Slip Differential on Vehicle Traction and Handling SAE 940875

9 NOMENCLATURE

a	acceleration	m/s^2
ABS	anti lock braking system	
closed loop	driving manoeuvre with steering corrections made by the driver	
CV	constant velocity	
F	force	N
high-μ	high friction surface	
low-μ	low friction surface	
LSD	Limited Slip Differential	
M	moment	Nm
open loop	driving manoeuvre without steering corrections made by the driver	
P	pressure	bar
T	torque	
T_b	torque bias ($T_b = T_{out, hi} / T_{out, lo}$)	-
v	speed	mph/kph
Veh. 1	vehicle with open or viscous differential	
Veh. 2	vehicle with open differential or POWR-TRAK	
YMR	yaw moment reduction	
α	slip angle	°
δ	steering wheel angle	°
Δn	speed difference between the output sides of a differential	rpm
ΔF	force difference between unequal forces	N

10 INDICES

B	braking
CoG	centre of gravity
f	front
high / hi	high (torque, friction coefficient etc.)
low / lo	low (torque, friction coefficient etc.)
l	left
MC	master cylinder
r (1st. Ind.)	rear
r (2nd Ind.)	right
S	steering
T	traction

C498/30/129/95

Effect of petrol and diesel engined variants on driveline durability

J D BARNES BEng, AMIMechE
GKN Technology Limited, UK

SYNOPSIS

This paper describes how differences between petrol and diesel engines affect the durability and service life of constant velocity joints. Road load data has been collected from two "C" Class vehicles, one with a diesel engine, one with a petrol engine, but with similar on-road performance. The data was then analysed using three techniques to allow direct comparison.

i) Static strength analysis
ii) CV joint life prediction analysis
iii) Torsional fatigue analysis

The results can be used to develop individual design and application rules for diesels to allow specification of more suitable drivelines.

NOTATION

$N_{(t,\beta)}$	driveshaft revolutions at torque/ angle class (t, β)
M_d	joint application torque
$T_{(t)}$	driveshaft torque at class t
T_{EJmax}	maximum joint torque
T_{Emax}	maximum engine torque
T_{Amax}	maximum adhesion torque
i_G	overall 1st gear ratio
Q_F	total vehicle weight
R_S	tyre rolling radius

1 INTRODUCTION

GKN, as a world leader in the supply of vehicle drivelines, is always seeking to enhance it's products with respect to their cost, weight and performance. To achieve this we continuously improve our application engineering capability, by reinforcing our knowledge of our products with a detailed appreciation of how they operate on the vehicle. By increasing our knowledge and understanding of each specific application, a better component selection can be made.

One area that GKN is studying is the effect of diesel engines on vehicle drivelines. Diesel engines derived from commercial vehicles have been fitted as options to passenger cars for years, although more recently, advances in diesel engine design have produced a generation of engines which have driving characteristics more like those of a conventional petrol engine. This trend coupled with lower fuel consumption and perceived "green" emissions, has lead to a wider public acceptance of diesel engined vehicles and their market share has been growing considerably over the past fifteen years. GKN Technology, GKN's UK R&D centre, set in motion a project to investigate how much the fundamental differences between diesel and petrol engines can influence the durability and service life of GKN's core product, the constant velocity (CV) joint.

1.1 Diesel Engine Characteristics

There are several differences between compression-ignition (diesel) engines and spark-ignition (petrol) engines that have implications to CV joint durability. The diesel combustion process is much more violent, and the subsequent in-cylinder pressures are considerably higher than in a petrol version of similar size and performance. This allows the diesel to produce significantly more torque, although it also leads to a reduction in engine speed due to the increased inertia of the stronger, more robust engine components. In some cases the diesel's peak engine torque may be over 30% higher and is often available at speeds as low as 2200 rpm (typical petrol engine torque peaks occur at well over 3000 rpm). Although diesels produce large torques, their power outputs are reduced compared to their petrol equivalents due to their lower operating speeds. This power deficiency often leads to a lack of vehicle performance, with the vehicle feeling sluggish and slow to accelerate compared to petrol rivals. This lack of performance is compounded by the extra weight a diesel carries due to it's heavier engine and ancillaries. This may add up to 100 kg.

1.2 Effect on Vehicle Driveline

The main contributor to CVJ ball track damage, which determines the life of the joint, is the torque level it experiences. It's operating angle and rotational speed are of less importance. The diesel's ability to produce large torques at low engine speeds automatically leads one to believe that the they are more damaging to CV joints than the equivalent petrol engine. GKN required up to date test data to support this idea and so it was decided to perform a specific back-to-back exercise to confirm the above theory.

Two equivalent vehicles from the same model range were set up to collect road load data under controlled conditions, thus allowing direct comparison of the analysed results. It was decided to compare the data from the vehicles using similar analysis techniques to those commonly used by application engineers to select joints for each application. This would produce a thorough understanding of the conditions the driveline experiences and make comparisons simpler. The comparisons were divided into three areas of study:-

(a) Study of the maximum driveshaft torques both on road and on test tracks.
(b) Calculation of CV joint lives using the Macielinski damage prediction model.
(c) Calculation of the torsional fatigue of the driveshaft.

2 TEST METHOD

Two "C" Class vehicles were chosen for the comparison. The petrol example was fitted with a 1.6 litre fuel injected engine, whilst the diesel was powered by a 1.8 litre turbocharged engine. These particular models were chosen for their similar on-road performance, ease of instrumentation and easy access to joint application information (driveshafts for both vehicles are manufactured and supplied by GKN).

A comparison of their maximum torque and power outputs and kerb weights is given in Table 1.

Table 1 Test Vehicle Comparison

	Max. Torque	Max. Power	Kerb Weight
Petrol	135 Nm @ 5200 rpm	110PS @ 6300 rpm	1100 kg
Diesel	180 Nm @ 2500 rpm	90 PS @ 4300 rpm	1200 kg

2.1 Data Acquisition System

Both vehicles were fully instrumented at GKN-T with the same eight channels, as given below, to allow data collection to be carried out simultaneously if required.

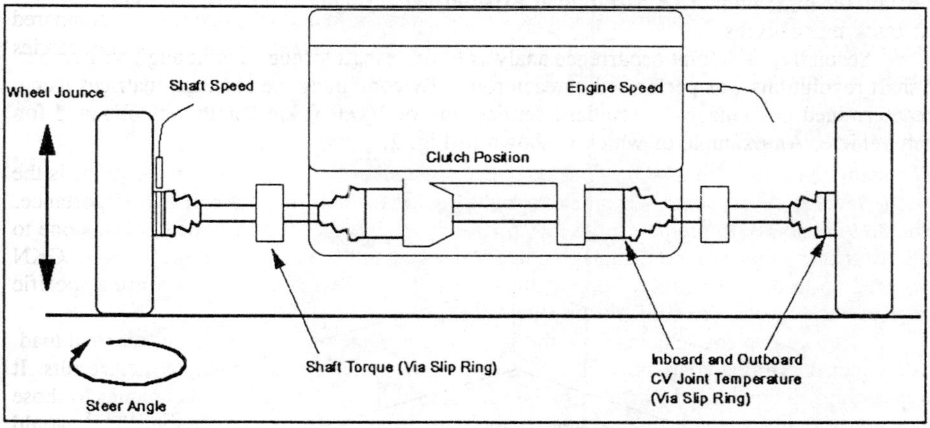

Fig 1 Measured Road Load Data Channels

Shaft torque was measured using a strain gauge full bridge arrangement, wheel jounce using a rotary potentiometer mounted on the lower suspension arm and steer angle obtained from a long stroke linear potentiometer mounted on the steering rack. Both speeds were measured using toothed rings and inductive proximity sensors, the clutch position sensor was a 50mm linear potentiometer, whilst both temperatures were measured by thermocouples and on-shaft conditioning.

Signals from the transducers were fed back, via slip rings if necessary, into the vehicle and recorded on VHS tape using a TEAC FM data recorder. After digitising the data onto a microVAX analysis computer three additional channels were calculated by computer manipulation of the recorded channels:-

(a) Inboard CV Joint Compound Angle - obtained from wheel jounce.
(b) Outboard CV Joint Compound Angle - obtained from steer and wheel jounce.
(c) Gear Position - obtained from shaft speed, engine speed and clutch position.

2.2 Data Collection

The vehicle testing was split into on-road and test track sections. The on-road test were performed over GKN-T's standard UK road routes. These consist of five separate routes, two routes on main roads, and one each on motorway, country, and urban road types. The same driver was used for both vehicles, with each route driven in dry conditions and at the same time of day. This aimed to reduce as much as possible differences due to weather and traffic conditions.

The test track work was carried out at a proving ground and was mainly tests aimed at generating high torques under controlled conditions that would be dangerous to perform on the road. These included forward and reverse hill starts, full throttle accelerations and driveline impact 'idiot' starts.

2.3 Data Analysis

The digitised data was analysed in three ways. Firstly simple signal statistics were employed to obtain the maximum driveshaft torques experienced during both the on-road routes and the test track manoeuvres.

Secondly, 3-D joint occurrence analysis of driveshaft torque vs joint angle vs number of shaft revolutions was performed on each route. By combining the resultant matrices using predetermined percentages, a standard service mix of 100,000 km length was obtained for each vehicle. An example of which is shown in Fig. 2.

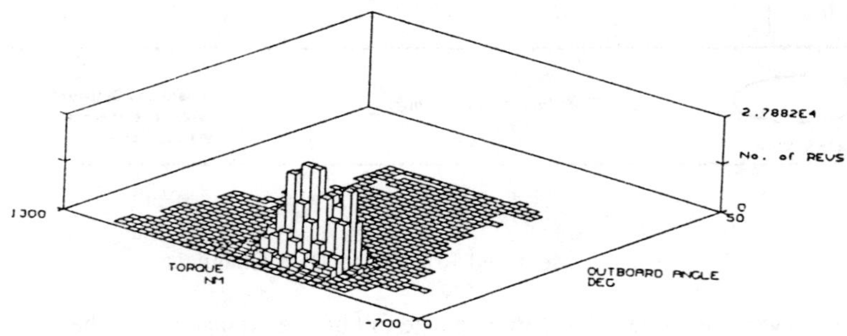

Fig 2 3-D Joint Occurrence Matrix

From this matrix, fatigue damage predictions were then made for the outboard joints using the Macielinski life prediction model (1) which is defined as (assuming for simplicity, a constant speed of 100rpm):

$$Total\ Damage = \sum \frac{N_{(t,\beta)}}{\left(\frac{Md}{T_{(t)}}\right)^3 \times \left(\frac{A}{Aa}\right)^3 \times 9{,}000{,}000}$$

where A = angle factor at angle class β = $(1 - \sin\beta)(\cos^2\beta)$
and A_a = angle factor at joint application angle

Using this damage value a prediction of the endurance life of the joint can be calculated as below:

$$Macielinski\ Life\ (km) = \frac{Service\ Data\ Distance\ (km)}{Total\ Damage}$$

Rainflow cycle counting analysis (2) was also performed on the torque signals to investigate the torsional effects on the driveline. From this analysis torsional damage values were calculated from each vehicle's torque signal for a 27mm diameter bar shaft (the interconnecting shaft between the two CV joints) as fitted to both vehicles.

3 RESULTS

3.1 Maximum Driveshaft Torque

The maximum torque a driveshaft experiences is obviously of prime importance to the application engineer when he makes his joint selection. Among the initial calculations made are static strength estimates that use vehicle data supplied by the customer to predict the maximum torque loading. Maximum joint torque, T_{EJmax} and maximum adhesion torque, T_{Amax} for the application are normally calculated as below:-

$$T_{EJmax} = \frac{T_{Emax} \times i_G}{2}$$

and

$$T_{Amax} = \frac{Q_F}{2} \times R_S \times \mu \times g$$

The lower of the two values is then multiplyed by a shock factor and compared with known allowable joint torques to select a suitably strong joint. Table 3 contains calculated maximum torque values from both the test vehicles, and the larger values obtained from the diesel show the effect of it's higher peak engine torque and higher weight. This difference may lead to a larger joint being specified for the diesel application based on purely theoretical calculations.

Table 3 Maximum Driveshaft Torques

Calculated Maximum Torque	Petrol	Diesel
Joint Torque T_{EJmax}	988 Nm	1157 Nm
Adhesion Torque T_{Amax}	1029 Nm	1129 Nm
Measured Maximum Torque		
Road Tests	911 Nm	895 Nm
'Idiot' Starts	1418 Nm	1425 Nm

Study of the peak driveshaft torques measured during the road tests and the 'idiot' starts (Table 3) shows that in practise these differences do not occur. The measured differences between the diesel and petrol are so small as to be negligible and it can be assumed that each vehicle can generate near identical peak torques. This similarity between the 'idiot' start peak torques may be due to these events being controlled more by the component inertias and clutch characteristics, than the engine torque. Both the clutches are of similar size although the diesel's has a slightly larger diameter.

3.2 Joint Life Prediction

In addition to the static strength calculations joint life prediction analysis is carried out using the Macielinski model to ascertain whether the joint's endurance life is sufficient for the application. The results of running this analysis on each vehicle's standard 100,000 km service mix are given in Table 4. A damage value of 1.0 indicates that fatigue failure of the joint would occur after 100,000 km. To allow direct comparison it was assumed that both vehicle drivelines were identical and equipped with the outboard CV joints as fitted to the petrol (the diesel's standard joints are larger than the petrol's). The results showed that the diesel engined vehicle is 19% more damaging to the CV joints than the petrol over the standard service mix.

Table 4 Joint Life Prediction Results

	Damage	Life (km)
Petrol	0.78	128,205
Diesel	0.93	107,527

These figures assume a constant shaft speed of 100 rpm. The calculated lives are therefore shorter than can be expected in service, however they do give a good comparative measure of the difference between the two vehicles.

Some of this increase in fatigue damage may be due to the diesel's heavier weight. The larger 1.8 litre turbo diesel with all it's associated ancillaries contributes to the overall vehicle test weight being 100 kg more than the petrol version. A consequent increase in power and hence drive torque is then needed to achieve the same on-road performance, with it's subsequent adverse effect on the durability of the diesel's CV joints. The diesel's average torque figures for each of the test routes were between 23 and 35 % higher than the petrol's values.

Values representing the vehicle's gear useage also have to be defined as part of the joint selection procedures, and they are a critical factor as changes can lead to selection of a different size of joint. The results in Table 5 show that the gear useage of the diesel is significantly different to that of the petrol. Although the number of revolutions spent in 1st and 2nd gear are comparable, the diesel spends much longer in 3rd and 4th and consequently less time in 5th. This is due to the inherent performance difference between the two vehicles. The diesel has to spend longer in the lower gears to achieve the desired vehicle speed and acceleration.

Table 5 Gear Usage Results - % Revs in Each Gear

	1st	2nd	3rd	4th	5th
Petrol	0.7	3.4	11.3	17.8	66.7
Diesel	0.8	3.5	14.2	23.2	58.3

3.3 Torsional Fatigue Analysis

Although not often used in joint selection, torsional fatigue analysis using rainflow cycle counting techniques can provide us with valuable information as to how the differing torque characteristics of the two engine types affect the driveshaft. The two rainflow plots of the vehicle's 100,000 km service mix showed very little difference between the two torque signals and this was confirmed by the calculated torsional damage values. These show that the petrol is slightly more severe on the interconnecting barshaft than the diesel, but the differences are so small as to be negilable. Here again a damage value of 1.00 would represent a failure of the barshaft after 100,000 km.

Table 6 Torsional Damage Results

	Torsional Damage
Petrol	0.081
Diesel	0.079

4 CONCLUSIONS

The fundamental differences that characterise diesel and petrol engines do not have a great effect on the environment the driveline experiences. Four main points have been clarified:

(a) The measured peak torques showed very little difference between the two vehicles compared with the torques predicted from basic theoretical calculations.

(b) Joint life prediction showed that the diesel is 19% more damaging to the outboard fixed CV joint than the petrol engined vehicle. This increase may necessitate the fitment of a slightly bigger joint (eg one size larger) for some diesel applications.

(c) Gear usage analysis showed that the diesel vehicle spends longer in 3rd and 4th gear and consequently less time in 5th gear than it's petrol equivalent. This is due to it's lack of power and inherent performance disadvantage. More time spent in a lower gear will lead to increased average torque levels and increased damage to the vehicle's driveline.

(d) Torsional fatigue analysis of the interconnecting barshaft showed that the two engine types produce very little difference in torsional damage.

In the past there was very little knowledge concerning how fitting a diesel engine to an application would affect the choice of driveline, and may have lead to oversized joints being specified due to concern over the diesel's higher torque levels. The data collected during this investigation has provisionally shown that although the diesel does have an adverse influence on driveline durability, the effect is much less than previously thought.

4.1 Further Work

Work is proceeding presently at GKN-T to confirm the findings obtained from the "C" Class cars tested by repeating the exercise with a pair of petrol and diesel larger "D" Class executive cars. These further tests will be performed with the two vehicles set at the same test weight condition, to remove the possibility of the weight difference affecting the measured torques.

5 ACKNOWLEDEMENT

The author gratefully acknowledges the permission of GKN Automotive Driveline Division to publish this paper, and wishes to thank colleagues who contributed to the activities reported.

6 REFERENCES

(1) MACIELINSKI, J.W. Propeller Shafts and Universal Joints - Characteristics and Methods of Selection. IMechE Proceedings, 1969-1970 Vol.184 Part 31 pp.516-543.

(2) DOWLING, N.E. A Review of Fatigue Life Prediction Methods. SAE Paper No. 871966. October 1987.

C498/30/144/95

Traction and handling safety synergy of combined Torsen differential and electronic traction control

PLATTEAU, S GUIDONI, and P SACCHETTINI
Zexel-Torsen SA, Belgium
JESSON CEng, MIMechE
Zexel UK, UK

ABSTRACT: This paper describes the technical development of a cost effective, practical synergy using the combined features of the new generation *TOR*SEN *Tor*que *Sen*sing differential and electronic Traction Control Systems (TCS). It utilizes the benefits of the differential to improve vehicle dynamics, leading to increased stability and safety in extreme handling conditions. At the same time a synergy between the differential and TCS is demonstrated achieving the optimum traction for a given vehicle and reducting drastically TCS duty cycle and hunting effects in winter operating conditions.

1. GENERAL INTRODUCTION

With the continuous technical evolution of automotive vehicles and the dramatic increase of traffic density, the vehicle's active and passive safety have become one of the car maker's main concerns. The on-going upgrade of engine power and vehicle speed have pushed the development for improved tyres, chassis & suspension, braking & traction management systems... (*active safety*) and protection systems such as shock absorbing body work, airbags... (*passive safety*) to prioritary levels. Thus, the development of such systems always have to meet severe cost constraints so a cost/efficiency compromise usually finds its way into serial production. The defined system is able to cover *most of the situations*.

However, in order to make it 100% efficient, the necessary add-up cost becomes too high to be competitive. For this particular reason *synergy* between systems is of utmost interest, as the characteristic of the combined system upgrades the sum of their single characteristics.

For instance, most of the car makers have developed electronic traction control systems governed by different principles. However, the electronic traction control systems advantage is to make use of existing equipment (ABS, brakes, throttle, ignition...), but have the basic disadvantage of increased sensors to become efficient. In addition, the typical operating mode is to *detect* an abnormal situation with the sensors and enable a corrective action with the logic control system.

In contrast, the fundamental characteristic of a *Torque Sensing* differential is the ability of instantaneous torque transfer *preventing* excessive wheel slip or spin instead of *correcting* it. This fundamental characteristic is particularly usefull in supporting electronic systems during the transitory phase of their response time. On the other hand, the *curative* action ensured by the electronic systems permits more design freedom for the *Tor*que *Sen*sing differential. The reality of having complex effects for each system's weakness permits us to simplify each of the systems with obvious cost reduction potential.

Here after we are going to consider three chapters describing the application of a *Tor*que *Sen*sing differential as a single axle differential, as an interaxle differential and the combination with electronic traction systems.

2. *TOR*QUE *SEN*SING IN A SINGLE AXLE

When fitted in the single driving axle of a vehicle, the *Tor*que *Sen*sing differential does not only improve traction, but also vehicle dynamics leading to a better stability and higher grip under difficult driving conditions. Before explaining how the *Tor*que *Sen*sing differential works in the field of vehicle dynamics, it is necessary, to clearly understand the two features of extreme operating conditions: the open differential and the rigid axle.

2.1. BASICS:

Input torque, weight distribution and tyre characteristics dictate the way the differential will work.

Weight distribution

During general driving conditions, a vehicle is subject to both longitudinal and lateral acceleration (Ax and Ay). In this case, each of the 4 wheels is loaded with a vertical force (Fz) that has to satisfy the equation of equilibrium which depends on:

- Total vehicle weight
- Static weight distribution (front/rear)
- Centre of gravity height
- Suspension roll stiffness distribution (front/rear)

For most of the passenger cars vertical load is supposed to be proportional to lateral acceleration. We can build the corresponding curves by driving a vehicle on a steering pad at different levels of longitudinal and lateral acceleration. The driver must adjust the steering wheel angle to keep the vehicle on a defined radius.

Figure 1 shows an example of weight distribution over Front and Rear Axle, during a steering pad manoeuvre at Ax=0g and Ax=0.2g.

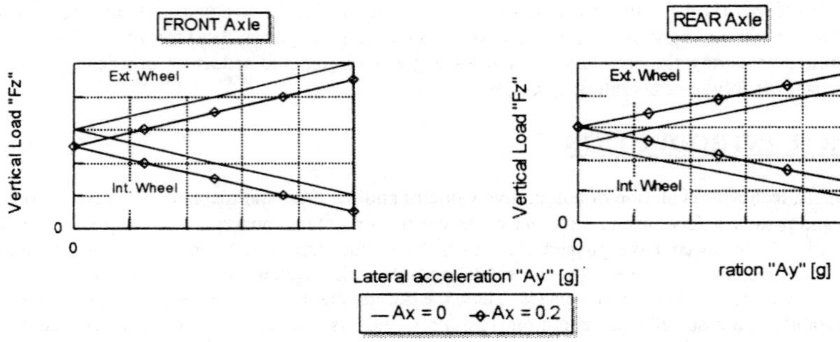

Figure 1

Tyre characteristics

As far as tyre characteristics are concerned, we will take into account longitudinal and lateral tyre friction coefficient (μ_x, μ_y) whose trends are shown by the non-linear functions of Figure 2.

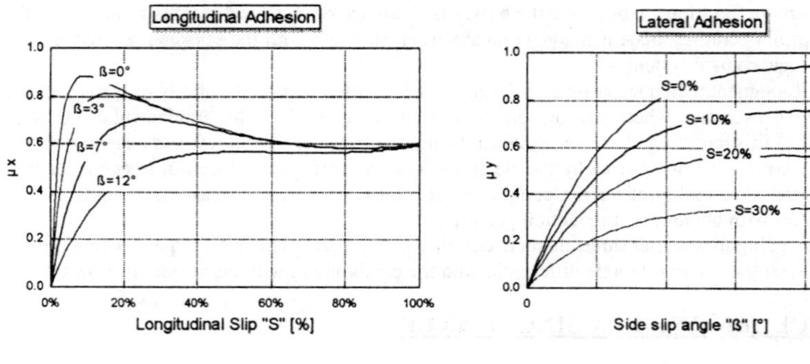

Figure 2

These coefficients will allow us later on to calculate longitudinal and lateral tyre forces (Fx, Fy).

2.2. SINGLE AXLE DIFFERENTIAL OPERATING MODES

DEFINITION OF TERMS

Trg	Ring Gear Torque	ωe	External Wheel Speed	Ax	Lateral Accel
Te	External Wheel Torque	ωi	Internal Wheel Speed	Ay	Longitudinal Accel
Ti	Internal Wheel Torque	ω_{cin}	Kinematic Wheel Speed	v	Vehicle speed
s	Wheel slip	ω_{el}	Elastic Wheel Speed	$\Delta\omega$	Wheel Speed difference
C	Track width	R	Circle radius	Rr	Rolling radius

GENERAL EQUATIONS

Wheel slip $$s = \frac{(\omega * Rr) - v}{v}$$ **Eq. 1**

Lateral Acceleration $$A_y = \frac{v^2}{R}$$ **Eq. 2**

OPEN DIFFERENTIAL:

For the OPEN differential the basic conditions are: $Trg = Te + Ti$ $\quad Te \cong Ti$

Let us now investigate what happens with the wheel speeds. The OPEN differential always dictates wheel torque (**None or very little wheels torques difference**) and allows the wheels to adapt their speed to the conditions described here under.
The wheel speeds will result from ***kinematic*** conditions (ω_{cin}) like vehicle geometry, steering pad radius (see Figure 3) and ***elastic*** conditions (ω_{el}) like tyre characteristics.

External Wheel Speed $\omega e = \omega e_{CIN} + \omega e_{EL}$ **Internal Wheel Speed** $\omega i = \omega i_{CIN} + \omega i_{EL}$

In practice it is more interesting to consider the wheel speed difference ($\Delta\omega$) rather than the speed of each wheel individually.

$$\Delta\omega = \omega e - \omega i = \Delta\omega_{CIN} - \Delta\omega_{EL}$$ **Eq. 3**

We find $\Delta\omega_{cin}$ thanks to Figure 3

$$\Delta\omega cin = \sqrt{\frac{A_y}{R}} * \frac{c}{Rr}$$ **Eq. 4**

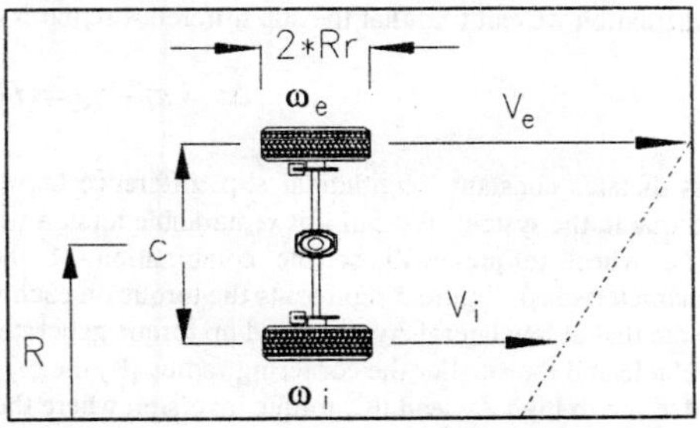

Figure 3

We find $\Delta\omega_{el}$ thanks to

$$\Delta\omega el = \frac{\sqrt{Ay * R}}{Rr} * (s_e - s_i)$$ **Eq. 1 + Eq. 2 + Eq. 4**

Torque and speed difference in function of lateral acceleration are shown in Figure 4
It is interesting to note that there is a range at low/medium Ay values where $\Delta\omega$ is positive: the external wheel is faster than the internal one. This is because, in that area, kinematic conditions (Pure geometric trajectory of the two wheels) play a bigger role than elastic conditions. At medium/high Ay values, elastic conditions have more influence than kinematic conditions because of the general torque increase and the vertical load decrease on the internal wheel. $\Delta\omega$ becomes then negative: the internal wheel is faster than the external one..

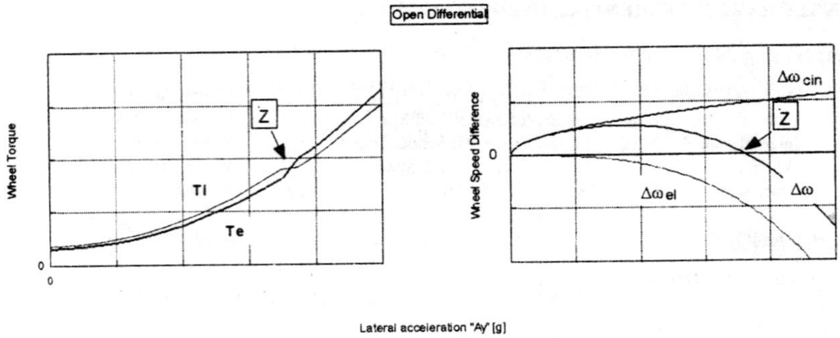

Figure 4

RIGID AXLE:

For the RIGID axle the basic conditions are: $Trg = Te + Ti$ $\omega_e = \omega_i$

Lets now investigate what happens with the wheel torques. The RIGID axle always dictates wheel speeds (**no wheel speed difference)** and allows the wheels to adapt their torque to the conditions described here under.
The "no wheel speed difference" condition will be compensated by a wheel slip difference. Indeed, at low input torques, the actual path of the outer wheel is shorter than the theoretical one. That means the actual wheel speed is slower than the theoretical one and that we have thus negative or reverse slip (s_e). It is the opposite for the inner wheel: the actual path of the inner wheel is longer than the theoretical one. That means the actual inner wheel speed is faster than the theoretical one and that we have thus positive or forward slip (s_i). By calculation we can find that the slip difference remains constant:

$$\Delta s = s_i - s_e = f(c, R) = \text{Constant}$$

Δs dictates constant longitudinal slip difference between internal and external wheels which will induce a torque in the system. We call this remarkable torsion within the axle **"wind up torque"** (at low input torques).
The wheel torques will be the combination of the basic conditions and the ***elastic*** conditions (tyre characteristics). Figure 5 represents the torque on each wheel in function of lateral acceleration.
Note that at low lateral Ay, the wind up torque generated within the axle gives an understeering behavior to the vehicle and the smaller the cornering radius (R) the bigger that negative effect!
Medium to high Ay lead to a torque inversion where the external wheel torque exceeds the internal one.

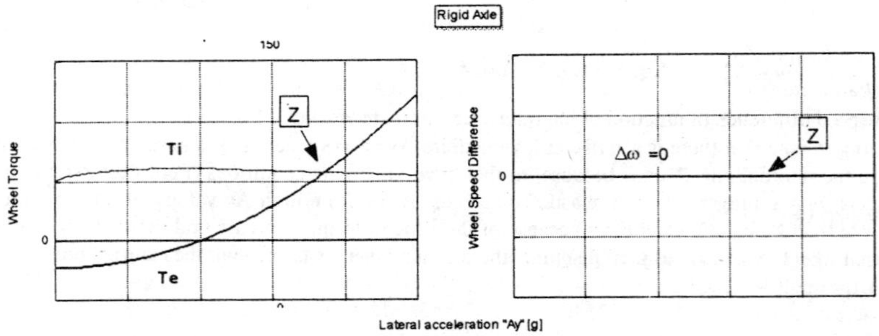

Figure 5

TORQUE SENSING DIFFERENTIAL

For the *Tor*que *Sen*sing differential the basic conditions are: $Trg = Te + Ti$
and one of the two following relationships dictated by grip and driving conditions:

$$For\ \Delta\omega \neq 0 \Rightarrow \frac{T_{max}}{T_{min}} = TBR$$

$$For\ \Delta\omega = 0 \Rightarrow \frac{T_{max}}{T_{min}} < TBR$$

In Figure 6 torque and speed difference of a ***Tor*que *Sen*sing** differential in function of Ay.

We can describe three different ranges of Ay:

Low Ay

The kinematic conditions force the outside wheel to turn faster than the inner; therefore higher torque is sent to the slower wheel; the *Tor*que *Sen*sing allows free speed difference and eliminates the adverse effect of wind up torque.

Medium Ay

The grip conditions are not strong enough anymore to unlock the *Tor*que *Sen*sing differential and the system will work as a rigid axle.

High Ay

Ay increases with additional input torque and the vertical load decreases on the internal wheel accordingly to Ay. The torque bias ratio value is not large enough to withstand the wheel torque difference, hence the differential unlocks and the inner wheel starts to spin.

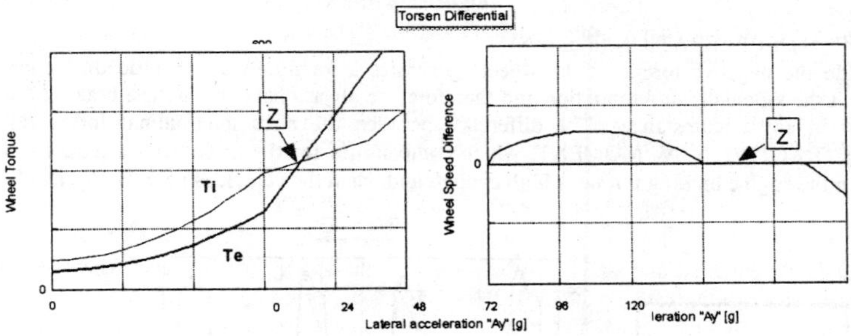

Figure 6

Observations:

We can display a remarkable **"Z"** point on the Torque and Δω curves.. It represents operating conditions that are identical for any kind of differentials where, for a defined Ay, Δω = 0 and Te = Ti. This **"Z"** point will be located at lower Ay with:

Decrease of
- Surface coefficient (μ)
- Roll stiffness of the non-driving axle

Increase of
- Roll stiffness of the driving axle
- R
- Ax

2.3. FRONT AND REAR WHEEL DRIVE: EFFECTS ON VEHICLE DYNAMICS

As can be seen, the *Tor*que *Sen*sing differential generates a torque difference between outer and inner wheels compared to the traditional OPEN differential, where the two output torques are almost equal. That torque imbalance induces a TRACTIVE YAW MOMENT (TYM) which is active in DRIVE as well as in COAST mode. Lets explain how the TRACTIVE YAW MOMENT acts on the dynamic behavior during a cornering condition. For this, let's consider the steering pad manoeuvre:

DRIVE MODE - ACCELERATION (Figure 7):

At low Ay the input torque level is low and the tyres are far away from their physical limits. The outer wheel is faster than the inner one and the inner torque is larger than the outer one. The difference between the tyres longitudinal forces results in an **understeering** TRACTIVE YAW MOMENT, which is slightly perceived by the driver.

At higher Ay the input torque level is high and the tyres are closer to their physical limits. The inner tractive wheel tends to go faster than the outer tractive wheel in consequence the outer torque will be larger than the inner torque. The difference between the tyres longitudinal forces results in an **oversteering** TRACTIVE YAW MOMENT, which is usually noticed by the driver.

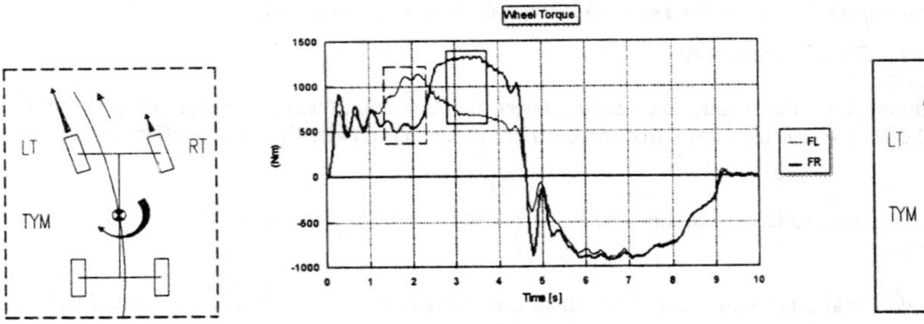

Figure 7

COAST MODE * DECELERATION (Figure 8)

On drop throttle the negative torque at the wheels generates a longitudinal slip ratio difference that is in accordance with the kinematic differentiation and therefore, we always have the highest braking torque on the external wheel in any circumstances. The difference between the tyres longitudinal forces results in an **understeering** TRACTIVE YAW MOMENT, which compensates for the drop throttle induced oversteering especially when the engine braking torque is high enough and when the vehicle is near its physical limit.

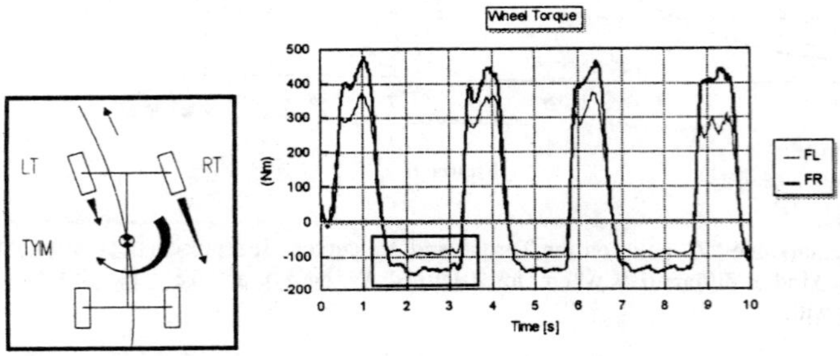

Figure 8

For a defined vehicle, steering pad radius and grip conditions, it is possible to predict the actual working condition thanks to the lines of the graph in Figure 9. This graph is build with the aid of the $\Delta\omega$=f(Ay) curves for different steering pad radii and at different longitudinal acceleration levels.

The lines represent the (Ax, Ay) pair when $\Delta\omega$=0.

TYM generates understeering below the line "$\Delta\omega$=0" and oversteering above.

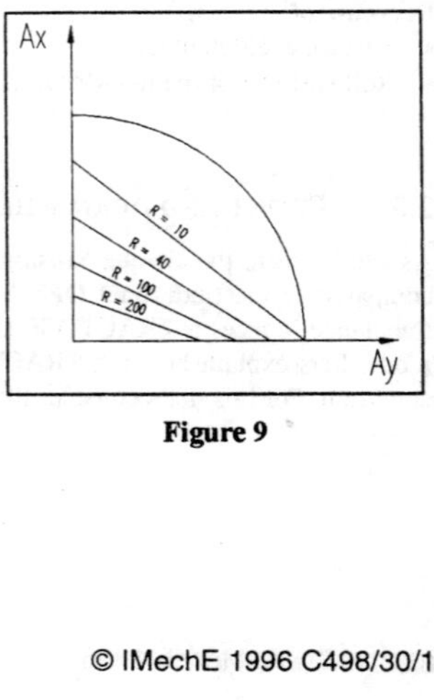

Figure 9

3. *TORQUE SENSING* IN A CENTREBOX

A *Tor*que *Sen*sing differential fitted as centre differential of an ALL WHEEL DRIVE vehicle distributes the torque to the front and the rear axles. Generally the basic distribution ratio is 50/50 front/rear in "normal" operating mode (straight road, low to medium wheel torques).

The necessity for a different torque distribution appears at medium to high wheel torques as soon as unstable driving situations occur such as cornering, sudden lane changes, split-μ road conditions or transitory DRIVE/COAST torque variations, causing increased tyre slip ratios. The aim of the *Tor*que *Sen*sing differential is to instantaneously distribute the torque to avoid increased longitudinal slip ratios of one of the driving axles in order to ensure the maximum potential of each axle's lateral stability.

Vehicle's dynamic stability and handling characteristics are determined by a list of parameters which could be divided in 4 main groups:

1. Vehicle dimensions & weight distributions
2. Vehicle elasto-kinematic suspension & steering characteristics
3. Tyre design characteristics
4. Torque distribution to the wheels

Today's vehicle designs optimize handling & dynamic stability in order to continuously improve active safety characteristics.

As far as ***Torque** Sensing Differentials* are concerned, the target is to control efficiently parameter "4", the Torque distribution to the wheels. The fundamental idea is to take advantage of the vehicle's intrinsic dynamic behavior parameters and their variations which dictate it's stability. The *Tor*que *Sen*sing centre differential passively reacts to the wheel's instantaneous torque withstanding ability, thereby regulating the longitudinal slip ratio characteristic. This last parameter is fundamental for the wheel's ability of lateral stability, as a wheel exceeding it's saturation level in terms of longitudinal forces (directly proportional to applied torque) will suddenly loose it's capability of lateral hold, with subsequent loss of lateral stability leading to vehicle loss of control.

This centre differential's *passive basic characteristic* actually turns out to be an *active dynamic characteristic*, as the parameters dictating it's dynamic torque distribution are actually *anticipating* the vehicle's loss of directional stability. This allows a particularly efficient and immediate corrective action during transitory situations at the limit of lateral acceleration. This proactive operating mode does actually *prevent* an eventual loss of stability instead of *correcting* it after it's occurrence, as electronic systems with accelerometers and yaw rate sensors do.

3.1. INTERAXLE DIFFERENTIAL OPERATING MODES

An interaxle centre differential operates in the following 4 basic modes shown in Figure 10 In reverse, DRIVE becomes COAST and vice-versa.

FORWARD			
MODE 1	**MODE 2**	**MODE 3**	**MODE 4**
DRIVE	DRIVE	COAST	COAST
Rear high axle torque	Front high axle torque	Rear high axle torque	Front high axle torque

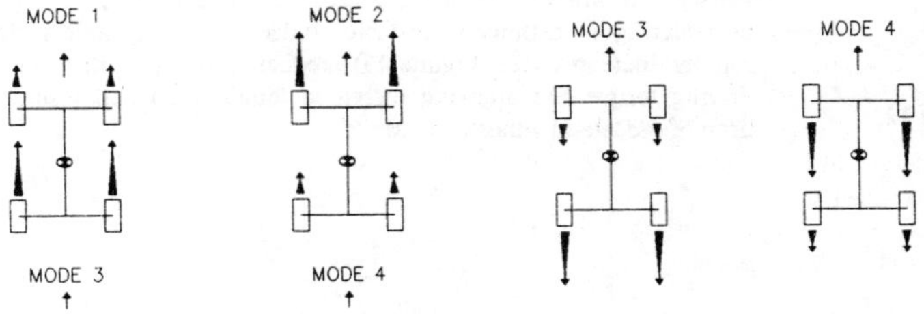

Figure 10

In DRIVE mode, a *Tor*que *Sen*sing differential will distribute the higher driving torque to the axle which tends to turn slower than the other one.
In COAST mode the higher braking torque will be distributed to the axle which tends to turn faster than the other one.

This shows that, in cornering at low speed, the centre differential operates following **mode 1** (forward driving) and **mode 3** (reverse driving), as the front wheels rotate faster than the rear one because of the larger path radius. For these modes, the differential's locking effect value should be low enough to avoid windup torque within the driveline causing excessive loading, unsmooth operation inducing tyre slip noise, stick-slip vibrations,...

Vehicle stability is not a prioritary argument for operating mode 1 and mode 3 however:
In case of too low locking effect value in mode 1 (DRIVE, rear high axle torque), the adverse consequence would be to let the front axle slip ratio increase to high values. This will cause increased understeering at the limit which is a situation that can be easily controlled.
In case of too low locking effect in mode 3 (DRIVE, reverse driving or COAST, rear high axle torque), the consequences are not relevant as far as vehicle stability is concerned. Vehicle dynamics is not involved in reverse driving and during drop throttle manoeuvre the front axle always tends to be the faster one (therefore corresponding to mode 4).
The driving situations for which the vehicle's dynamic stability becomes a serious matter of concern are **mode 2** (power oversteering) and **mode 4** (drop throttle oversteering):

3.2. POWER OVERSTEERING

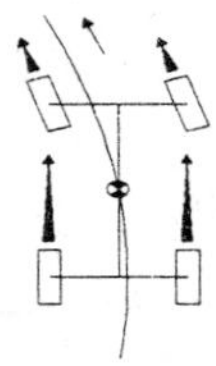

Figure 11

During a cornering manoeuvre at low speed and low torque (Figure 11), the higher driving torque will be biased to the rear axle (kinematic condition: front axle turns faster than rear axle):
When more input torque is added, vehicle speed rises, the rear axle slip ratio increases (elastic conditions catch up kinematic conditions) until the rear axle reaches the speed of the front axle. At that time there is no differentiation at the centre differential which operates as a rigid axle. From this neutral "steady state" condition, there are two possible dynamic behavior evolution's.
a. The front axle slip ratio increases causing vehicle understeering.
The centre differential will react by biasing the surplus torque to the rear axle, thereby correcting the understeering. Should this correction be insufficient, the driver will naturally react by releasing the throttle that will reduce the understeering. This manoeuvre does not need driving skills and can be done by any driver.

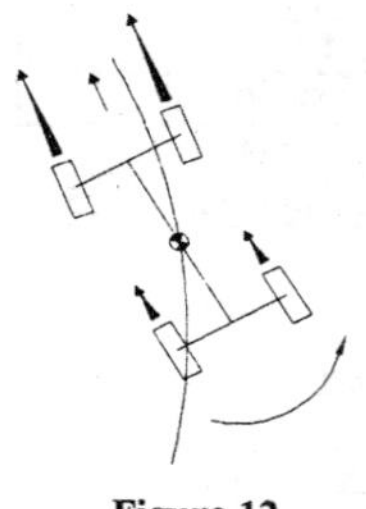

Figure 12

b. The rear axle slip ratio increases causing vehicle oversteering.
The centre differential will react by biasing the surplus torque to the front axle, thereby correcting the oversteering (Figure 12):

In this situation the driver will require consistent assistance from the centre differential because his instinctive reaction will be a throttle release that will worsen the initial oversteering instead of correcting it. We know that oversteering is extremely unstable. Indeed, the rear axle's side slip angle rapidly increases (see Figure 13), reducing it's potential to withstand the driving torque and allowing increased longitudinal slip resulting in a rapid drop of the lateral adhesion capability.

The front axle's side slip angle will normally not increase, as steering correction always tends to keep the front wheels in the correct trajectory.
Therefore the centre differential must ensure enough torque biasing capability to relieve the rear wheels from longitudinal driving forces, even for increased side slip angles causing a very high traction potential difference between front and rear axles.

However, in a DRIVE (power ON) manoeuvre the dynamic weight shift to the rear helps the rear axle to keep it's traction potential, unlike in a COAST manoeuvre (power OFF) for which the dynamic weight shift to the front worsens the rear axle's traction potential by reducing the vertical load. Therefore, the centre differential's locking effect value must be higher in coast than in drive mode.

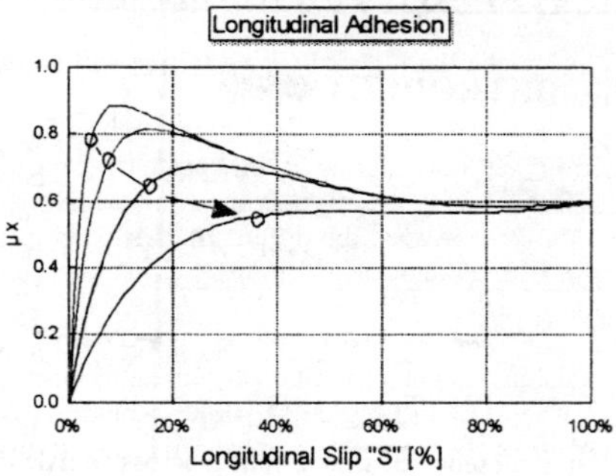

Figure 13

3.3. DROP THROTTLE OVERSTEERING

The most critical situation for vehicle stability during cornering is the COAST mode (power OFF), as this generally corresponds to a "panic" manoeuvre (too high vehicle speed or sudden decrease of curve radius...). In addition, the dynamic weight shift further decreases the rear axle's vertical loads, reducing its potential adhesion capability.

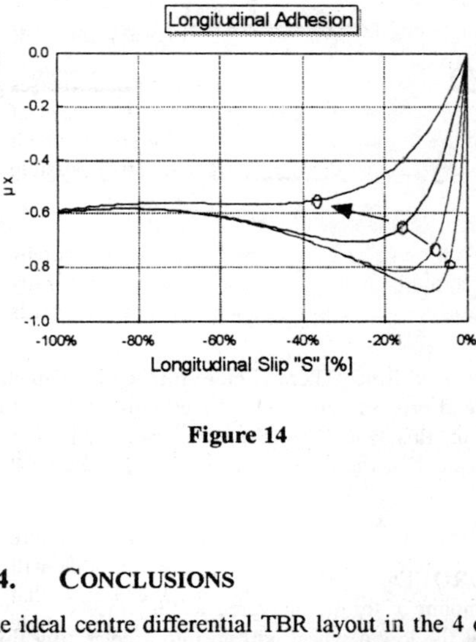

Figure 14

Furthermore, the sudden variation from positive (DRIVE) to negative torque (COAST) causes a critical transitory phase, as it generates a variation of several dynamic parameters. In this case, the centre differential has to bias the necessary amount of surplus braking torque to the front axle in order to relieve the rear axle in terms of torque.
As vehicle oversteering starts, the rear axle's side slip angle increases, reducing the longitudinal and subsequently the lateral grip capability (See Figure 14). The centre differential has to further remove braking torque from the rear axle and bias it to the front axle to ensure the rear axle's lateral stability. The necessary TBR (Torque Bias Ratio) of the centre differential can reach high values, as the ratio between front and rear torque raises rapidly in this particular case where rear torque has to decrease while front torque has to increase.

3.4. CONCLUSIONS

The ideal centre differential TBR layout in the 4 operating modes is a function of vehicle dimensions (wheel base, track width, centre of gravity height...), suspension elasto-kinematic design (stiffness front/rear, angular variations, ...) and engine torque characteristics for given road conditions. Therefore the ideal design characteristics for a centre differential can be determined after a great deal of subjective vehicle tests. The optimization will be a compromise between different setups, depending on the surface conditions (dry asphalt, wet asphalt, snow, ice...)..

4. SYNERGY: *TORQUE SENSING* + TRACTION CONTROL SYSTEMS (TCS)

4.1. INTRODUCTION

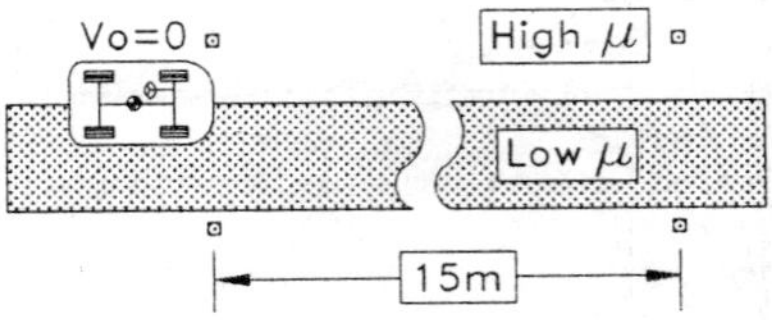

Figure 15

In adverse weather conditions, the coefficient of friction between tyre and road can vary greatly. We are daily facing situations when the friction coefficients for the left and right wheels are not the same, we have what we call a "split μ" situation. A simple split μ acceleration test is described Figure 15. The vehicle is aligned on the test surface and accelerates at full throttle from 0 to 15 meters

We can represent that simple split μ across an axle (Figure 16)

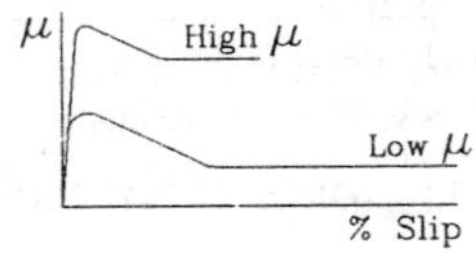

Figure 16

These curves describe the relationship between the road surface and the tyres. Actual operating conditions can fall anywhere on these curves. Maximum available traction can be described by the sum of the peaks. Various traction control strategies are described Figure 17. The Tyre/Road μ-Slip curves for the two surfaces are shown, along with the expected operating conditions.
% SLIP describes how much faster a wheel is spinning than it should for a given vehicle speed.

CONTROL STRATEGIES

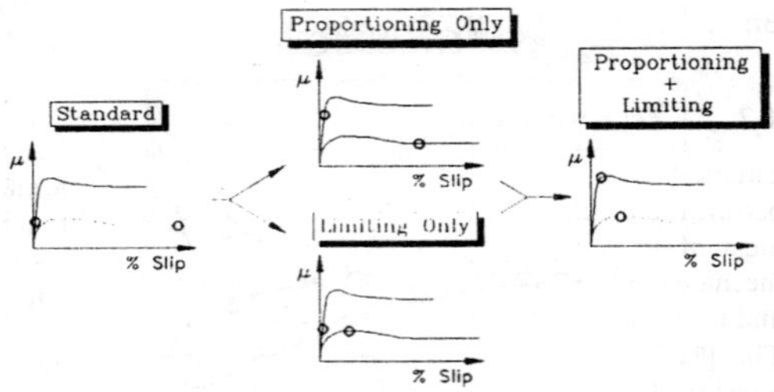

Figure 17

STANDARD:
OPEN DIFFERENTIAL with NO TCS
The low μ wheel spins up, reducing its coefficient of friction. The amount of torque delivered to the high μ wheel will be approximately equal to that of the low μ wheel and thus it will utilize only a portion of the available traction.

PROPORTIONING ONLY:
***TOR*QUE *SEN*SING DIFFERENTIAL with NO TCS**
One way to improve the use of available traction is to *proportion* it. Since There is more traction available at the high μ wheel, more torque should be sent there. In the case of proportioning only, drive torque still has the potential to exceed that available at the low μ wheel, causing this wheel to spin. However, due to the proportioning, total traction utilization has increased. This means that the amount of drive torque the wheels can accept has increased.

LIMITING ONLY:
OPEN DIFFERENTIAL, with LIMITED DRIVE LINE TORQUE
Another way to improve the use of traction is to *limit* the amount of torque delivered to the wheels. Drive torque in excess of that sustainable by the available traction at the low μ wheel will go into accelerating that wheel. By limiting the torque, the effective low μ can be optimized at or near the peak. The open differential provides a torque balance between the two axles, and thus the effective μ at the high μ wheel will be increased as well.

PROPORTIONING + LIMITING:
BRAKE TCS + OPEN
+ *TOR*QUE *SEN*SING
ENGINE TCS + *TOR*QUE *SEN*SING
From the last graph we see that in this situation, optimal use of available traction requires *both proportioning and limiting.*

Thus generally,

- LIMITED SLIP DIFFERENTIAL can ***proportion***, but cannot ***limit.***
- ENGINE TCS + OPEN DIFFERENTIAL can ***limit***, but cannot ***proportion.***
- BRAKE TCS + OPEN, BRAKE TCS + *TOR*QUE *SEN*SING AND ENGINE TCS + *TOR*QUE *SEN*SING are capable of ***proportioning and limiting.***

The efficiency of the different systems can be easily shown by actual test results of a simple Split μ acceleration (Figure 18).

In this specific case,

- STANDARD configuration is the OPEN differential with NO TCS.
- PROPORTIONING is provided by the *TOR*QUE *SEN*SING differential.
- LIMITING is provided by IGNITION TIMING control.
- PROPORTIONING and LIMITING is the combination of *TOR*QUE *SEN*SING differential and IGNITION TIMING control.

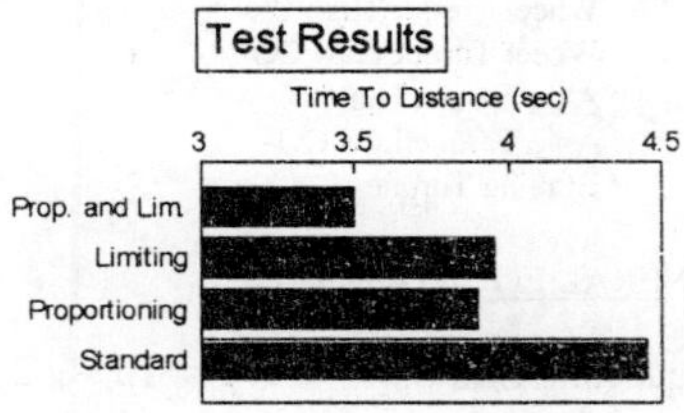

Figure 18

As can be seen, taken individually, PROPORTIONING and LIMITING each help reduce time to distance. Because each approach attacks a different portion of the traction control problem, **SYNERGY** exists between the two. The ***Superior System*** incorporates ***Both Proportioning And Limiting***
We are now going to review the combination of the *Tor*que *Sen*sing differential combined with Engine and Brake TCS.

4.2. *TOR*QUE *SEN*SING + ENGINE TCS

Engine based Traction Control Systems (ETCS) all perform the same basic ***function;*** they provide a mean of limiting torque. The way they differ is in the mechanism (i.e., sub-throttle, ignition timing, etc.) and in the algorithm that performs the control.
The performance advantage of adding the *Tor*que *Sen*sing differential to an Engine Traction Control System is readily apparent in the Split μ acceleration actual data shown Figure 19.
The *Tor*que *Sen*sing equipped vehicle puts more torque to the ground, the benefits of which are twofold:

- Improved vehicle control and stability (there is less wheel spin)
- Improved tractive effort.

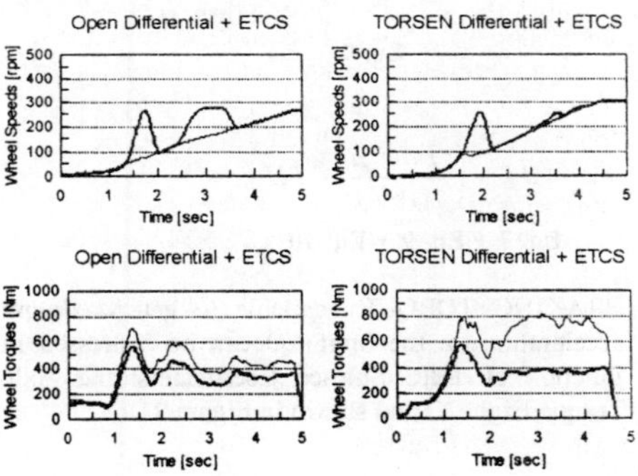

Figure 19

On a flat surface, improved tractive effort will mean improved acceleration. On a slippery grade it could mean the difference between continuing on your way or getting stuck.

4.3. *TOR*QUE *SEN*SING + BRAKE TCS

As mentioned previously, Brake Traction Control Systems (BCTS) ***are*** capable of providing both proportioning and limiting. So how can the *Tor*que *Sen*sing differential help?.
Since the *Tor*que *Sen*sing differential begins working ***before*** wheel spin occurs, the response of the system will be improved. Also since much of the torque proportioning will be taken care of by the *Tor*que *Sen*sing, *less* brake effort will be required.

4.3.1. Theoretical Approach

The calculations here after show general equations of the needed ***Braking Torque*** for a BTCS with *Tor*que *Sen*sing or OPEN differential

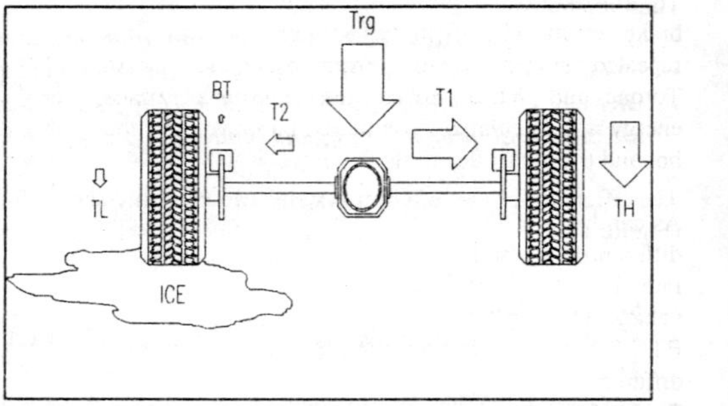

Figure 20

<u>DEFINITION OF TERMS (See Figure 20)</u>

Trg	Ring Gear Torque
TH	Wheel Torque High CO
TL	Wheel Torque Low CO
T1	<u>Axle</u> Torque High CO
T2	<u>Axle</u> Torque Low CO
BT	Braking Torque

<u>GENERAL EQUATIONS</u>

Torque Ring Gear $$Trg = T1 + T2$$ **Eq. 5**

Torque Bias Ratio $$TBR = \frac{T1}{T2}$$ **Eq. 6**

Axle Torque Low CO $$T2 = BT + TL$$ **Eq. 7**

We know $$TH = T1$$ **Eq. 8**

<u>BRAKING TORQUE CALCULATION</u>

Wheel Torque Low CO can be expressed as a function of Wheel Torque High CO, taking into account the Low μ / High μ ratio

$$TL = x \times TH \qquad \text{with } x = \frac{\text{Low } \mu}{\text{High } \mu}$$

Eq. 9

$$T2 = BT + \left(\frac{Low\ \mu}{High\ \mu} \times TH\right)$$

Eq. 7 + Eq. 9 + Eq. 10

BRAKING TORQUE available to get Maximum acceleration on the Split CO can be expressed in function of TBR, Surface Coefficients and Axle Torque High CO and shown in Figure 21.

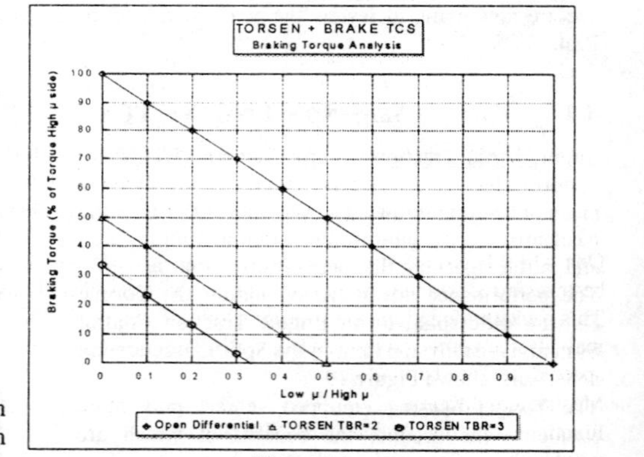

Figure 21

$$BT = \frac{TH \times \left(1 - \left(\frac{Low\ \mu}{High\ \mu} \times TBR\right)\right)}{TBR}$$

Eq. 6 + Eq. 8 + Eq. 10

4.3.2. Actual Data

<u>*Asphalt / Ice Split μ*</u>

As an example, we show actual Split μ acceleration data for a vehicle having the *Tor*que *Sen*sing differential with BTCS (Figure 22). In the wheel torque data, notice the *Tor*que *Sen*sing differential begins proportioning at 0.4 seconds, while the brake torque does not begin to rise until 0.9 seconds. Thus the *Tor*que *Sen*sing differential begins working a full half second ***before*** the BTCS actuates. Also, notice that with the *Tor*que *Sen*sing differential, the required brake torque is cut in half.

To illustrate the significance of the reduction in brake effort, the Split μ acceleration test was repeated several times. From measured Brake Torque and Wheel Speed, the cumulative brake energy was calculated. Eventually, the brake gets too hot and the controller shuts down the BTCS.

As can be seen Figure 23, with the *Tor*que *Sen*sing differential the brake energy accumulates at a slower rate. Thus the *Tor*que *Sen*sing differential equipped vehicle was able to complete 17 runs before the BTCS shut down, compare to only 11 with the open differential.
This represents an ***improvement of over 50%.*** These actual data confirm the theoretical approach.

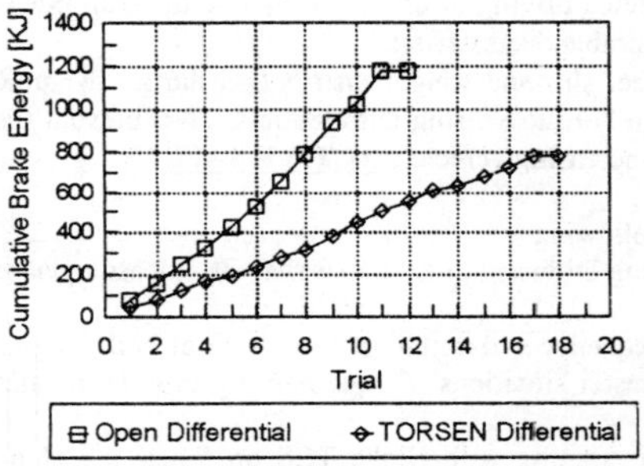

Figure 23

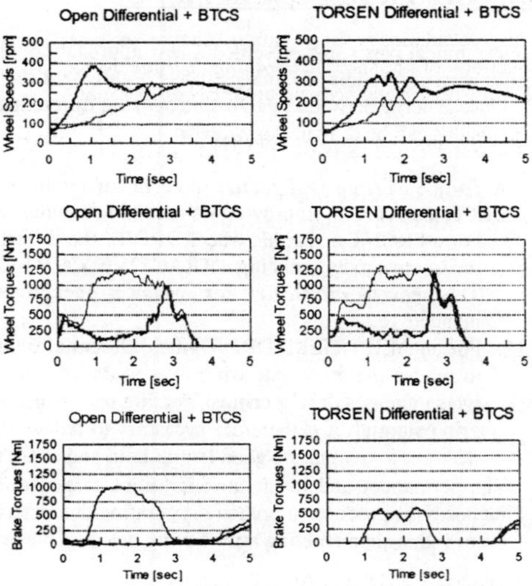

Figure 22

<u>*Equal μ or slight Split μ difference: HUNTING*</u>
Under hard acceleration on a true equal μ surface or under mild acceleration on a Split μ surface with a slight difference in coefficients of friction below the tyres (Ice-snow), Brake TCS will typically experience "hunting" across an axle. Hunting occurs when the two brakes across an axle get out of phase.
One wheel begins to spin away from the other and its brake tries to slow it. This then causes excess torque to be transferred to the other wheel and, as the first wheel comes back under control, the second begins to spin. This back and forth process is accompanied by objectionable NVH (Noise Vibration Harshness) in a vehicle. It may also be perceived by the driver as a lack of proper control. Hunting, by definition, cannot occur without a speed difference between the wheels.
Mechanical differentials will have a ΔT threshold, below which they will not differentiate. This threshold is a function of the mechanical friction within the differential, and thus it is directly proportional to total axle torque (ΣT). Therefore, for a given differential and ΣT, there will be a given ΔT required before differentiation occurs (and thus before hunting can occur).
Under most circumstances, a Torque Proportioning differential will have a high enough differentiation threshold that hunting will be reduced or even eliminated.
This is illustrated by the test results given Figure 24. For this example, we have a SNOW-ICE Split μ and the driver tries to get maximal acceleration with minimal wheel spin.

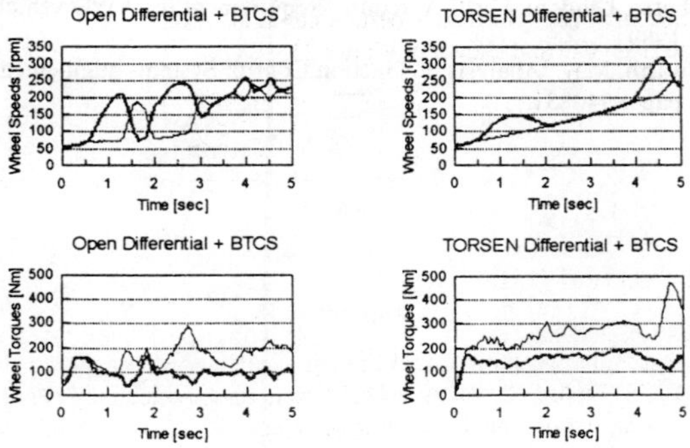

Figure 24

5. GENERAL CONCLUSIONS

This technical paper has presented a new approach for the use of *Tor*que *Sen*sing differentials and in addition to the normal traction advantages it has highlighted the investigation into vehicle dynamic benefits including the close synergy between this *Tor*que *Sen*sing differential and Traction Control Systems. The following major conclusions can be drawn:

- A *Tor*que *Sen*sing Differential fitted in car modifies vehicle dynamics in ACCELERATION as well as in DECELERATION, improving its natural behaviour in two ways:
  1. For a FRONT or REAR wheel DRIVE, the differential generates directly a wheels torques difference at the tractive axle, inducing a TRACTIVE YAW MOMENT.
     This reduces drastically, for instance, unexpected oversteering after power OFF during a cornering manoeuvre.
  2. For an ALL WHEEL DRIVE, the differential transfers more torque to the higher grip side axle in order to relieve the other axle when it is at its limit of adhesion. For instance, the side slip angle of the rear tyres increases during critical steering pad manoeuvres such as power ON or power OFF, diminishing its grip potential. It is therefore necessary to reduce the applied driving or braking torque to the rear axle if we want to maintain a good lateral hold and avoid undesirable oversteering.

  The permanent and instantaneous torque response to wheel slip and weight distribution during dynamic manoeuvres produces a proactive operating characteristic to *Tor*que *Sen*sing Differentials. They prevent the event rather than reacting after having detected the event, increasing vehicle's **ACTIVE SAFETY**.

- As far as TCS synergy is concerned, we can conclude the following:
  1. In all the cases, the performance of TCS + *Tor*que *Sen*sing Differential is greater than or at least equal to the performance of TCS + OPEN.
  2. *Tor*que *Sen*sing Differentials reduce the duty cycle (Frequency + duration of actuation) of both Engine and Brake TCS systems in split μ or lateral weight transfer situations. *Tor*que *Sen*sing even eliminates the need for TCS actuation in some cases
  3. *Tor*que *Sen*sing Differentials reduce or even eliminate hunting with Brake TCS on Equal μ and on specific Split μ situations (Slight difference in coefficient of friction like ICE / SNOW Split μ). This translates to reduced and improved driver perception of system control.

  All this shows that an extremely advantageous **SYNERGY** exists between *Tor*que *Sen*sing and Traction Control Systems (Engine and/or Brake TCS) and is now being used by automotive manufacturers in production vehicles.

6. REFERENCES

Tamburro, Guidoni, Ferro, Visconti "Requirement for AWD vehicles".
CRF 1989
Holzwarth, May "Analysis of Traction Control Systems augmented by limited slip differentials"
SAE paper 940831

C498/30/164/95

A P Suretrac – accelerated rig durability testing of a torque biasing limited slip differential through vehicle road load simulation

J P CHIPPENDALE MIMechE, **M J WILSON,** and **A J YOUNG** BSc, CEng, MIMechE
Automotive Products plc, UK

SYNOPSIS

Test techniques for the A.P. Suretrac Torque Biasing Limited Slip Differential have been developed through the extensive use of data analysis from representative vehicles - covering Urban, Rural and Motorway Routes. By recording individual wheel speeds and drive shaft torques, a complete spectrum of transmission driveline and differential duty levels has been built up for a variety of styles of driving over these typical routes.

The paper includes a description of how the data was collated into a form suitable for building up an accelerated rig test schedule. This data has given confidence in matching the rig test to real life durability requirements. A description is also given of the new fully automated test rig designed for this purpose.

1 INTRODUCTION

In the development of any new, innovative product such as Suretrac, there is a need to obtain reliable feedback as to how the product performs under 'real world' situations. It is also necessary to be able to obtain quickly, indications of the validity of different test configurations. To achieve this aim requires the translation of the conditions under which the unit is expected to operate into a comprehensive test schedule, which is capable of being used in a test rig environment. A further objective is to accelerate any test work to minimise development time and to complement vehicle proving tests.

Initial test work had centred on a four-square rig which imposed a fixed speed difference across the differential. The power unit/drive-line was a conventional car engine and transmission, while a Heenan and Froude eddy current dynamometer provided the power absorption. The limitations of this system were clearly apparent, prompting the need to upgrade the test to reflect the duty cycle during the life time of the differential. This inferred a requirement to produce a rig capable of being programmed to suit a variety of test needs. A library search, alongside discussions with various sources within the industry, revealed little in the way of a complete rig durability schedule for the testing of limited slip differentials. This initiated a programme of work where instrumented vehicles were driven round a variety of test routes to obtain Road-Load Data. This was subsequently analysed to produce a rig test schedule. Previous to this exercise work had started on specifying, designing and commissioning a test facility capable of providing the degree of flexibility and control necessary for the comprehensive testing of the Suretrac differential.

2 DESCRIPTION OF SURETRAC

Suretrac has been described in detail elsewhere - (1) and (2). However, for completeness, a brief description is included here.

Suretrac is a torque biasing limited slip differential in which the usual bevel gear set is replaced by an arrangement of cams and followers. The general arrangement is shown in Figure 1, while the principle of operation is shown in Figure 2.

Internal friction between the cams and followers, and between the followers and the mating drive spline in the cage, enables the unit to transmit whatever unequal torques the tyre contacts can withstand, up to a limiting torque ratio set by design. The ability to generate static friction at the component mating surfaces before relative motion occurs, gives a friction based torque biasing unit such as Suretrac the advantage of inhibiting the onset of wheel spin before any build-up. This improves directional stability for normal road use.

Figure 1 shows a diagrammatic overview of the Suretrac differential. The drive comes through from the main transmission to the cage via the pinion and crown wheel in the normal way. The followers are mounted in splines in the cage, taking the drive. From there the drive is split to the two cams attached to the left and right drive shafts, and hence to the wheels.

For straight-line driving no internal relative motion occurs, so both cams rotate at the same speed. In this condition there are no internal friction losses within the differential.

When a speed difference does occur, the followers move along the cage spline, maintaining contact with both cams. This motion is opposed by frictional forces on each sliding face, the combination of these forces providing a torque ratio between the outputs.

Figure 2b shows the forces acting on a single follower during differential motion. The follower is moving to the left, with the left hand cam moving faster than the right. The friction in the cage spline (μL) opposes the motion of the follower and provides part of the slip-limiting effect. This is done by increasing the magnitude of the right hand Normal Reaction force (N2) and decreasing the left hand force (N1). The remainder of the slip limitation is provided at the interface between cams and followers, where the frictional forces (μN1 and μN2) combine with the Normal Reaction contact forces (N1 and N2) as shown, to give low and high resultant drive forces on the respective cams.

What is less obvious is that statically, when conditions at the tyres dictate it, they can

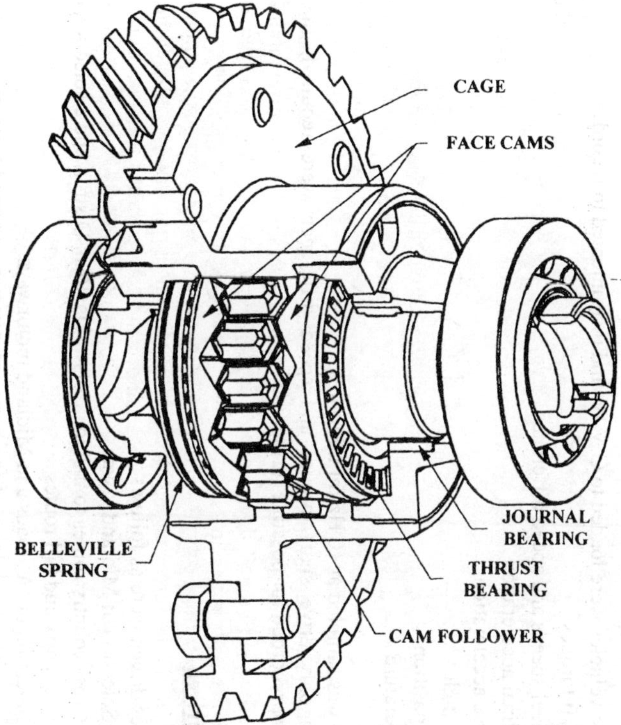

FIGURE 1 GENERAL ARRANGEMENT

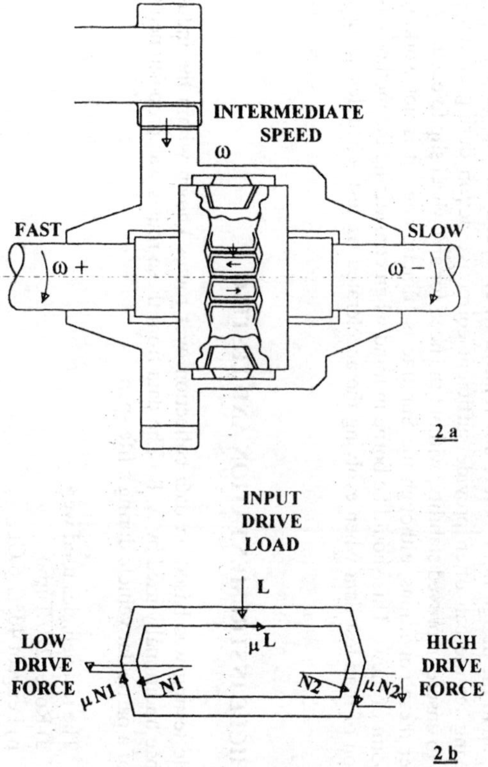

FIGURE 2 THE ENGINEERING PRINCIPLE

cause drive shaft wind-up. This wind-up results in a build-up of the internal friction forces from a balanced straight ahead condition to any required level up to the point at which the unit starts to break out into relative motion. This produces any intermediate torque ratio demanded, up to the design limit, before any relative wheel speed (or slip) takes place. This torque biasing feature can often provide sufficient corrective action during changing road conditions to ensure improved stability, without the need for any wheel slip to occur.

Under these conditions, although the Suretrac is 'functioning', it is not working from the wear point of view. This should be borne in mind when considering the methods used for the collation of road test data when evolving the accelerated rig test schedules, as described below.

3 VEHICLE INSTRUMENTATION AND TESTING

The vehicle testing was based on data collection over routes known within the industry as well as those traditionally used by A. P. The data from these routes was proportioned to that expected of a particular vehicle during a life span.

The test vehicles used were :-
a) Rover 220 Turbo
b) Peugeot 405 1.6 GL.

The vehicles were loaded to G.V.W. and instrumented to record :-

- Drive shaft torques
- Front wheel speeds and speed difference
- Longitudinal acceleration
- Transverse acceleration
- Steering angle
- Throttle position
- Oil temperature

The data was sampled at 10 Hz for all routes.

Although only drive shaft torques and speed difference were required, the other parameters were recorded for possible further analysis.

3.1 Routes driven

The routes driven were as follows :-

A) Selection of Midland towns
B) Cross country routes to include Cotswolds, Peak District and a selection of cross country and hill routes.
C) Selection of 'A' roads and Midland motorways.

Table 1 shows the breakdown of the basic information obtained from each route along with the calculations to derive the equivalent number of route cycles to reach 160 000 kilometres (100 000 miles). These calculations are allied to vehicle and engine size involving proportions of total usage spent on the various routes and the average vehicle speed for each route. 160 000 Kilometres is generally accepted as the initial minimum durability target expected by vehicle manufacturers.

Table 1 Derivation of route proportions.

ROUTE	TEST TIME -Sec.	DIST.-- Kms.	AVER. SPEED KPH	NUMBER OF ROUTES IN 160 000 Kms.
TOWN 1	1621	8	17.7	3024
TOWN 2	2585	19.2	26.7	1896
COTSWOLD	6377	97.6	55.1	77
XCOUNTRY	7153	129.6	65.2	114
PEAK	7153	129.6	65.2	114
HILL	1982	37.3	67.7	82
A ROADS	1811	33.6	66.79	451
M'WAY	5356	173.6	116.8	305

4 DATA ANALYSIS

For each of the routes a matrix of torque / speed difference was derived with each cell showing the 'time-at-level' in tenths of seconds. An example of the type of information gained is shown in Table 2.

An examination of the matrices was made in order to determine the maximum speed difference seen across the differential. From this a philosophy was adopted that the rig test would impose a value no greater than 95% of the maximum speed difference seen during the appropriate route. The selected rig speeds were based on the average speeds for the type of route, plus a compromise between the need to keep overall test times down and the upper limit of the transmission speed. There was also a need to ensure that overall temperatures within the transmission were maintained within manufacturers' specified limits.

Each of the 'time at level' cells were then converted to a time equivalent to that being run at the selected rig speed difference. Each torque level was then summed to obtain the equivalent 'time at torque level'.

Further time reductions were obtained by selecting fewer torque bands for the rig running. Adjustments for wear life damage were made by using the non-linear torque dependency factor obtained from previous rig testing. An exponential type of characteristic was evolved for the relationship between unit fatigue life and applied torque by testing at a series of fixed torque levels.

For each of the routes the new times were grouped together and an overall timing schedule obtained representing the work done during the 160 000 kilometre life. To obtain a mixture of types of route the schedule was split into 10 sections with each section comprising of a run through each of the route groupings. Using this modular approach, any desired combination of routes and overall durability target can be derived to suit an individual manufacturer's requirements.

As two different types of vehicle had been analysed, the information gained enabled an interpolation to be done, such that the work done for any engine rating between the two chosen could be deduced.

In order to validate the rig test results Automotive Products have concurrently run a number of durability vehicles. These have been subject to a variety of duty cycles and loading situations. Analysis of the componentry from these, together with the calculations of total work done (as derived from the road load data), have all been amassed to give a comprehensive set of correlated data. Suretrac is currently being used in several successful motor sport applications both in this country and abroad also providing valuable information on the validity of the design.

Correlation between the rig work and general road running has been good. The prediction of wear characteristics over a wide range of conditions has improved significantly also enabling design changes to be confidently tested under realistic conditions.

5 DURABILITY TEST RIG

A new test schedule involving the simulation of a wider spectrum of output speed, torque and output speed difference raised the need for a more versatile durability test rig. For greater flexibility an electric motor drive was chosen. The arrangement ensured the ease of making complete transmission changes and facilitated the use of different drive-line configurations with the minimum of adaptation. Separate output dynamometers, to achieve variable control over the output speed difference as well as the total torque, were deemed necessary. To achieve full flexibility on the control side it was seen this could only be done under computer control with the necessary interlocks to enable 24 hour running.

Fig 3 presents a schematic of the durability rig showing the main features.

The development of the rig has enabled the simulation of a wide range of Road-load data parameters. This has given A.P. the ability to test quickly design variants, confident that the results feed across into real world situations. As well as the durability considerations the following test schedules may be simulated :-

1) Split μ Starts
2) Compact Spare tests
3) Noise Signature tests
4) Tow tests (Alpine climb)

6 CONCLUSIONS

In the cost conscious Automotive industry the need to reduce the time from inception to implementation highlights the need for a comprehensive yet flexible test facility such as the one described in this paper. The ability to translate road load data and customer vehicle test schedules into a rig based test has been a major advance for the A.P. Suretrac Project. This has resulted in significant improvements, both in the validity of test, and time of test to be achieved. The vehicle testing of the Suretrac Torque Biasing Limited Slip Differential over 160 000 Kilometres can now be represented by an accelerated rig schedule in 100 hours. This enables the effect of design changes to be evaluated in the minimum time.

REFERENCES

(1) JARVIS, R.P. and YOUNG, A.J. A New Form of Limited Slip Differential. S.A.E. Paper 930672. Also printed in Automotive Transmissions and Drivelines (SP-965)

(2) CHIPPENDALE, J. Sure and Simple - the A. P. Suretrac Differential. Racecar Engineering Vol. 5 No. 1

Table 2:- Matrix of speed difference v Axle torque range - Town1

MEAN SPEED DIFF. - Kph	overdrive torque <-------- -400 -> -300	-300 -> -200	-200 -> -100	-100 -> 0	--------> drive torque 0->100	100 -> 200	200 -> 300	300 -> 400	400 -> 500	500 -> 600	600 -> 700	700 -> 800	800 -> 900	900 -> 1000	1000 -> 1100	Total time 1/10ths secs.
-2.5	0	0	0	0	0	0	0	0	0	0	0	0	0	0	0	0
-2.3	0	0	0	0	0	0	0	0	0	0	0	0	0	0	0	0
-2.1	0	0	0	15	0	0	0	0	0	0	0	0	0	0	0	15
-1.9	0	0	0	26	17	0	0	0	0	0	0	0	0	0	0	43
-1.7	0	0	0	40	17	0	0	12	0	0	0	0	0	0	0	69
-1.5	0	0	16	47	23	16	0	0	0	19	0	0	0	0	0	121
-1.3	0	0	17	39	37	27	15	11	15	13	15	0	0	0	0	189
-1.1	0	0	32	75	46	14	15	0	0	0	11	0	0	0	0	193
-0.9	0	0	41	94	59	22	15	0	15	14	17	0	0	0	0	277
-0.7	0	15	52	94	78	23	19	14	22	13	0	0	0	0	0	330
-0.5	0	18	75	143	87	37	21	21	27	23	24	0	0	0	0	476
-0.3	0	33	131	189	111	79	58	35	44	35	25	28	17	0	0	785
-0.1	12	51	248	312	172	113	81	93	74	44	46	36	20	14	0	1316
0.1	0	94	415	360	236	182	130	143	89	73	56	50	33	16	0	1877
0.3	0	77	374	307	199	145	104	112	81	67	47	48	51	26	0	1638
0.5	0	41	199	190	129	75	68	48	38	37	43	21	23	19	0	931
0.7	0	11	81	129	89	70	33	35	26	18	30	16	12	13	0	563
0.9	0	0	62	88	63	44	22	17	23	0	14	15	0	0	0	348
1.1	0	0	43	45	29	20	13	15	14	16	13	14	0	0	0	222
1.3	0	0	27	36	19	18	15	0	0	0	11	0	0	0	0	126
1.5	0	0	0	21	0	0	0	0	0	0	0	0	0	0	0	21
1.7	0	0	0	14	13	0	0	0	0	0	0	0	0	0	0	27
1.9	0	0	0	0	0	0	0	0	0	11	0	0	0	0	0	11
2.1	0	0	0	0	11	0	0	0	0	0	0	0	0	0	0	11
2.3	0	0	0	0	0	0	0	0	0	0	0	0	0	0	0	0
2.5	0	0	0	0	0	0	0	0	0	0	0	0	0	0	0	0
Total time-1/10ths secs.	12	340	1813	2264	1435	885	609	556	468	383	352	228	156	88	0	9589

SURETRAC LIMITED SLIP DIFFERENTIAL DURABILITY TEST FACILITY

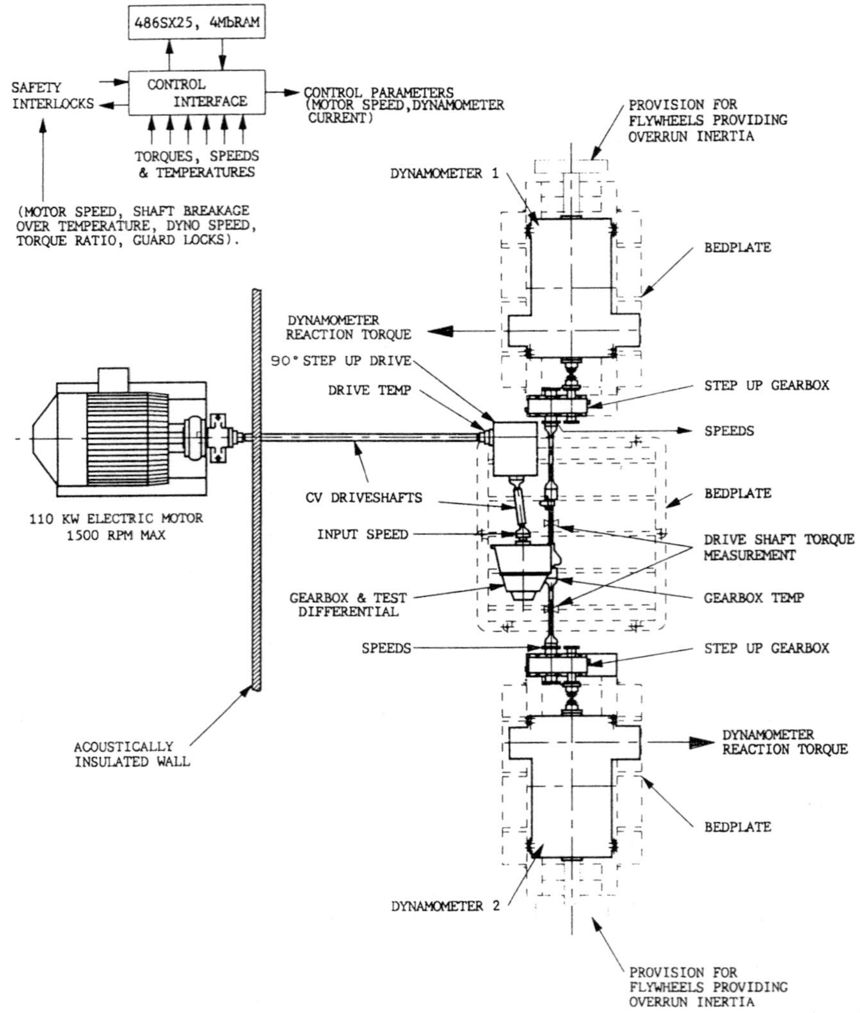

FIG 3

Emissions

C498/28/104/95

Reproducible determination of permeation rates of fuel system components

SCHÄDLER and J MÜLLER
MS Chemie AG, Switzerland

SYNOPSIS

The use of plastics in automotive fuel systems has become general practice. However it involves the problem of meeting stringent permeation resistance requirements and proving that these requirements have been met.
A test instrument, designed to measure permeation, is described in this paper.

1. INTRODUCTION

Reasons of cost and weight require the use of thermoplastics in automotive fuel systems worldwide. At the same time the hydrocarbon emissions from a vehicle are drastically limited by the applicable legal requirements (2g HC per vehicle per day in the so-called SHED test, according to the California authority for the prevention of air pollution). Plastics components involved in future developments will be required to meet automotive specifications of permeation resistance.

Current developments in this field comprise of multilayer fuel lines connected with units such as fuel injection rails with integrated coupling technology. The consistent use of variety of thermoplastic multilayer tubings permits a lower number of interfaces within the fuel system. Further developments in highly integrated fuel system components must meet even stricter permeation and cost requirements.

Comparison and assessment of the permeation resistance of fuel system components require a quantitative determination of the hydrocarbon (and alcohol) emissions under defined ambient conditions. For this purpose a Swiss raw material producer has developed a permeability testing method that meanwhile has become established worldwide.

The system consists of testing a fuel system component e.g. a fuel line, through which a heated and pressurized fuel is passed and which is introduced into a closed container (recipient). The fuel permeating through the test specimen is transferred by the carrier gas (N_2) into a container with activated carbon. This activated carbon absorbs the permeated fuel vapours. The fuel permeation can be measured via the weight gain of the activated carbon.

Based on the experience gained with optimized laboratory scale units, a high quality measuring device has been developed; the Permeation Tester (P-Tester) for determing the permeation rates of various fuel system components. The aim is to contribute to the standardization of permeation measuring methods and equipment industry wide. The development of the P-Tester enables the automotive industry to develop cost effective fuel system components.

2. THEORETICAL BACKGROUND

Permeation is the penetration of one medium through an other. This process is complex and must be subdivided in four steps. (Fig. 1)

Fig.1. Mechanism of permeation - divided into four steps

2.1 Adsorption: The molecules bond on the surface of the polymer due to intermolecular interactions. (Fig. 1a)
The Adsorption is described by the Henry's law:

$$c = s \cdot p$$

c: Concentration of the permeating medium in the polymer
s: Solubility of the permeating medium in the polymer
p: Partial pressure of the permeating medium at the polymer surface

2.2 Absorption: The adsorbed molecules are dissolved in the plastic that means they penetrate into the material. (Fig. 1b) The Absorption is described by the Henry's law too.

2.3 Diffusion: The fuel molecules move irregularly in the polymer. Due to the concentration gradient a molecule flow in the direction of the smaller concentration results. This means that there is a diffusion rate from inside a fuel line to the outside (Fig. 1c). The diffusion is described by the first Fick's law. For a polymer tube with wall thickness d the following mass flow rate results:

$$\frac{dm}{dt} = - D \, A \, \frac{\Delta c}{d} \, \rho$$

m: Mass of the permeating medium
D: Diffusion coefficient
A: Diffusion surface
Δc: Concentration difference from inside to outside
ρ: Density of the permeating fuel

2.4 Desorption: Fuel molecules reaching the outside of a tube are retransferred into vaporous condition. The description is the same as for the diffusion. (Fig. 1d)

3. DEVELOPMENT OF THE MEASURING PRINCIPLE

At the beginning of the development work permeation measurements for determing the barrier properties of plastic fuel system components were carried out *statically.* Hereby a test specimen with defined dimensions is filled with test fuel and sealed to be gas tight. The weight loss resulting from permeation is determined directly. The component to be tested is kept in a temperature controlled air conditioning cabinet. The pressure in the test specimen is a result of the test fuel's vapour pressure at the selected temperature and thus cannot be influenced. With components differing in their permeation rate, the test fuel may change its composition unless regularly replenished. This effect has a great influence on the measuring results when using methanol-containing test fuels.

With the more recent dynamic measuring principle the fuel continuously circulates through the test specimen. Compared with the static measurement, this offers substantial advantages.

- The pressure and temperature of the fuel can be set independently and thus more exactly.
- Moreover the dynamic principle ensures an almost constant fuel composition during the measurement.

4. THE DYNAMIC MEASURING PRINCIPLE

Fig. 2 shows the schematic design of a modern permeation tester. The unit consists of a fuel circuit (primary circuit) and a carrier gas circuit (secondary circuit).

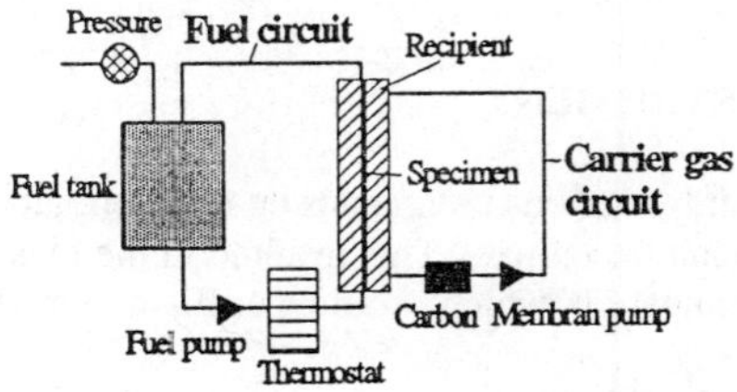

Fig.2 Schematic design of the permeation tester

In the *primary circuit* the test fuel is passed through a *heating bath* and the specimen by a *gear pump.* Pressure is generated by means of nitrogen. This circuit is connected with a *fuel tank* of 4 litres capacity. Compared with the circuit's volume the tank's capacity is big. In spite of components differing in their permeation rate, the overall composition of the fuel remains approximately constant. A fuel line or another part of the fuel system is kept in a gas tight test cell, called *recipient.*

In the *secondary circuit* dry, inert carrier gas (nitrogen) passes the permeated fuel vapours to a filter of *activated carbon* to be absorbed. The flow rate of the carrier gas circuit can be controlled by a reducing valve.

For the calculation of the permeation rate the weight gain Δm is plotted versus the time t. The resultant curve can be basically subdivided into three areas (Fig. 3):

- The length of the *adsorption phase* is highly dependent on the material. The better the barrier properties of a material, the longer this phase.
- The *equilibrium phase* is characterized by a constant desorption rate. The slope of the linear regression straight line during this phase is the measure of the permeation rate. The length of the equilibrium state should be at least 200 hours for one permeation test.

- The *saturation phase* characterizes the area following the actual measuring phase during which the slope of the linear regression straight line decreases as a result of the activated carbon being saturated or the test fuel having changed its composition. This especially applies to heavily permeating systems. The saturation phase is not to be used for the calculation of the permeation rate.

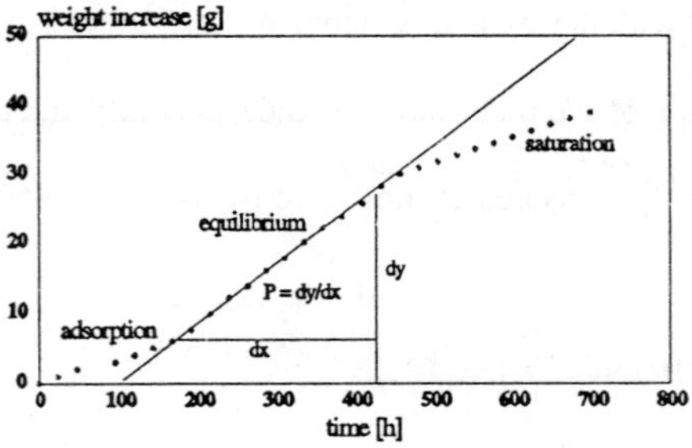

Fig. 3 Schematic permation curve

5. THE MEASURUNG INSTRUMENT

The measure series carried out by various companies on different laboratory scale units during the recent past were very difficult to compare. This resulted in the idea to design a standardized unit for measuring fuel permeation which would permit a reproducible determination of permeation rates.

The new P-Tester developed for this purpose is characterized by the following improvements:

- *Recipients* made of steel provide higher safety

- An additional *heating system* for the recipients permits the setting of different temperatures on the surface of the test specimen and prevents deposits of dissolved plasticizer on the wall of the recipient.

- The *carrier gas* is transported by means of high-performance *membran pumps*; the flow rate can be exactly set and controlled.

- The *activated carbon* is kept in aluminum cylinders. An *on line weighing-system* facilitates handling and makes the determination of the permation rates more accurate. The calculation of the permeation rate out of the weight increase of the activated carbon is performed by means of an in-built industrial personal computer. A data sheet with all parameters and measure points is made for every permeation test. This allows a perfect control of the test influencing parameters.

- The temperatures of the fuel and the recipient are electronically monitored and controlled.

- Simultaneous measurement of the *change in length* of a pipe as caused by swelling and/or loss of plasticizer.

- A *safety system* with electronic control unit and hydrocarbon sensor enables the instrument to be operated in "Ex-Zone 2" proof rooms.

- All operations are program-controlled

- To ensure the fuel composition's constancy one's attention is drawn to change the test fuel at a 5% loss of the original volume (4 liters) by permeation.

Measuring series were carried out on unplasticized monolayer tubings to determine the accuracy of the permeation tester. They resulted in a scattering range of 2% of the permeabilities measured.

The absorbance of the used activated carbon resulted from measurements of mass balance expressed by the so-called adsorption factor, and was roughly 98%. The permeation rates shown by the permeation tester are to be corrected by this factor.

The zero-calibration of the P-Tester was obtained by means of a steel tube which allows no permeation.

6. INFLUENCES ON THE MEASURING RESULT

6.1 The fuel temperature is the most significant influential parameter. As a rule of thumb, a temperature increase by 10 to 15 K can be assumed to double the permeation rate. It is therefore of the utmost importance that the fuel temperature is exactly controlled and kept constant.

6.2 The influence of the fuel pressure is less significant. An increase in the pressure was found to result in a slightly higher permeation rate for multilayer fuel lines. For monolayer tubings the fuel pressure influence is neglectible. When measuring the permeability of parts like connectors, a higher pressure can result in a higher permeation rate. In this case the mechanism which leads to emissions is microleakage rather than permeation.

6.3 The recipient heating permits defined test conditons and can serve to simulate different situations in the practical operation of fuel system components. Rising recipient temperatures lead to an increase in the permeation rates.

6.4 With plasticized materials a higher carrier gas flow rate can lead to higher permeation rates. This effect is particularly marked with materials having poor barrier properties. For in this case the tendency to dissolve plasticizer out of the material increases, which is due to the high volume of permeating fuel.

6.5 The type of fuel, i.e. the composition of the test fuel and particularly alcohol (methanol) content is a significant parameter of influence on the permeation through plastics. A comparison of measuring results is therefore only possible when standardized fuels are used.

The test parameters given in Table 1 have proved to be standardized test conditions with relevance to general practice. With these conditons the "worst case" is simulated. It is important to know that a test with the dynamic measuring principle under the worst case conditions is only suitable for the development work of plastic fuel system components such as fuel lines, fuel rails and connectors. The permeation rates determined by the dynamic system differ to the SHED test's results. Reasonable correlations are not possible.

Parameter	Value
Test fuel temperature	60 °C
Test fuel pressure	4 bar
Recipient temperature	60 °C
Carrier gas flow rate	1.5 l/min
Test fuel	FAM B

Table 1. Parameters in the operation of the permeation tester which represent the worst case

7. CONCLUSIONS

The described permeation measuring unit is a great help for the development of plastic fuel system components to meet future requirements. By using the same measuring principle and testing under the same test conditions permeation results of different laboratories will be comparable. One of the biggest advantages of the P-Tester is the automatic calculation which will allow very accurate and reliable permeation results even for laboratories with little trained staff.

C498/28/249/95

A review of the problems associated with diesel exhaust particles and the available methods for the extraction of the adsorbed hydrocarbon fraction

A J SMITH BSc, **D H TIDMARSH** BSc, PhD, CEng, FIMechE, MBIM, MIMgt, and **H DAVIES** PhD
University of Central England, UK
M WILLCOCK BEng, PhD
A E Goetze Limited, Bradford, UK

SYNOPSIS

Due to the concern about exhaust emissions from diesel powered vehicles and the effect these have on health and the environment, there is a need to assess both the quality and quantity of these emissions. Particulate emissions are of special interest because of the specific health problems associated with them. Analysis of diesel particulates should include the weight, chemical and physical composition and ultimately the size of the particulate matter collected.

This paper will review the issues associated with diesel particulate emissions and the methods available for extraction of the adsorbed hydrocarbons. It will go on to discuss the relative merits and failings of each method and draw conclusions as to the most appropriate method for routine analysis.

1. INTRODUCTION

1.1 Background

One of the major differences in exhaust emissions between those from spark ignition (petrol) engines and those from compression ignition (diesel) engines is that of particulate emissions. The particulate matter emitted in the exhaust of a vehicle powered by a compression ignition engine is legally defined as, "any material collected on a specified filter medium after diluting diesel exhaust gases with clean filtered air so that the temperature does not exceed 325°K (52°C)."[1]* The emissions of particulate matter from compression ignition engines are much higher than that from spark ignition engines. Van den Hout and Rijkeboer [2] found that light duty diesels emit between 4-7g particulates/litre of fuel, with 7-14g particulates/litre of fuel for heavy duty diesels, compared with < 1g particulates/litre of fuel for spark ignition vehicles.

* Numbers in parenthesis denote references at the end of the paper.

Particulates as defined above are formed from a combination of elemental carbon and adsorbed hydrocarbons (organic fraction), water, sulphur compounds and inorganic compounds (Figure 1). Both the elemental carbon and organic fraction are derived from partial combustion of both fuel and lubricating oil. Other components, water, sulphur compounds and inorganic compounds are derived from impurities in the diesel fuel.

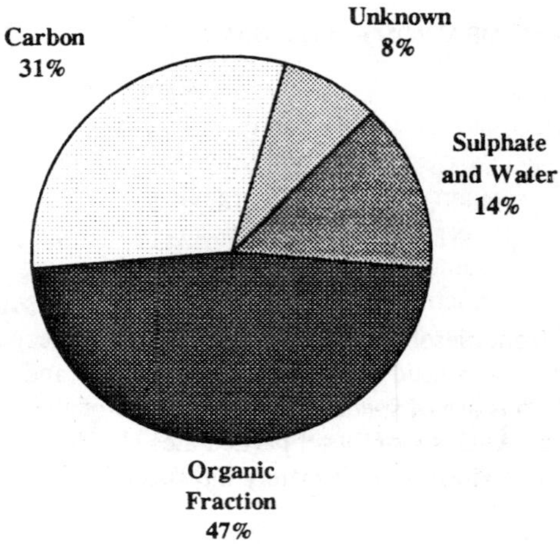

Figure 1. Typical Composition of Direct Injection Diesel Exhaust Particulate.

Haynes [3] proposed a two stage mechanism for the formation of elemental carbon. The first stage is particle formation where particles are formed by pyrolysis and oxidation. The initial particles typically consists of unsaturated hydrocarbons and polycyclic aromatic hydrocarbons, and have a diameter of less than 2nm. The second stage, particle growth, results from material condensing onto the particles and agglomeration with hydrocarbons.

Combustion bomb studies [4] have shown three different paths to soot formation depending on the temperature of combustion and the fuel composition. At temperatures of less than 1700°C only aromatics and highly unsaturated aliphatic compounds with a high molecular weight produce solid carbon by pyrolysis. At temperatures typical of diffusion flames (~1800°C), all hydrocarbons normally used in fuel will react to produce solid carbon if burned under sufficiently rich conditions. At temperatures higher than that found in diesel engines a nucleation process utilising carbon vapour is the primary particulate mechanism.

At the high temperatures associated with combustion, the components of a diesel particulate, i.e. carbon and hydrocarbons, remain distinct. However, as the exhaust gases are released into the atmosphere and diluted with ambient air, the temperature of the exhaust gases drops. This allows high molecular weight compounds in the exhaust, such as gas phase hydrocarbons, condense from the gas phase to the liquid phase. The elemental carbon then acts as a seed for the formation of a mature particulate as the heavy hydrocarbons condense on to its surface. This results in the surface of the elemental carbon being coated in a layer of condensed hydrocarbons, sulphur compounds and inorganic compounds.

In order to fully assess the impact of diesel particulates on both health and the environment the two major components, elemental carbon and organic fraction, must be analysed individually. Some form of extraction method is therefore required to separate the particulate into its individual components. There are two main types of extraction method; chemical and physical. Chemical methods use some form of solvent, e.g. dichloromethane (DCM), to dissolve the organic fraction. The solvent containing the dissolved organic fraction can then be analysed, both for quality and quantity, using gas chromatography. The most common of these methods is soxhlet extraction (see 3.1). Physical methods use heat alone or heat plus a vacuum to extract the organic fraction. These methods rely on heat energy to volatilize the organic fraction back into the gas phase (see 3.2). The organic fraction can then be quantified by the loss in mass of the particulate sample paper. If the gas phase organic fraction can be captured it can then be injected directly into a gas chromatography column for qualitative analysis.

Newkirk [5] pointed out that these different methods do not measure exactly the same thing. The chemical method of using solvent extraction measures soluble organic fraction (SOF), which can be defined as, "any material, independent of boiling point, that is soluble in the test solvent under the test conditions." Physical methods using purely heat desorption measure the volatile organic fraction (VOF), which can be defined as, "organic material that responds on a flame ionisation detector (FID) and has a boiling point less than approximately 600°C". Methods that use heat and a vacuum measure the total volatile fraction (TVF) which can be defined as, "all material on a filter paper that can be volatilized at 225°C and 5 torr (absolute)."

2. PROBLEMS ASSOCIATED WITH DIESEL PARTICULATES

2.1 Health Effects

There are a number of concerns about possible deleterious effects of diesel particulates on health. Pope *et al* [6] published results that suggest that there is a correlation between mortality and morbidity with atmospheric concentrations of particulates of a size less than 10mm (PM_{10}). This work showed an apparently linear relationship between atmospheric concentrations of PM_{10} and both daily mortality from all causes and respiratory morbidity. The author went on to suggest that a 1% increase in mortality from all causes could be associated with a 10mg/m^3 rise in atmospheric PM_{10} concentration although no biological mechanism to explain this phenomenon was put forward.

Perhaps the most major health concern with diesel particulates is that associated with the organic fraction. As previously explained, this is the fraction of the particulate made up of unburnt and partially burnt fuel an oil. As such it can potentially contain a number of carcinogenic compounds, which, when found on small particulates (e.g. PM_{10}), can penetrate deep into the lungs. The International Agency for Research on Cancer (IARC) has three classifications for compounds that are, or potentially, carcinogenic to humans.[7] These are; chemicals that are proven to be human carcinogens (IARC Group 1); probable human carcinogens (Group 2), this group is split into those that have at least limited evidence of carcinogenic effects on humans (2A), and those that have evidence of carcinogenic effects on animals (2B).Table 1 lists a number of compounds that have been found in the exhaust of vehicles and their IARC classification. The IARC have also classified diesel exhaust itself as group 2A.

Table 1. IARC Classification of Carcinogenicity of Some Particulate Associated Exhaust Compounds [7]

Compound	IARC Group
anthracene	3
benzo(a)anthracene	2A
benzo(a)pyrene	2A
benzo(e)pyrene	3
benzo(j+k)fluoranthene	2B
2-nitrofluorene	2B
1-nitropyrene	2B
1,8-dinitropyrene	2B
1,6-dinitropyrene	2B

The elemental carbon portion of the particulate is itself generally regarded as being non-toxic. However, there is some evidence to suggest that the physical properties of particulate elemental carbon may cause damage to the lungs. Work Done by Nikula *et al* [8] used rats to show that prolonged exposure to high concentrations of diesel exhaust particulate could cause an increase in lung tumours. It was suggested that the accumulation of particulate matter deep in the lungs may retard the pulmonary clearance mechanisms, thus clogging the gas exchange areas deep in the lung. These particles may also act as seeds for other compounds to become adsorbed on to, causing locally raised concentrations of potentially carcinogenic organic material. This problem is compounded by the fact that the lungs clearance mechanism is impaired, thus potentially carcinogenic compounds are trapped.

2.2 Environmental Effects

The carbon content of diesel exhaust is the main contributor to another major concern about diesel exhaust particulates, that of soiling and visual nuisance. Diesel vehicles were estimated to produce 42% of the black smoke in the UK in 1990 [9]. In urban areas the problem is more acute, as a greater proportion of the total black smoke is from vehicles. Ball and Casswell [10] estimated that two thirds of smoke in London was from road traffic.

While of less overall significance than that of health problems, the problem of soiling of buildings is still of great importance. In 1987 £110 million was spent cleaning the exterior of buildings in urban areas [11]. A survey conducted by McCrae and Williams found that 40% of respondents were concerned about blackening of building walls, with around 20% targeting road traffic as the primary source.[12]

Diesel exhaust particulates are also a major contributor to reduced visibility caused the scattering and absorption of light. Hamilton and Mansfield [13] estimated that diesel exhaust particulate matter is responsible for 25-45% of visibility reduction in urban areas. While difficult to quantify such a subjective nuisance it still desirable to reduce or remove it if possible.

With these problems in mind, governments around the world have set increasingly higher standards for diesel exhaust emissions and more are on the way. This will increase the need for an easy, accurate and precise method for the analysis of diesel particulate matter.

3. EXTRACTION METHODS

It is important when discussing the methods available for the extraction of the organic fraction of diesel particulates to highlight the distinction between quantitative and qualitative analysis. Quantitative analysis measures what proportion of the particulate mass was made up of organic material. Qualitative analysis measures the composition of the organic material matrix. If the two are combined, one gets the quantity of organic fraction as a percentage of the mass of the particulate, and also the composition of the organic fraction, i.e. what organic compounds make up the organic fraction.

The former, quantitative analysis, is carried out on a weight basis. The sample filter paper is conditioned under temperature and humidity controlled conditions before particulate sampling for at least 4 hours [14]. The particulate sample is then taken and the sample filter paper re-conditioned in precisely the same way as before and re-weighed. The difference in weight is the quantity of particulate. The same method can then be used to quantify the organic fraction, i.e. the weight before and after extraction is the quantity of organic fraction.

Qualitative analysis can be carried out using some form of chromatography, either High Performance Liquid Chromatography (HPLC) or Gas Chromatography (GC). Chromatography is used to separate the sample matrix into its individual components. These can than be identified by the properties they exhibit during analysis, e.g. retention time, mass or fluorescence. If internal standards are used chromatography can also quantify the individual components.

3.1 Chemical Methods

This group of methods use a chemical solvent to extract the soluble organic fraction. In the case of diesel particulates the most commonly used solvents are cyclohexane, acetonitrile and methylene chloride [14]. This can applied in two main ways; soxhlet extraction or sonication.

Soxhlet extraction uses a condenser apparatus as shown in figure 2. The principle of the method is as follows:

1. The solvent in the reservoir (C) evaporates as its heated and the vapours rise into the condenser (D).
2. The vapours then cool and condense back into a liquid. The droplets of solvent then run down the walls of the extractor (B) into the thimble (A) containing the sample filter paper.
3. This continues until the level of solvent in the thimble reaches the return siphon (F).
4. Liquid solvent then returns to the solvent reservoir to be reheated.
5. This process is repeated continuously for at least 8 hours.
6. The solvent with the sample in is then dried under a dry nitrogen atmosphere until all the solvent has evaporated.
7. The dried sample is then re-constituted with 1ml of solvent (carbon disulfide).
8. This mixture can then be injected as a liquid into a gas chromatograph for qualitative and quantitative analysis.

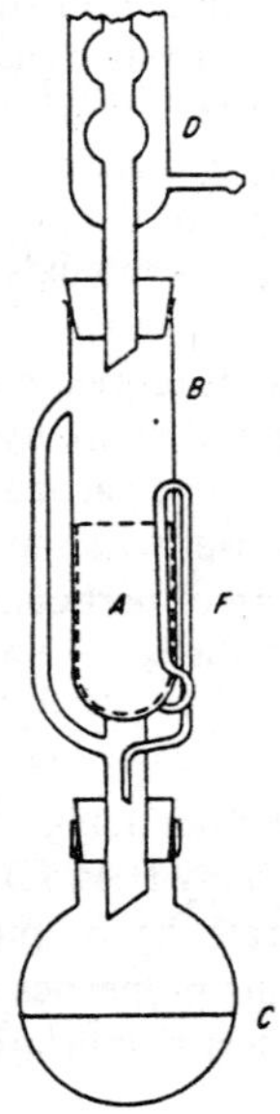

Soxhlet extractor.
D **= reflux condenser,**
B **= body of extractor,**
A **= extraction thimble,**
F **= liquid return siphon,**
C **= extracting solvent.**

Figure 2. Soxhlet Extraction Apparatus [15].

Extraction using ultra sound is similar to soxhlet extraction in that it uses the same solvents. It differs however, in that rather than continually washing the sample with heated solvent, the sample and solvent are exposed to high frequency sound in an ultrasonic bath. This has the effect of vibrating the organic fraction into the solvent. The sample can then be prepared and analysed as for soxhlet extraction.

3.2 Physical Methods

This group of methods rely on the volatility of the organic fraction to separate it from the elemental carbon fraction. There are a number of methods for achieving this, thermogravimetirc analysis, vacuum oven volatilisation and solid sample injection.

Cuthbertson *et al* [17] used a thermogravimetric analyser to analyse diesel exhaust particulate samples. The principle of the method is as follows: the conditioned sample filter paper is weighed and placed in an inert atmosphere furnace, i.e. nitrogen. A temperature program is then run, for example, 20°C-550°C at a ramp rate of 50°C/minute, to volatilize the organic fraction to a gas. The quantity of organic fraction can then be calculated by re-conditioning and re-weighing the sample filter paper and subtracting this weight from the pre-extraction weight. The difference in weight before and after is the quantity of organic fraction present in the sample. This method can be used in conjunction with a Heated Flame Ionisation Detector (HFID) to achieve crude qualitative results. The HFID responds to the organic material to create a peak on a chart recorder. Cuthbertson found that this method could be used to distinguish dodecane (C_{12}) and eicosane (C_{20}), but the addition of cetane (C_{16}) almost completely removed any distinction between the peaks. This method therefore a primarily quantitative one which offers too little resolution for adequate qualitative analysis of the extracted material. Little sample preparation is required and a result is achieved in around thirty minutes.

Vacuum oven volatilisation works on a similar principle to that of thermogravimetric analysis. It differs in that the final oven temperature is lower (225°C) and a vacuum (5 torr) is used to lower the boiling point of the organic fraction [5]. The organic fraction is quantified in the same way as thermogravimetric analysis, and again little qualitative analysis can be carried out.

Solid sample injection was first developed by Cuthbertson and Shore [18] of Ricardo. This method attempted to bring together the best parts of soxhlet extraction and thermogravimetric analysis. Rather than extracting the organic fraction with a solvent prior to injection into a gas chromatograph, the sample filter paper was put directly in to a modified gas chromatograph injector. The sample was flash heated to volatilize the organic fraction on to the cooler gas chromatograph column. The organic fraction then re-condensed on to the end of the column. A temperature programme was then run to sequentially volatilize the organic fraction and allow it to pass through the column to the detector. This method offers both quantitative and qualitative analysis with little sample preparation required.

4. CONCLUSIONS

The main advantages and disadvantages for each method are listed in table 2. Until the advent of solid sample injection there were very distinct applications for the different chemical and physical methods of extraction. Chemical methods offered an extract that could be further analysed for composition, at the expense of a very complex sample preparation routine and time consuming extraction process. Added to this, the effect of the solvent on the sample is very difficult to quantify and could produce misleading results. If the composition of the organic fraction was required, in Research and Development for example, these were the only methods available. Physical methods on the other hand offered relatively quick quantitative results, at the expense of detail as to how the sample was made up. They also did away with any interference associated with the solvent used in chemical extraction and required little sample preparation and as such were ideal for routine analysis of large numbers of samples. Solid sample injection offered potentially the best of both worlds, the speed and simplicity of a physical method with resolving power of a chemical method. This makes this method suitable for both routine and analytical testing and is used by Ricardo as a standard method but has yet to be established industry wide.

Table 2. Comparison of the Methods Available for the Extraction of Particulate Organic Fraction

Extraction Method	Advantages	Disadvantages
Soxhlet Extraction	Can analyse the extract for content	Complex sample preparation, lengthy extraction process, possible interference from sólvent, possible error fromthermal decomposition, o measures molecules soluble in the test solvent.
Sonication	Can analyse the extract for content	Complex sample preparation, lengthy extraction process, possible interference from solvent, health and safety regulations associated with ultr sound, only measures molecul soluble in the test solvent.
Thermogravimetric Analysi	Quick, easy method, no solvent interference	Poor resolution for qualitative analysis.
Vacuum Volatilisation	Quick, easy method, no solvent interference	Poor resolution for qualitative analysis.
Solid Sample Injection	Quick, easy method, no solvent interference, ca analyse extract for cont	Not yet widely used but has potential to become industry standard.

One could conclude then that solid sample injection offers the best compromise between ease of use and quality of data, and as such has a brighter future than that of chemical or other physical methods. With further development this method could become the industry standard for routine testing and research and development work. This may become more important in the future if prospective emission regulations take into consideration the quality of the particulate as well as the quantity.

REFERENCES

[1]. EEC Directive 91/542/EEC. Official Journal of the EC 1991, L295 Vol. 34, Daughter Directive of .

[2]. Van den Hout K.D. and Rijkeboer R.C. TNO. 1986, Road Transport Research Institute.

[3] Haynes, B.S. and Wagner, H.G. Soot Formation. Prog. Energy Combustion Science, 1981, Vol. 7, p. 229-273.

[4] Amann, C.A. and Siegla, D.C. Diesel Particulates - What they are and Why. Aerosol Science Technology, 1982, p.73-101.

[5] Newkirk M.S. Application of Methods for Determining Total Organic Contribution to Diesel Particulates. Transactions of the ASME, 1992, p. 578-589, Vol. 114.

[6] Pope, C.A. III *et al.* Daily Mortality and PM_{10} Pollution in Utah Valley. Arch. Environ. Health, 1992, Vol. 47, p. 211-217.

[7] Harrison, R.M. *et al.* Diesel Emissions and Urban Air Quality. 2nd Report of the Urban Air Review Group, 1993, Dpt. of the Environment.

[8] Nikula, K.J. *et al.* Influence of Soot Associated Organic Compounds on Pulmonary Carcinogenicity of Diesel Exhaust in Rats. Toxic and Carcinogenic Effects of Solid Particles in the Respiratory Tract. 4th International Inhalation Symposium, Hanover, 1993.

[9] Dpt. of Environment 1992. Digest of Environmental Protection and Water Statistics No. 15., 1992, HMSO.

[10] Ball D.J. and Caswell R. Smoke from Diesel Engined Road Vehicles - an Investigation into the Basis of British and European Emission Standards. Atmos. Environ., 1982, vol. 17, 169-181.

[11] Mansfield T. Sources of Building Soiling and a Review of the Stone Cleaning Industry 1991. International Conference on Stone Cleaning, 1992, Edinburgh.

[12] McCrae I.S. and Williams I.D. Road Traffic Pollution and Public Nuisance. Proceedings of the 4th International Highway Pollution Symposium, 1992, Paper No. III.B.3., Madrid.

[13] Hamilton R.S. and Mansfield T.A. Airborne Particulate Carbon: its Sources, Transport and Contribution to Dark Smoke and Soiling. Atmos. Environ., 1991, vol. 23, p. 1591-1599.

[14] Warner M.A. Filter Extraction Procedures and Results for Various EPA/ECTD Particulate Samples. 1982, EPA 460/3-82-008.

[15] Shugar G.J. and Ballinger J.T. Chemical Technicians Ready Reference Handbook. 1990, McGraw-Hill Inc.

[16] Schaus F. Principles of Gas Analysis an Particulate Analysis. Monitoring and Control of Industrial Air Pollutants, 1994, Sunderland.

[17] Cuthbertson R., Stinton H.C. and Wheeler R.W. The Use of a Thermogravimetric Analyser for the Investigation of Particulates and Hydrocarbons in Diesel Engine Exhaust. 1979. SAE Paper 790814.

[18] Cuthbertson R., Shore P., and Sundstrom L. Direct Analysis of Diesel Particulate Bound Hydrocarbons by Gas chromatography with Solid Sample Injection. 1987. SAE Paper 870626.

C498/28/260/95

₋vap sub-systems with reduced evaporative ₃missions –a support to environmental protection

) WEBER
)ELPHI, Luxembourg

In the last decades the exhaust gas emission levels have been decreased rapidly. With regard to the value of hydrocarbons consisting of exhaust-gas emissions and volatile components, the importance of evaporative emissions has been increased. As a result the more stringent US-96 evaporative legislation and the now proposed EC-2000 Evap requirements seem to be obvious.

This publication deals with parameters lowering the evaporative emissions. After classifying the different sources the topic is focused on the tank system including carbon canisters. Therefore the following emissions are considered: vapour escaping the vehicle during trips and soaking (US and Cal. Enhanced Evap-test) and the hydrocarbons occurring when the cars are filled up (Onboard Refueling Vapour Recovery, ORVR).

1 Introduction

Evaporative emissions of a vehicle may have different sources. Most of the components are hydrocarbons evaporating during vehicle driving or soaking (figure 1). Comparing the importance of the HC-sources, the hydrocarbons evaporating from solid materials like plastic have a significant influence when the vehicle is new. In addition all fuel ducting lines and hoses are to be designed to prevent HC-permeation. But the main target of investigations during vehicle development should be the tank vapours, the exhaust emissions during the so called "Running Loss" and the emissions when the car is being refueled.

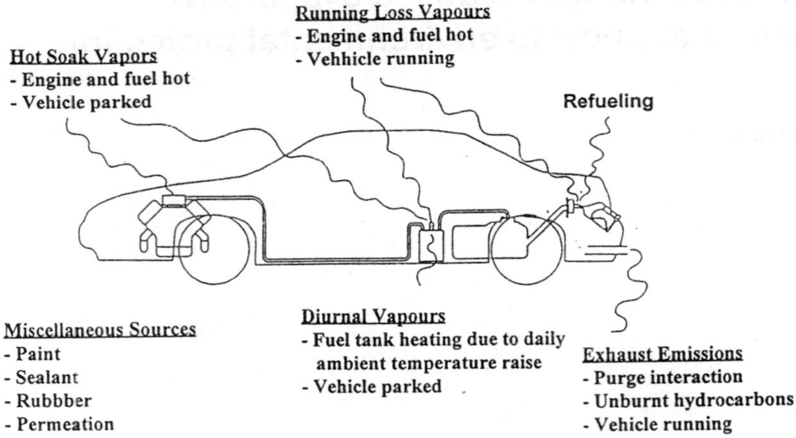

Figure 1: Classification of Evaporative Emissions

As the emission requirements on the exhaust side have been decreased rapidly during the last decades and will be more stringent in the future, the value to reduce evaporative emission levels should be discussed (figure 2). On the left hand side the US-TLEV and LEV limits in the exhaust tests are shown. The order of magnitude of the LEV-limit as well as the improvement from TLEV to LEV lies in the range of 0.05g/mi.

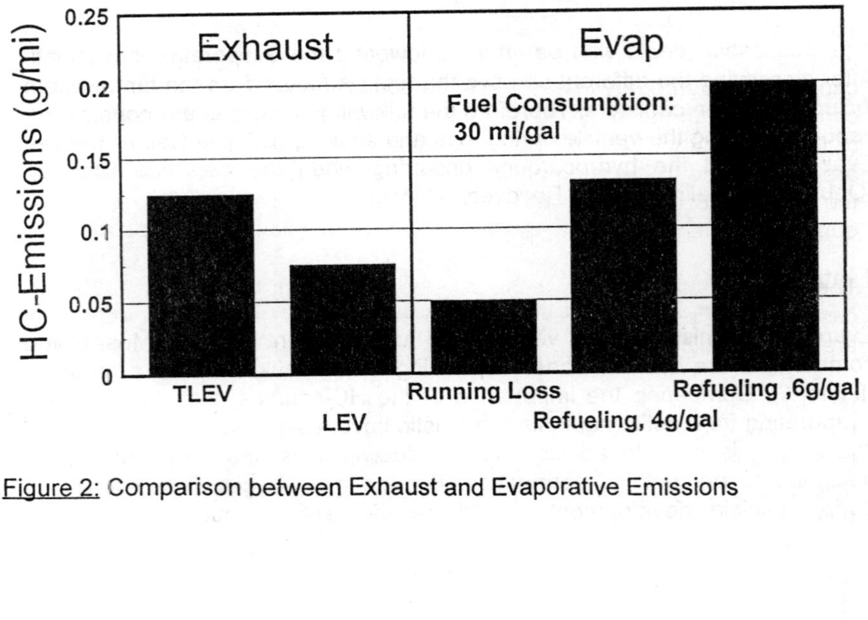

Figure 2: Comparison between Exhaust and Evaporative Emissions

ORVR-vapor generation rates are in the area of app. 4-6 g of vapour/gal (app.1-1.5 g/l) of fuel dispensed. With the assumption that a vehicle's fuel consumption is app. 30 mi/gal, the refueling emissions will have values between 0.13 and 0.2 g/mile. If these emissions are prevented, the benefit will be higher compared to the improvement on the exhaust side described above. The Running Loss limit is also in the same of magnitude. As a conclusion the evaporative legislation is useful with regard to exhaust emission levels, although different boundary conditions like test cycles, fuel composition and temperatures were not taken into account.

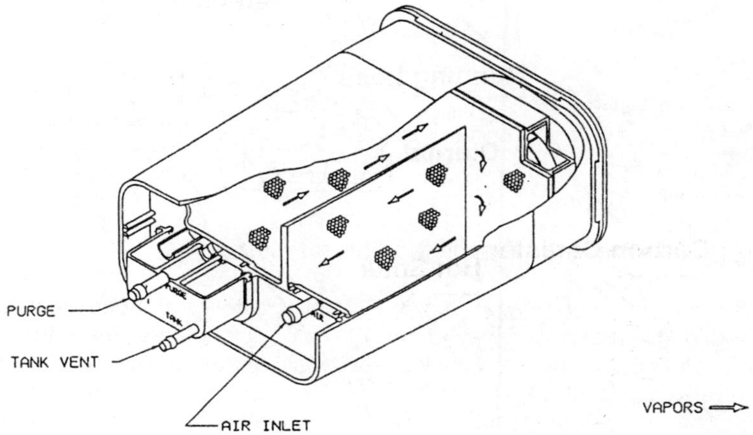

Figure 3: Example of a Carbon Canister

To prevent hydrocarbons from escaping a tank system carbon canisters (example: figure 3) are used. The mixture of fuel vapour and air enters the canister is absorbed by the activated carbon inside and pure air can leave the canister through the air tube. To purge the canister, air flows from the air tube to the purge tube. On their way through the HC molecules can desorb and are transported to the internal combustion engine where they are burnt. Both the exotherme adsorption and the endotherme process depend on carbon quality, type of hydrocarbon, ambient pressure, humidity and temperature.

The determination of the canister size according to the legislation and customer test procedures is a real challenge as a lot of boundary conditions are to be considered:

i.e.:

fuel composition[1]	purge volume flow
tank volume	geometry of carbon canister
flow resistance of hoses	carbon working capacity
test cycle	temperatures

2 Enhanced Evap Test Results

To simulate the behaviour of the whole tank system including fuel and carbon canister a calculation programme[2] (figure 4) has been created which allows accurate predictions (figure 5). The programme consists of several modules working independently, so that all conceivable test sequences can be simulated. For example, a comparison between the test results and the predicted results is presented in figure 5.

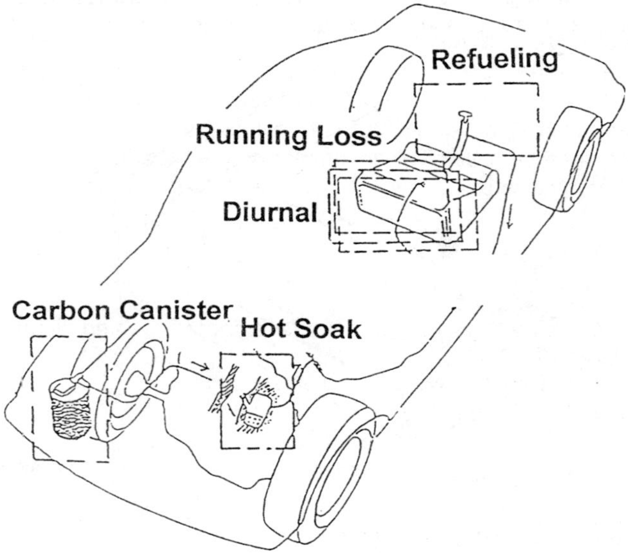

Figure 4: Various Mathematical Modules in Calculations Programme "EVAP"[1]

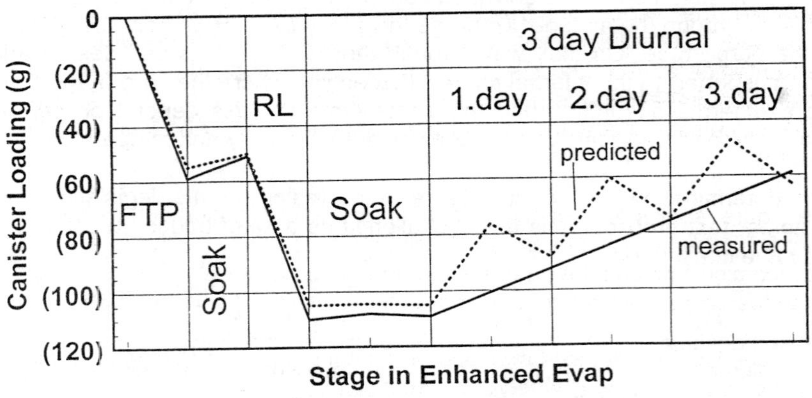

Figure 5: Enhanced Evap Results - Calculation and Test Results

The figure shows the canister loading in the Enhanced Evap procedure. Before the test starts the canister is filled initially. Therefore the full loaded canister has the weight of 0g. During all stages of the Enhanced Evap there is only a small deviation between testing and calculation when the canister is loaded as well as it is purged. The so called back purge can also be observed when the tank system is cooled down and the recondensation effect in the tank sucks air through the canister.

The canister calculation module also gives the results of the breakthrough emissions but they can't be compared in this case because only the total vehicle evaporation emissions have been measured. As a result these kind of calculation programme saves a lot of money and costs during the development process by predicting necessary purge rates, canister volumes, tank peak temperatures etc..

3 ORVR

The well known ORVR test procedure defines temperatures and fill rates:

filling rate: 4 - 10 gal/min
tank temperature ca. 28°C
dispensed fuel temp: ca. 19 °C

emission limit: 0.053 g/l of fuel disp.
9 RVP fuel

To prevent the hydrocarbons from escaping the filler neck a liquid seal is used due to its advantages concerning warranty and suitability to different fuel nozzle geometries.

The necessary ORVR tests are run in an ORVR shed where the refueling emissions as well as the canister loading are determined. Similar to the Enhanced Evap procedure the canister size should be as small as possible. Therefore the vapor generated during the refueling event has to be decreased. The main parameters influencing these vapor generation rates are the temperature in the bulk fuel and the fuel dispensed as well as the air entrainment impacting the partial pressure of the fuel components. Using the EVAP calculation model the following dependencies could be found (figure 6, bottom)

The figure shows the canister loading or vapour generation rates as a function of the fuel tank temperature with the heat transfer efficiency between the incoming fuel and the inner tank vapour as a parameter. The temperature of the fuel dispensed is held constant at the legislation value of 19.3°C.

The incoming fuel replaces the vapor in the tank dome. In parallel this vapour can be cooled down when the tank temperature is higher than the temperature of the fuel dispensed. In addition the bulk fuel will heat up the cooler incoming fuel. These two effects can be found on the right hand side of the figure (tank temperature > incoming fuel temperature). If there is no heat transfer between tank vapour and incoming fuel (0%), the incoming fuel will replace the tank vapour but is heated up by the bulk fuel. That generates vapour and the canister loading increases. The better the heat transfer efficiency, the higher the influence of the recondensation effect caused by the vapour cooled down. Therefore the vapour generation rates can decrease with increased fuel tank temperatures at high heat transfer efficiencies. As a result high heat transfer efficiencies should by chosen to decrease canister size under legislation conditions.

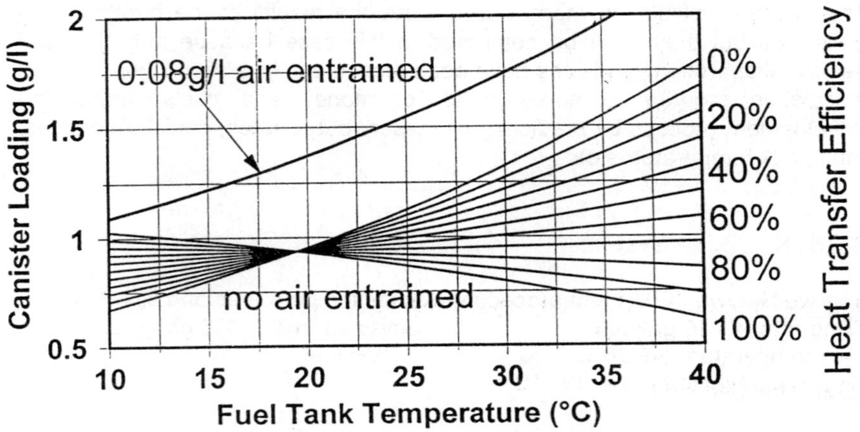

Figure 6: ORVR Canister Loading as a Function of Temperature Difference, Heat Transfer and Air Entrainment

In reality tank temperatures can be significant higher than those of the fuel dispensed. Then the tank system especially the bulk fuel is heated up by the incoming fuel and no recondensation will occur. The vapor generation rates will be so high that the fillabilty is not acceptable. In these situations the heat transfer mentioned above has to be small. In the development process both effects are taken into account.

All the results presented in figure 6 in the bottom don't include the air entrained during the refueling event. An air entrainment of app.0.08g/l fuel dispensed will increase the canister loading by app. 0.5 g/l. By this the air flow to the tank influenced by the fuel flow should as small as possible.

The development process can be accelerated using all the background knowledge discussed. The improvement of the refueling system will lead to smaller vapour generation rates and therefore smaller and less expensive canisters. Figure 8 contains the refueling results with a base system (left hand side) and the developed tank (right hand side). Canister size can be decreased by more than 20%: better package conditions and lower process can be achieved. In parallel the emissions must meet the legislation limit of 0.2 g/gal (0.053 g/l)

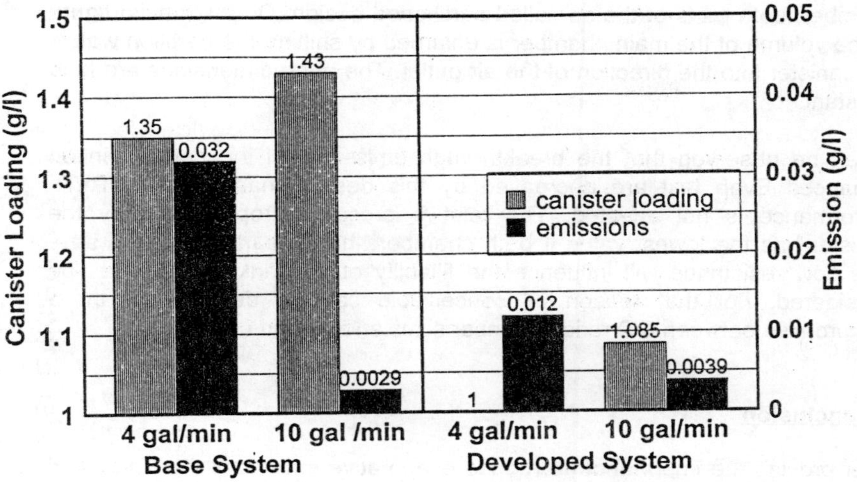

Figure 7: Non-Working and Working ORVR-System

4 Canister Design

The improvement on the tank system concerning vapour generation rates is only one possibility to let smaller canisters work. The canister design also has impact on its performance (figure 8). Beside the well known influence of the L/D ratio of canisters other design parameters are to be taken into account.

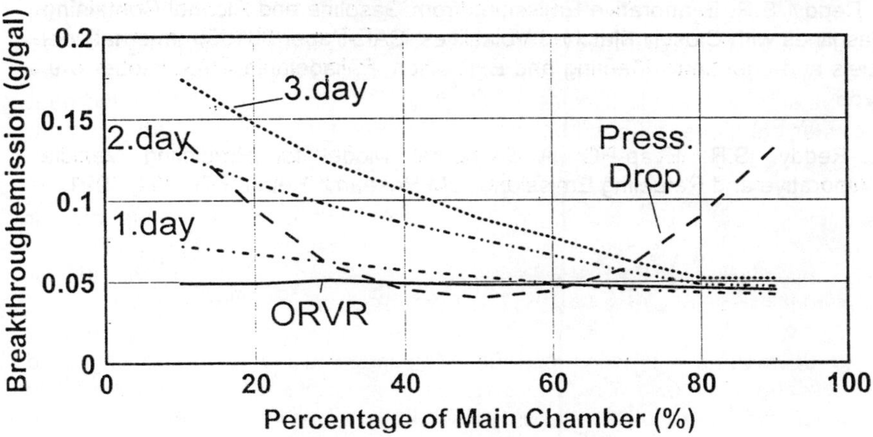

Figure 8: Influence of Canister Design on Breakthrough Emissions

The canister (see also figure 3) consists of two chambers: the main chamber to be entered by the vapour flow coming from the tank and the second chamber, both parts of the so called partitioned design. On the x-axis (figure 8) the volume of the main chamber is enlarged by shifting the partition wall in the canister into the direction of the air outlet. The outer dimensions are held constant.

It can be observed that the breakthrough emissions of the canister in an Enhanced Evap test are decreased by this design change. The ORVR-performance is not affected. The relative pressure drop caused by the canister has the lowest value if both chambers have nearly the same size. This flow resistance will influence the fillabilty of the tank and has to be considered. For that reason all conceivable canister designs will be a compromise between L/D ratio, chamber sizes and carbon used.

5 Conclusion

After proving the importance to reduce evaporative emissions the Enhanced Evap and ORVR test conditions and significant influencing parameters are discussed. A computer simulation model helps to meet the emission levels and customer requirements as well as to ensure a short and cost effective development process. In addition the knowledge and test experience in ORVR-systems and canister design supports the future improvement of Evap Sub Systems.

6 References

1. Reddy, S.R. Evaporative Emissions from Gasoline and Alcohol-Containing Gasolines with Closely Matched Volatilities, SAE-Paper 861556, International Fuels and Lubricants Meeting and Exposition, Philadelphia, PA, October 6-9, 1986

2. Reddy, S.R. Evap-PC: A Computer Model for Predicting Vehicle Evaporative and Refueling Emissions, GM Research Report F7L-764, 1991

C498/34/109/95

Reduction in petrol engine exhaust emissions by the use of a positive displacement throttle loss recovery turbine

R W DRIVER CEng, MIMechE
PDT Engineering Technology Limited, UK

SYNOPSIS

A novel rotary positive displacement turbine is described that has the unique ability to efficiently recover petrol engine throttle losses. It allows engines to reduce exhaust emissions and speed by half at idle. Engines optimised to use this turbine reduce fuel and exhaust emissions in town and city driving by over 20%. The device's novel mechanical arrangement makes it uniquely applicable to the low flow, low speed regime of petrol engines. The turbine has a mechanism that solves the friction problems that beset traditional devices. An enhancement to the turbine simultaneously provides compressed air for cold start fuel atomisation.

NOTATION

A	=	leakage area
Cd	=	vehicle coefficient of aerodynamic drag
Cr	=	vehicle coefficient of rolling resistance
cp	=	specific heat at constant pressure
cv	=	specific heat at constant volume
Fa	=	vehicle frontal area
g	=	gravitational acceleration
ml1	=	air mass leakage from '1' to 'f'
ma	=	inlet air mass at temperature Ta
$m$al	=	air mass leakage from inlet to '1'
n	=	cp/cv
nc	=	number of vane compartments
P1	=	total pressure in compartment '1'
Pa	=	ambient pressure
Pf	=	total pressure in final compartment
r/sec	=	revolutions per second

R = gas constant
$T1$ = total temperature in compartment '1'
Ta = inlet temperature
Tf = temperature in final compartment
Tfi = ideal total temperature in final compartment
V = vehicle velocity in miles/hour
W = vehicle weight

1 INTRODUCTION

The handling of the intake air in a petrol engine and the mixing of the fuel with the induced air is a complex process and the source of both inefficiencies and emissions that are unacceptable to our pollution-aware generation.

At full throttle, fuel atomisation can generally be optimised to give efficient combustion with low unburned fuel emission. The part throttle situation is more complex, to turn down the power output from the engine, the fuel input and the mass flow must be reduced. This is normally done by closing down the area through which the air is induced and creating a partial vacuum in the inlet manifold by a largely irreversible process. The engine under these conditions expends a significant amount of energy to draw the flow through this restrictor, and a high proportion of this energy is then lost as the gas expands to the inlet manifold pressure. At idle conditions, as much as 50% of the energy is taken simply to overcome this throttle-loss. If this energy could be extracted and returned to the crankshaft, engines would idle at half conventional speeds and energy consumption and emissions could also be halved. In heavily populated areas, such as the LA basin in the USA, where traffic congestion is characterised by engines operating at part-throttle, the potential exists to make a dramatic impact on fuel usage and output of pollutants.

The cold-start cycle produces the worst-case pollution scenario. A heavily fuel-rich condition must be imposed on the flow to ensure that a critical air-to-fuel ratio can be achieved for ignition to be successful. This is achieved at the expense of leaving partially-burned fuel in the exhaust. If an air compressor, that was very effective down to engine idling speed, could be used to generate pressure and temperature conditions that would atomise injected fuel to less than 10μ droplet size, the normal overfuelling that causes unburned fuel pollution could be eliminated.

Although modern engine technology with catalytic converters, stratified-charge combustion, fuel injection and computerised ignition has found a way to minimise some of these problems, the compression/expansion-control devices in this paper offer the potential to further improve the situation. Indeed, in many of the less regulated parts of the world the use of these simpler and more fundamental techniques could completely obviate the need for expensive devices such as catalytic converters.

Conventional air-moving systems, such as compressors or turbines, that would be considered for such applications fall into two classes: turbo-devices, which rely on high speed for gas sealing and for efficiency, and positive displacement devices which rely more on close mechanical fits for sealing.

In a typical engine the variation between idle and maximum speed is about 6:1 whilst the mass flow has a variation of 16:1. A radial or axial turbine with variable geometry inlet and meeting these speed and mass flow variations would have a maximum speed and efficiency of 70 000 r/min and 60% respectively and have an even worse efficiency at low engine speed.

Sliding vane turbines have been used for many years and are well known. They can run at engine speed and can be thermodynamically arranged to cover all intake conditions. The problems associated with these devices are friction, sealing and lubrication. Adequate vane tip sealing at idle conditions results in excessive tip loading and wear at high speed. Inertia and pressure loads produce friction losses from the faces of the vanes as they slide in their grooves. Oil lubrication contaminates the intake air and dry lubricants have relatively high coefficients of friction and short lives under these conditions. Turbine isentropic efficiency is likely to be less than 50% with very poor mechanical efficiency. Roots type machines are unsuitable as a turbine since there is no internal expansion between inlet and outlet. They also have poor compression efficiency above a compression ratio of about 2:1.

This paper describes a unique positive displacement device, previously applied only as a heat pump, that is inexpensive and has all of the desired characteristics for throttle-loss recovery and simultaneously providing compressed air for cold start fuel atomisation, but in which the seemingly inevitable efficiency and rotor tip wear problems are eliminated by the use of a novel mechanical arrangement.

2 POSITIVE DISPLACEMENT TURBINE

2.1 Turbine construction

The device has six cam-shaped vanes which rotate inside a static circular casing and avoid contact with the casing by pivoting as they rotate (Figure 4). The vanes are connected via crank arms to connecting rods arranged like the spokes of a wheel (Figure 5). The little end of the conrod is centred on line with the vane tip and the big end centred on the centre of the static casing. This makes the vane tips rotate in a circle concentric to, and nearly coincident with the bore of the static casing. Close to the side ends of the vanes are discs which rotate with the vanes and in which the vanes pivot. These side discs extend along the sides of the static casing beyond the vane tip diameter. The spoke conrods and pivoting shafts are all carried on plain journal bearings. Lubrication for the bearings is outside the plane of the turbine and divorced from it. The bearing loads are relatively light which allow the bearings to have the optimum length/diameter ratio for minimum friction losses.

Inefficiencies in this positive displacement device are caused by both mechanical friction and leakage of air between adjacent vane compartments. All the moving parts use standard plain bearings and the device has one location ball bearing.

Intake mass flow variation is provided by sliding a portion of the casing in a circular movement about the centre of rotation, thus exposing the required intake cross-sectional area.

Apart from bearings, all constituent parts of the throttle loss recovery turbine and air compressor are able to be made from pressure diecast magnesium or glass/graphite polymer composite. The weight of the turbine assembly and drive is approximately 5 kg.

Unique characteristics of the turbine allow it to run at engine speed and have the same life as the engine. A two litre engine at an idle speed of 500 revs/min and 50 kPa inlet manifold pressure provides about 0.3 kW power to the turbine, and when running at 3000 r/min with the same intake manifold pressure provides about 1.3 kW of power to the turbine. An extension to the camshaft drive belt would probably be the most convenient way of transferring this amount of power from the turbine to the crankshaft.

The value of a throttle recovery turbine at high engine speed is marginal, therefore to reduce stresses and increase the turbine's life it is suggested that a freewheel device is provided that operates when the throttle is open and the engine speed is over say 3000 r/min.

2.2 Air atomisation

If injected fuel is further atomised by an air jet, the resultant droplet size can be reduced to less than 10μ. At this droplet size the fuel will ignite in a cold engine as easily as in a hot one, thereby obviating the need for a conventional cold start system.

The throttle loss recovery turbine can retain a proportion of the intake air and compress it sufficiently to provide compressed air for additional fuel atomisation (Figure 4).

3 TURBINE EFFICIENCIES AND PERFORMANCE

3.1 Assumptions for the rotary positive device

With a knowledge of predicted bearing loads and industry accepted bearing theory the friction losses can be calculated.

Turbo machinery and roots compressor isentropic efficiency can be accounted for if it is assumed that air expands isothermally in accordance with Joule's law when leaking from high to lower pressure. Air leakage between adjacent vane compartments in the present case can be calculated using well known theory. The final fluid conditions are found when stable conditions prevail and can be used to calculate the isentropic efficiency of the turbine.

In the proposed turbine there are two vanes between inlet and outlet. The equation for the mass of ambient pressure air leaked ($m$al) past one vane to adjacent compartment '1' is:

$$m\text{a1} = \sqrt{\frac{\text{n g R}}{\frac{\text{n - 1}}{2}} (\text{Ta} - T1)} \quad \frac{P1 \text{ A}}{T1 \text{ R (r/sec nc)}}$$

and for compartment '1' the air leaked (m1l) past one vane to the outlet 'f' is:

$$m\text{1l} = \sqrt{\frac{\text{n g R}}{\frac{\text{n - 1}}{2}} (T1 - T\text{f})} \quad \frac{P\text{f A}}{T\text{f R (r/sec nc)}}$$

After a number of iterations a solution is found when Tf becomes constant.

$$\text{Turbine isentropic efficiency} = \frac{\text{Ta} - T\text{f}}{\text{Ta} - T\text{fi}}$$

Gross power = power requirement/engine mechanical efficiency.
Engine indicated power = gross power plus power absorbed by throttle loss.

3.2 Turbine performance

With Tf and bearing loads calculated the turbine efficiencies can be found.

The gross engine power can be calculated from the assumptions in Table 1 and using $[(Cr\ W\ 2.2) + (Cd\ Fa\ V^2.028)]\ V/(375\ \ \text{engine mechanical efficiency}) = \text{gross engine power}$.

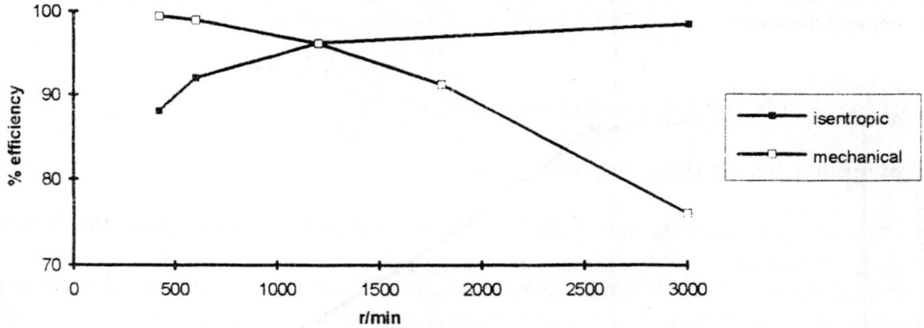

Figure 1 Turbine isentropic efficiency and mechanical efficiency.

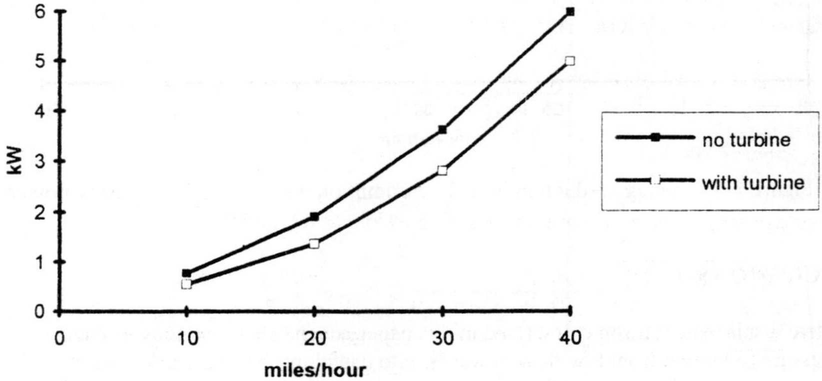

Figure 2 Engine indicated power requirement.

Table 1 City driving assumptions

vehicle weight	1000 kg
road gradient	level
vehicle coefficient of rolling resistance	.01
vehicle coefficient of aerodynamic drag	.38
vehicle frontal area	1.86 m
petrol engine capacity	2 litres
engine mechanical efficiency variation	80% to 88%
engine idling % of city driving time	50
engine acceleration % of city driving time	25
engine under steady power % of city driving time	25

From calculations of turbine and engine power the reduction in fuel consumption can be calculated.

If the engine is optimised to use the throttle recovery turbine the average reduction in fuel consumption and exhaust emissions under steady load conditions between 10 and 40 miles/hour varies between 40% and 20% respectively. Under the driving conditions in Table 1 the average reduction between 0 and 40 miles/hour is over 21%, and 50% at idle by virtue of a lower engine idle speed.

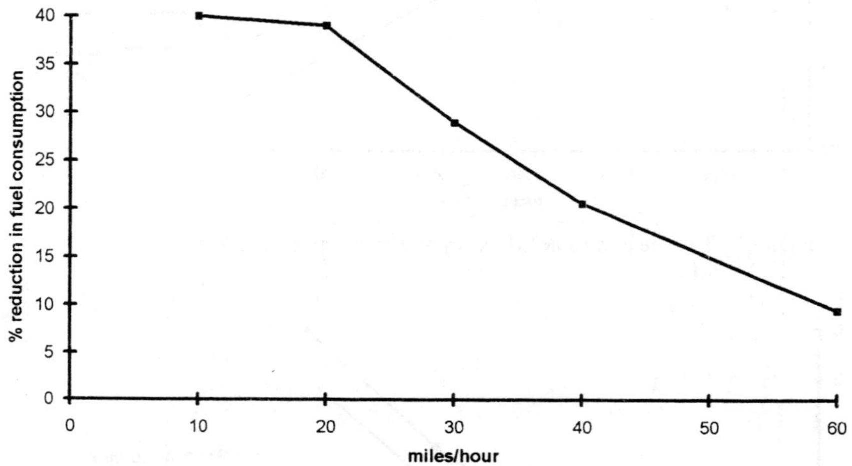

Figure 3 Percentage reduction in fuel consumption on level road and steady power.

4 CONCLUSIONS

The positive displacement turbine described in this paper has the ability to convert engine intake pressure reduction from low flow upwards, into useful mechanical work at rotational speeds several orders of magnitude slower than conventional turbines. This allows work to be returned to the crankshaft by a simple mechanical coupling giving a dramatic improvement in efficiency and halving of the idle speed.

The average reduction in fuel consumption and exhaust emissions between 0 and 40 miles/hour in city driving, where half the time is at idle and the remaining half is equally shared between acceleration and steady power, is calculated at over 21%, whilst at idle it is 50%.

The device's novel mechanical arrangement makes it uniquely applicable to throttle loss recovery.

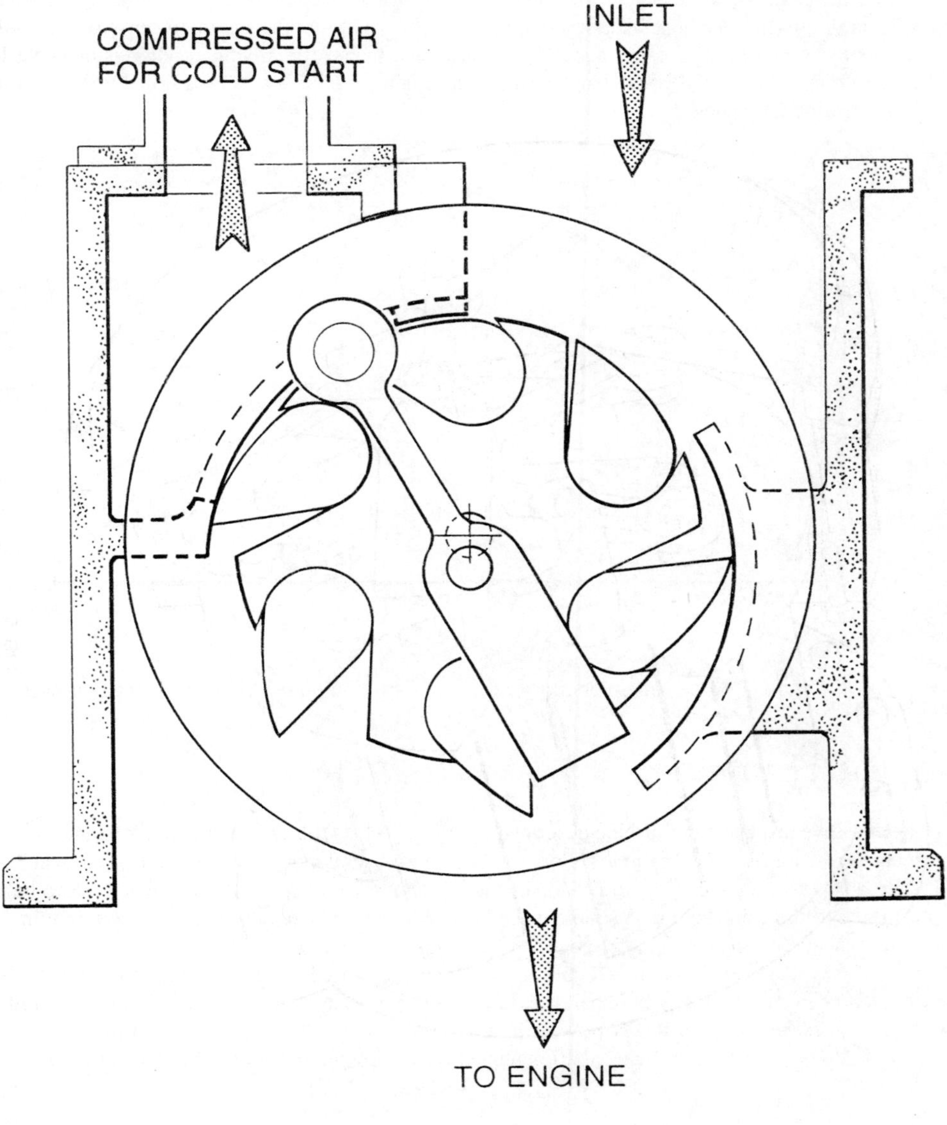

FIG 4

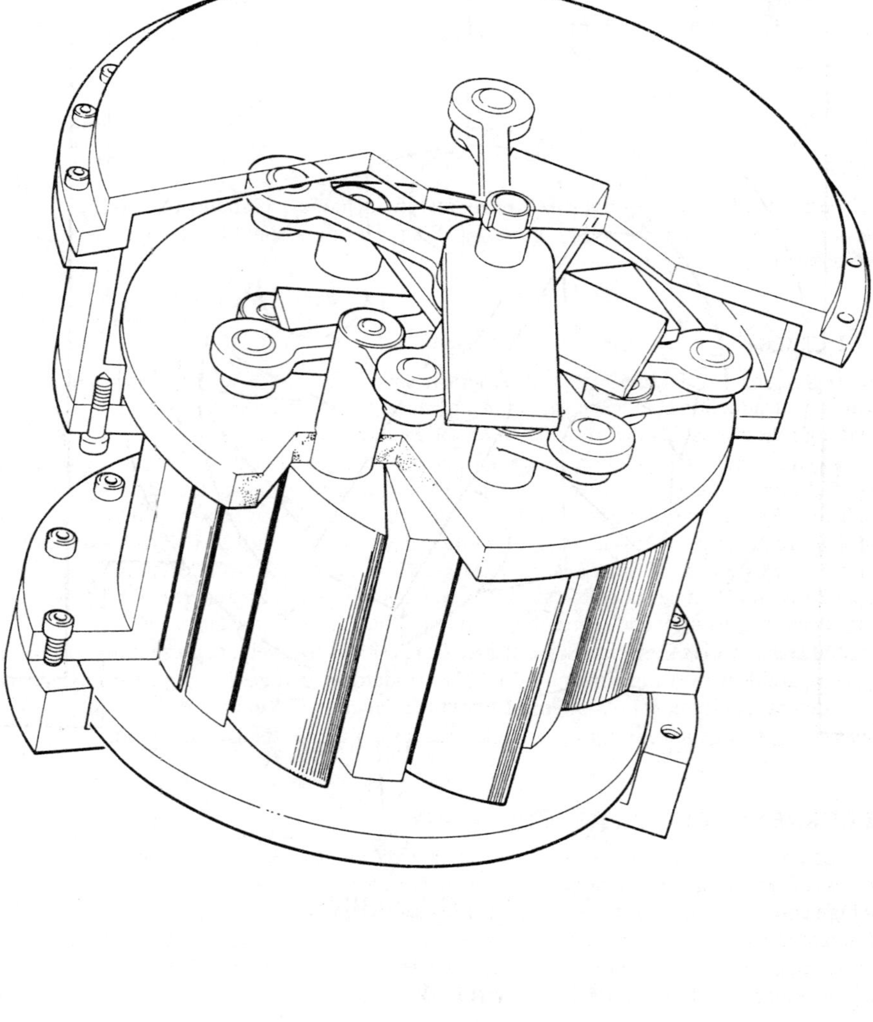

FIG 5

C498/34/132/95

Electrically heated catalyst system for low emission standards

F W KAISER, W MAUS, and R BRÜCK
EMITEC, Germany

SYNOPSIS

Design Criteria for EHC systems are explained. Catalytic behaviour is shown by computations and emission test results. ULEV and proposed EC Stage III limits could be achieved with low amp EHC systems.

INTRODUCTION

It is well known that the United States of America, and the state of California in particular, began with the introduction of exhaust gas emission limitations during the 1960's, thereby taking on the leading role in this field worldwide. With the Federal Clean Air Act of 1970, Congress paved the way for a national regulation program for the environment. Because of unusually severe pollution, California alone was given the authority to introduce tougher legislation for new vehicles. The intensified limitations of 1975 required the introduction of catalytic converters. Since 1994, the requirement comprises four categories of „clean" vehicles: TLEV (Transient Low Emission Vehicle), LEV (Low Emission Vehicle), ULEV (Ultra Low Emission Vehicle) and the ZEV (Zero Emission Vehicle) in which an electrical engine is driven by batteries. The ZEV are to be introduced increasingly and represent the only obligatory level in California.
The so-called Exhaust Gas Emission Stage III, EURO 2000 respectively, at present in proposal form, are comparable to the Californian LEV/ULEV standards with regard to engine and exhaust gas aftertreatment requirements. As compared with the Federal USA standards, with Stage III, Europe has also taken on a leading role, not least because of higher loads in the field.

1 THE CURRENT STATUS OF TECHNOLOGY

Future emission limitations can be achieved by means of substantial improvement in the effectiveness of exhaust gas aftertreatment. Since the present catalyst systems have already made possible high conversion rates with the catalyst in operating condition (above about 300 °C), the greatest potential by far for distinct improvement in pollutant reduction lies with the cold start. Appropriate changes to the catalyst system and supporting engine modifications become necessary to realize this development.

At present, three systems are receiving the majority of attention in the solution of the cold start problem: the electrically heated catalytic converter, a variety of burner and exhaust gas ignition systems and so-called hydrocarbon traps. Of these, the electrically heated catalytic converter (EHC) has currently the highest developmental and test status as well as the recognition and support of such important authorities as CARB and EPA [16, 17]. The examinations which follow, therefore, concentrate on the EHC. To achieve ULEV values in the USA FTP test, at least one part of the catalytic converter must be catalytically active shortly following engine cranking, i.e. during the first 20 s idle phase of the test [23]. Legislation in Europe will become substantially more stringent if the EC stage III is adopted, as has been suggested, with an idle/warmup phase of just 11 s (Fig.1).

In the USA and Japan the trend can already be noticed towards positioning the converter in closer position to the engine rather than in the underfloor position which is unfavourable for the cold start. Whether through this measure alone „clean" engines can fulfill LEV, EURO 2000 or even ULEV standards remains to be seen. Definite advantages have been proven by the geometrical and physical assets of metal supporting catalysts [1-7], which are being more frequently installed closer to the engine as so-called „Maniverters" by leading Japanese automobile manufacturers [8, 9]. Further increases in effectiveness can be achieved through the cell canal structure to improve mass and heat transport [10, 11, 12, 15].

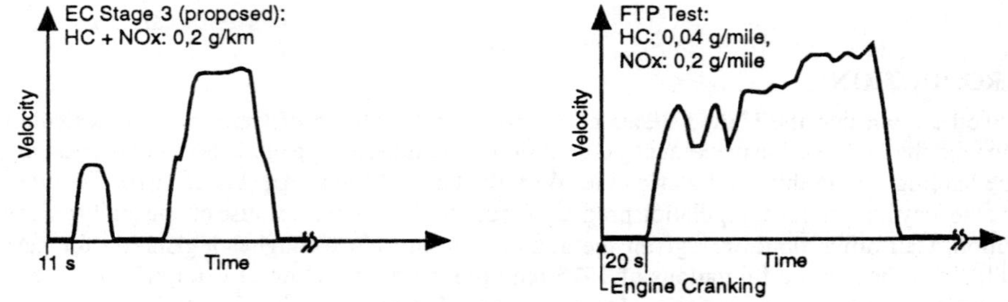

Fig. 1: Future development of emission standards: Europe, USA

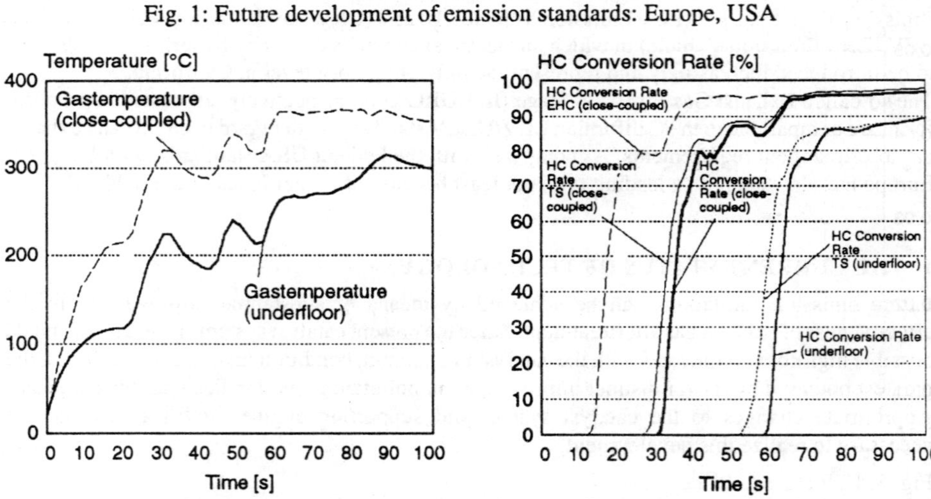

Fig. 2: Temperatures and HC conversion rate dependent on position and system of catalyst (evaluation results)

Dependent on the heat range in front of the converter, the engine-out emission, the position of the catalyst and the catalyst mass to be heated, the change to active heating will be necessary. This applies to the LEV and ULEV standards as well as in certain circumstances as an intermediate step to ZEV, the Near Zero Emission Vehicles (NZEV). Considering that in highly populated areas particularly, under extreme conditions a very high HC and NOx air pollution level can prevail, the first consideration is that e.g. in a tunnel atmosphere a ULEV or NZEV can emit less pollutant as is taken in by a combustion engine. Such an approach, while appearing excessive in the light of today's technological status, should serve as motivation towards continuing reduction of pollutant automotive emissions.

Thus in the future, the successful introduction of the EHC could lead to a new valuation of fuel and battery power for vehicles, since electrically powered vehicles in the situation mentioned above make no positive contribution towards reducing air pollution. In addition, the overall status of the energy and ecology of all polluting gases in the system (CO_2, HC, NOx, CO) needs to be reviewed. As was shown frequently in the past the EHC is a possible tool to reduce cold start emissions. Numerous earlier publications have shown that the heated catalytic converter is an effective tool in achieving future limitations [16, 24]. That heating after engine cranking is sufficient or even optimal was first proven in 1991 [22] which led development towards lowering electrical output. As this development progressed, modifications to the heated catalyst, exhaust system and engine management made the reduction of power consumption to within an area of less than 1.5 kW possible, which the utilization of only one battery in a conventional 12 V vehicle system allowed [24, 29, 30].

Figure 3 shows a lively development in the reduction of electrical heating power of electrically heated catalytic conversion systems has taken place, and within the time of only 6 years, numerous applications have already reached the target area of HC limits at a heating power of less than 1500 watts. Important progress up to the production stage of the heated catalyst system was achieved with regard to durability and provision of the energy requirement from the vehicle systems. The service life under field conditions is being tested by a number of automotive companies worldwide.

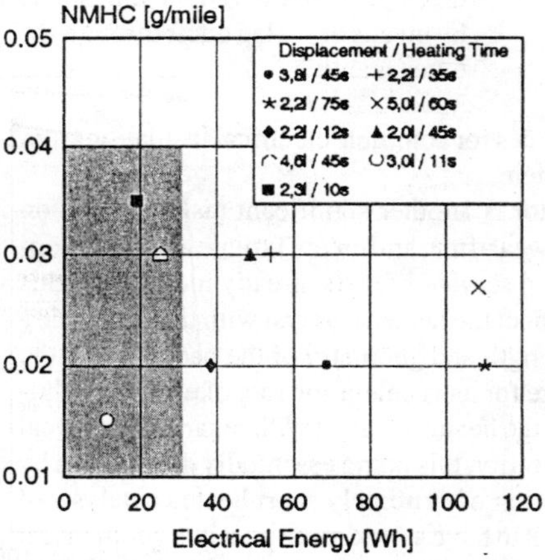

Fig. 3: FTP HC results dependent on the heating power of various heated catalytic converter applications [24]

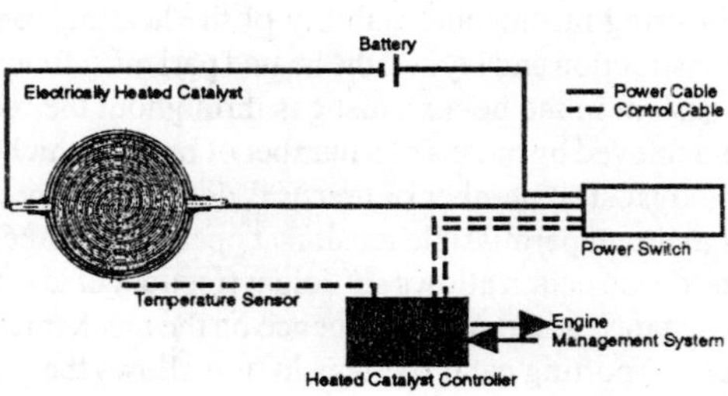

Fig. 4: Electrically Heated Catalytic System

Built on the experience of the foil metal catalyst development, test methods for the EHC were developed to simulate the various loads required for customer operation. The objective of the following work is to review the methods and results of durability tests as well as demonstrate an application of the EHC both in fresh and aged condition by means of the FTP and EC Stage III Test. For the tests, the new generation Emitec Series 6 heated catalytic converter, the EMICAT, was put into operation. The heated catalytic converter is integrated into a system (Fig. 4) comprising heated catalyst control and a power switch (semiconductor or relay).

2 SERIES 6 HEATED CATALYST

The Series 6 heated catalyst comprises a heated and an unheated part, i.e. the heated element and the „supporting" catalyst. The supporting catalyst is joined to the heated part with brazed, electrically insulated braces. Both catalysts are wound in the proven Emitec S design construction (Fig. 5). The material underwent development over a period of many years to mature to the present production stage suitable for high temperature applications in automotive catalytic converters [25] and modified for the S and SM shaped metal honeycomb structure METALIT. These metal supports have undergone millionfold testing in operating production vehicles, motorcycles and racing cars [13, 14]. Through variations of the geometric influencing variables, electrical resistance between 0.35 and 0.06 ohms (34 A-200 A/12 V) can be achieved [27]. The conduction package thus designed is rolled into an S-form, whereby common insulation is ensured by an air gap.

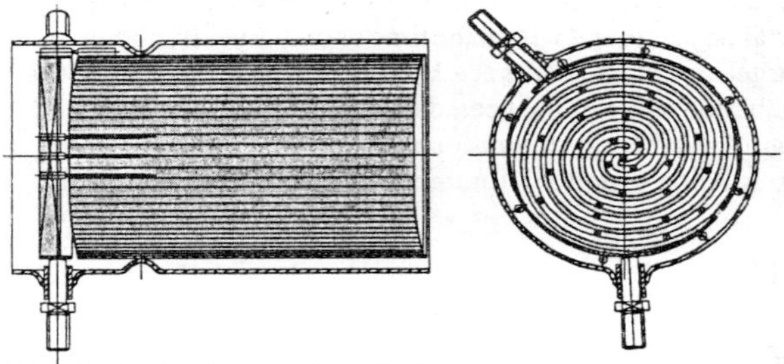

Fig. 5: Series 6 heated catalytic converter

The realization of insulation between the metal foil electrical conductors of the honeycomb is one of the most critical construction characteristics. Air gap insulation guarantees wear-free insulation, even during increased vibration loads on the system. Contact within the layered package takes place via two opposing shells by means of sealed ducts through the outer mantle.

The construction of the heated catalytic converter provides for complete electrical insulation from the supporting catalyst and thus from the exhaust system.

Ensuring mechanical stability of the heating conductor is another significant task imposed on construction and layout; the heated part must function wear-free, and creep, fatigue and vibration-resistant in the hot exhaust gas throughout the required service life. As already mentioned, this is achieved by means of a number of braces which connect the heated catalyst with the supporting catalyst. The number of braces is determined by the length and geometry of the package and the maximum permissible tension at operating temperature for an acceleration rate of e.g. 60 g. Thus the construction allows sufficient freedom of layout of the heated catalyst with regard to electrical resistance without any influence on the mechanical stability, this being essentially determined by the supporting catalyst. This in turn allows the realization of extremely short heated catalysts of widely varying diameters -Ø 60 to Ø > 127 mm.The task to have a maximum possible geometrical and/or catalytic surface with a minimum of possible heat capacity (m x cp) in order to heat up to operating temperature quickly for optimal effectiveness, applies even more strongly to the electrically heated catalyst. In this case the rational use of vehicle electrical systems is essential, with simultaneous realization of fast heating up times (< 20 s and/or ~ 11 s. See Chap. 1).

These criteria determine, therefore, not only the technical contention on the question of ceramic versus metal supports [13, 14], but also of the metallic sintered counterparts.

3 DURABILITY TESTING

In order to prove durability, the heated catalysts must be subjected to the following standard tests at Emitec, which, with the exception of the special electrical tests, must also be successfully carried out on production metal supports. A combination of field experience and simulated testing is crucial.

- Vibration test
- Inner thermal cycling test (ITC)
- Hot shake test (HST)
- Outer thermal cycling test (OTC)
- Engine thermal cycling test (ETC)
- Electrical inner thermal cycling test (EITC)
- Vehicle durability test

4 APPLICATION PRINCIPLES

The significance of the heated and main catalysts with regard to overall emissions can be differentiated in that, as mentioned earlier, the electrically heated catalyst is effective almost exclusively during the cold start. For phases under load, during which the catalyst is at a chemically highly active thermal niveau (Bag I after ca. 125 s and during Bag II of the FTP test), the subsequently connected main catalyst assumes most of the conversion activity. In Bag III, after the 10 minute cooling phase, the EHC is not heated, so that here the converter begins anew with passive conversion. Accordingly, in the following the emissions during the first 125 s are examined. The heating-up period should - as has been indicated - lie within the idle phase directly following engine cranking.

4.1 Electrical Energy, Mass to be Heated, Geometrical Surface.

All currently known EHC constructions make use of metal. In the case of actively heated honeycomb constructions it becomes immediately clear that ceramic supports would be at a disadvantage with regard to the higher specific heat capacity (cp) since valuable electrical power is now needed for heating. Differentiation can be made between metal foils and sintered metal honeycomb processed from extruded powdered metal. With the inherently thicker walls – from 110 to 160 µm - , these present a greater mass to be heated (m_M), as if the identical geometric surface is to be achieved with foil EHCs with wall thicknesses of 50 to 65 µm. As we are aware, catalytic effectiveness depends also on the geometric surface presented. The connections are represented in Figure 6. It must be noted that the washcoat mass ($m_{w/c}$) on the EHC exerts further significant influence on the light-off behaviour [23]. The necessary electrical energy to heat up the EHC is:

$$Q_E = (m_M + m_W) \cdot cp \cdot \Delta T \qquad (1)$$

Thus the objective of the EC Stage III and FTP tests must be to bring the greatest possible catalytic surface up to catalyst actuation temperature with an available amount of power during the idle, and, together with the resulting exothermics, similarly to actuate the subsequent passive catalyst. The electrical energy is carried directly via the metal honeycomb structure into the washcoat.

(The energy need not, as in a combustion system, be carried into the coating and honeycomb body via convection loss of the heat transport coefficients). By heating the EHC from the inside, the heat converted from electrical power is effective in the boundary layer through which a minimal energy loss operation is achieved for the desired catalytic exothermics.

4.2 EHC Diameter, Electrical Power

The following table represents the tested EHC, which were heated with 1000 and 1500 watts under steady air conditions without coating:

Table 1:

Size (mm)	Cell density [cpsi]	Mass [g]	Amperage [A]
Ø 70 x 10	300	26	100
Ø 90 x 11	160	36.3	150
Ø 90 x 11	160	35,6	100
Ø 90 x 11	160	63	150

With identical cell density and metal foil thickness, the heating-up behaviour is approximately proportional to the mass and/or the cross-section. The mass to be heated in the Ø 118 mm heated part has already become too large to become active within the required 10 to 15 s with 1000 Watts; a higher power input of 1500 watts (150 A, 10 V) is necessary (Fig. 7).

Exact determination of the heating curve of catalysts under normal conditions during engine operation is complicated. Owing to the difficult thermoelement arrangement in the honeycomb body, only qualified differentiation is possible between the temperatures of the catalyst structure and the gas. An FEM program developed by Emitec for the computation of catalytic converter structure temperature as well as coating and temperature-dependent conversion is being initiated. The results compiled indicate concurrence with those from actual operation.

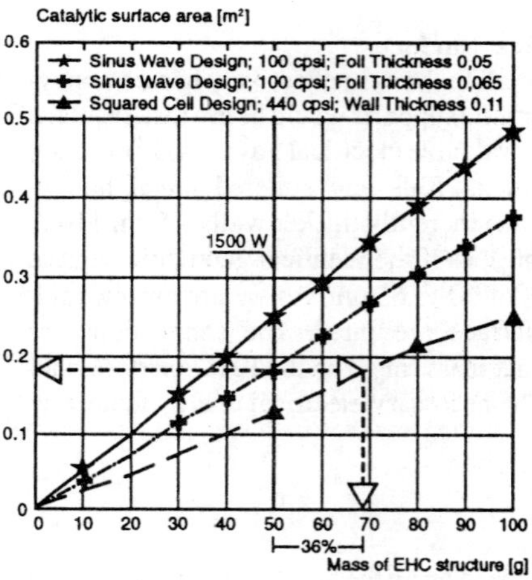

Fig. 6: Geometric surface, EHC mass, heating-up power of 1500 W to achieve 400 °C in 10 s

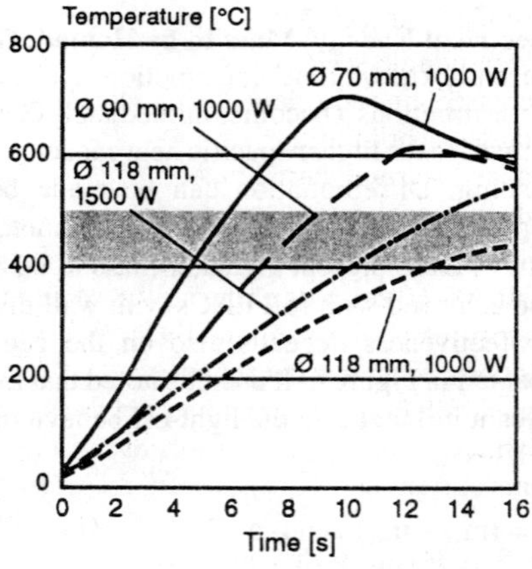

Fig. 7 Heating curve Series 6 EHC, Ø 70, 90, 118 mm with 1000 watts and Ø 118 mm with 1500 watts, uncoated, steady air conditions

5 PERIPHERAL CONDITIONS

5.1 Catalytic Converter Configurations

Research included heat cascade formations [28], formed as previously mentioned from heated parts of varying diameter, supporting catalysts of identical diameter as the heated parts, followed by main catalytic units of Ø 118 x 150 mm. Table 2: Catalyst configurations tested

No.	Catalyst	Dimensions [mm]	Cell density [cpsi] Amperage [A]	Catalyst volume [l]
1	EHC 1. Monolith 2. Monolith	Ø 70 x 10 Ø 70 x 50,8 Ø 118 x 150	300/100 400/ – 400 TS/ –	0.04 0.19 1.64 1.87
2	EHC 1. Monolith 2. Monolith	Ø 90 x 11 Ø 90 x 50,8 Ø 118 x 150	160/150 400/ – 400 TS/ –	0.07 0.32 1.64 2.03
3	EHC 1. Monolith 2. Monolith	Ø 90 x 11 Ø 90 x 50,8 Ø 118 x 150	160/100 400/ – 400 TS/ –	0.07 0.32 1.64 2.03
4	EHC 1. Monolith	Ø 118 x 11 Ø 118 x 150	160/150 400 TS/ –	0.12 1.64 1.76

5.2 Vehicle and Engine Modifications

For examination of emissions, a Porsche 968 (3 l, 4-cylinder, 4 valve engine) was used as a test vehicle. The vehicle was modified for EHC application [24], achieved in particular by

- catalyst in close coupled position to the engine
- reduction in heat loss before the catalyst engine management optimized for faster rise in exhaust gas temperature before the catalyst

and for the cold start by

- increased idle speed
- increased air mass flow during idle
- retarded spark timing
- reduced warmup enrichment
- reduced acceleration enrichment

Further, suitable controlled secondary air was injected into the exhaust manifold, whereby during the first 40 - 60 seconds of the cold start, lambda = 1 could be fixed. The 115 A series alternator was substituted with a minimally stronger one with optimized current delivery at lower engine revolutions. With this configuration the FTP and EC tests were carried out on a Porsche exhaust emission roller type dynamometer in the test centre for environmental protection. In addition to the exhaust emissions, exhaust gas temperatures and engine control data as well as vehicle systems current and voltages were recorded during the tests.

5.3 Aging of Catalytic Converters

On an engine test bench with a 5 l, 8-cylinder engine, the catalyst components described in 5.1. were subjected to exhaust gas via a five-way distributor. One method to guarantee a consistent mass flow distribution and load on all test variants was to compensate the pressure loss differences with orifices. In addition the positions of test components were alternated. Aging was done with a mass flow of 750 kg/h at a maximum catalyst inlet temperature of approximately 860 °C and lambda = 1. The aging process was carried out over a period of 96 h, whereby a fuel cut-off phase was interjected in the overrun 5 times every 5 minutes for 5 s respectively.

6 FTP AND EC STAGE III EXHAUST GAS TEST RESULTS

In order to evaluate the influence of the heated and unheated catalyst systems on the emission results, the first 125 s, decisive for the cold start, were examined (see Chapter 4). From about 60 to 100 s the significant share of conversion is taken on by the main catalyst. Table 2 contains the more important data for differentiation between the tests. All converters have a main catalyst of Ø 118 x 150 mm, 1.64 l.

Table 3: FTP results, fresh condition

Cat.-No.	Type of catalyst	Heating-Time	Emissions for second 0-125		
			HC (g)	CO (g)	NOx (g)
2	Ø 90 x 50,8 150 A	0-16	0.352	4.590	0.356
2	Ø 90 x 50,8 150 A	–	0.455	5.458	0.438

Table 4: FTP results, aged condition

Cat.-No.	Type of catalyst	Heating-Time	Emissions for second 0-125		
			HC (g)	CO (g)	NOx (g)
2	Ø 90 x 50,8 150 A	0-16	0.599	7.041	0.492
2	Ø 90 x 50,8 150 A	–	0.853	10.369	0.329
3	Ø 90 x 50,8 100 A	0-16	0.659	8.263	0.387
1	Ø 70 x 50,8 100 A	0-16	0.540	6.540	0.432
1	Ø 70 x 50,8 100 A	–	0.748	9.840	0.364
4	Ø 118 x 150 150 A	0-16	0.697	10.292	0.536

Table 5: EC III results, aged condition

Cat.-No.	Type of catalyst	Heating-Time	Emissions for second 0-125		
			HC(g)	CO(g)	NOx(g)
2	Ø 90 x 50,8 150 A	0-16	0.250	3.422	0.119

Table 6: EC III results, fresh condition

Cat.-No.	Type of catalyst	Heating-Time	Emissions for second 0-125		
			HC(g)	CO(g)	NOx(g)
2	Ø 90 x 50,8 150 A	0-16	0.526	7.244	0.234
2	Ø 90 x 50,8 150 A	–	0.656	14.770	0.226

7 DISCUSSION ON EMISSION RESULTS

7.1 FTP Tests, Cold Start

For the EHC heated with 150 A, Ø 90 x 11 mm (test components No.II, Table 2), Fig. 9 shows the accumulated HC emissions over 125 s. Through catalyst aging, the cold start emissions increase during the first 125 s by 89 %. The influence of EHC heating in a fresh catalyst is relatively minor. The improvement in this case in HC emission is about 29 %. Heating in the aged condition, however, effects a reduction of about 45 %. This shows that the electrically heated catalytic converter takes on an increasingly important function with regard to adherance to limitations during the service life of the catalyst.

Influence of aging:

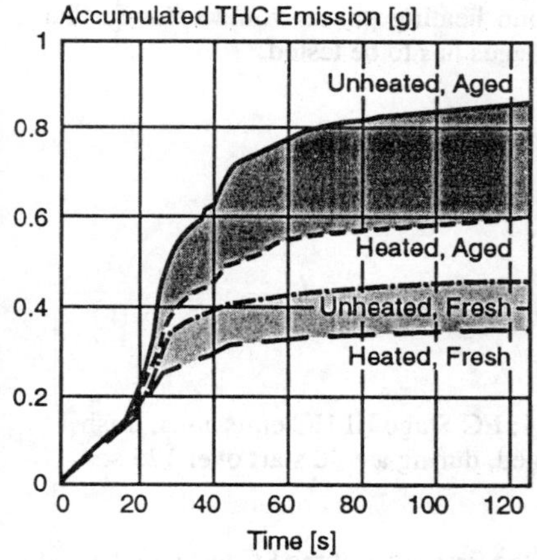

Fig. 8: Accumulated HC emissions during 125 s of FTP Test heated and unheated EHC, fresh and aged

Influence of the EHC and supporting catalyst diameters:

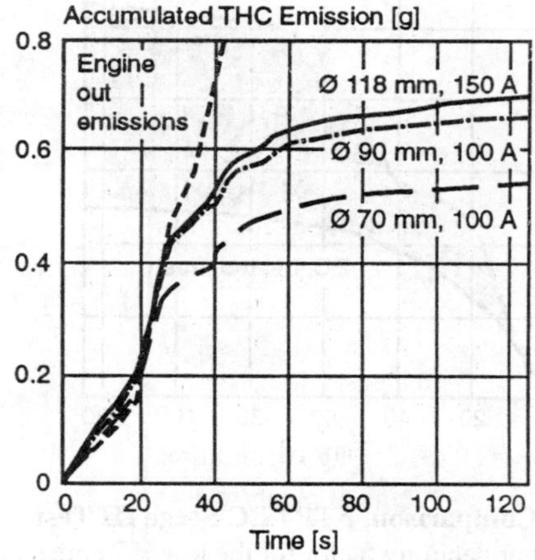

Fig. 9: Influence of the catalyst diameter on HC emission, aged during the first 125 s of FTP

The heat cascade [28] with the smallest EHC and supporting catalyst diameter shows the most favorable cold start behaviour. The faster heating of the lower mass and sufficient geometric surface provide this advantage. Accordingly, despite the 50 % higher heating power, the system with Ø 118 mm is less effective (Fig. 9).

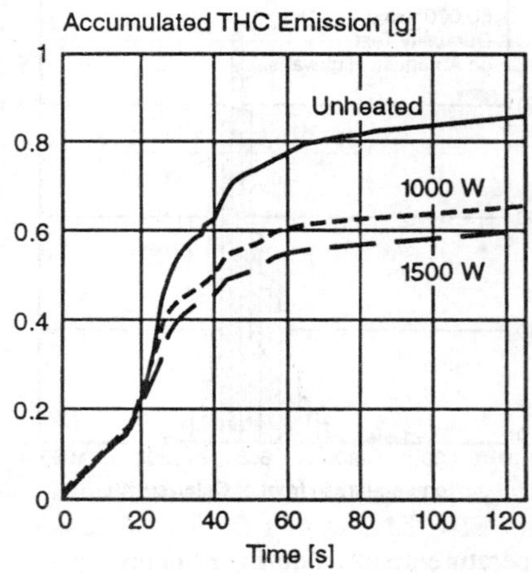

Influence of amperage:
Taking the example of a catalyst 90 mm in diameter, the influence of heating energy between 0 - 1500 Watts is emphasize. The heating improves the cold start emissions by 23 % with 1000 Watts and 30 % 1500 Watts. The slightly better results of the 150 A catalyst can be explained by more active catalyst volume during hill 1 in Bag 1 which gives a higher conversion rate in the acceleration phase (Fig. 10).

Fig. 10: Influence of heating energy after ageing. Accumulated HC emission during first 125 s of FTP

7.2 EC Stage III Tests, Cold Start

The EHC system with Ø 90 x 11 mm heated catalyst and 150 A heating indicates a 53 % higher cold start emission after 125 s when aged (Fig. 11) than when fresh. In an aged condition the influence of the second idle phase on the emissions can be clearly recognized. Whether an EC Stage III directed heating strategy with a second heating phase would offer further advantages has to be tested.

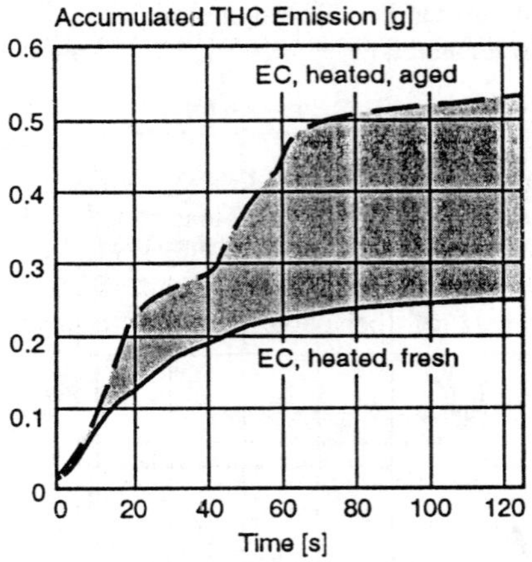

Fig. 11: EC Stage III HC emissions, fresh and aged, during a cold start over 125 s.

7.3 Comparison, FTP / EC Stage III Test

Another deciding factor for the low HC emission during the first 125 s of the EC test with heated catalyst Ø 90 x 11 mm and 150 A, are the approximately 35 % lower engine out emissions in comparison to the FTP test. When making the comparison, it must also be noted that EHC as well as main catalyst aging were carried out under temperature and load conditions characteristic for Germany.

Under American conditions, lower aging rates can be expected, as shown in Fig. 12. However, this factor is not valid for Light Duty Truck, which can be operated at a high load niveau.

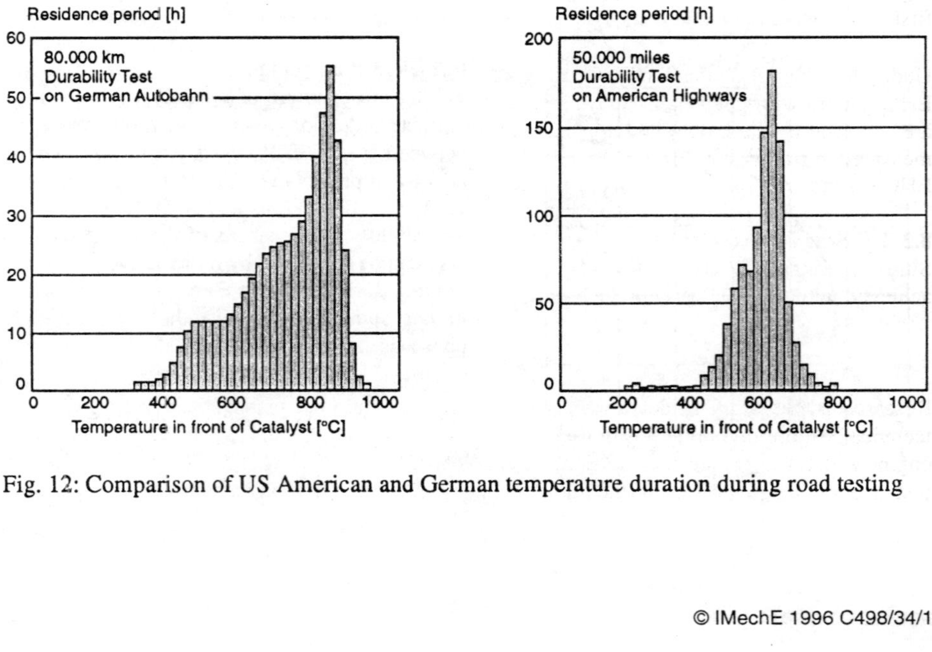

Fig. 12: Comparison of US American and German temperature duration during road testing

8 FTP AND EC STAGE III OVERALL RESULTS

8.1 FTP tests Bag III

During the inactive heated first 125 s of the Bag III warm start, the smaller EHC supporting catalyst diameter, designed according to the heat cascade principle, also shows decreased emissions. This can be explained by the thermodynamically comparable behaviour of the active cascade in Bag I (Fig. 13).

8.2. FTP overall results

The entire catalyst volume, with 1.87 l and 2.03 l at 62 to 68 % of engine displacement, is insufficient even with the EHC to achieve ULEV values (Table 7). During the load phases with higher mass flow (space velocity) so-called „breakthroughs" occur frequently. With an identical aged main catalyst of Ø 127 x 74.5 + 120 mm (2.46 l), the overall results can be improved by factor 2. Thus it can be shown that with an improved cold start behaviour, the entire catalyst volume can be reduced (Fig. 14).

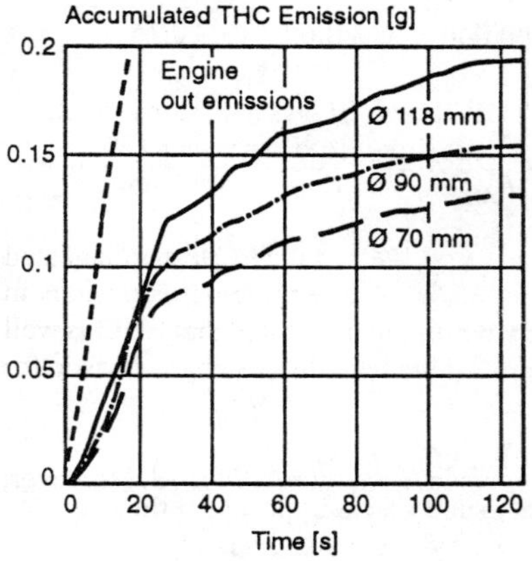

Fig. 13: Bag III HC emissions during the first 125 s, accumulated unheated

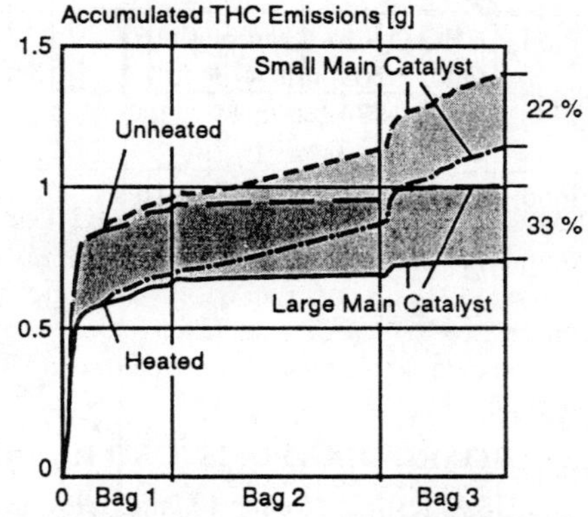

Fig. 14: Influence of catalyst concepts for reduction of cold start emissions and emissions in warm condition on the FTP overall result

Under the assumption that the EHC influences the first 125 s of Bag I only, the test results from EHC systems with small main catalysts can be transposed to larger main catalysts. The best results are achieved by the EHC Ø 70 mm 100 A. The THC result was calculated as 0.036 g/m. The measured result was 0.031 g/m. These values indicate that ULEV standards can be met with an EHC system with minimal electrical power.

8.3 EC Stage III results

Similarly to the FTP test the proposed EC Stage III limits of HC + NOx = 0.2 g/km could be achieved with the EHC system Ø 70 mm

		HC + NOx [g/km]
THC [g/km]	0.045	0.088
NOx [g/km]	0.043	
CO [g/km]	0.69	

Research is necessary to determine whether with optimal heat management by means of a technically more favorable position in closer position to the engine, cascade formation and an engine with lower engine out emission, Stage III could be achieved even without the EHC. In this instance, rapid aging owing to the higher temperatures in the field must be taken into account.

Table 7: FTP overall results with differing main catalyst volumes and heated/unheated EHCs

Cat.-No.	Convertertype	Heating-Time	Condition	THC [g/m]	CO [g/m]	NOx [g/m]
2	EHC Ø 90 x 11 mm Ø 90 x 50,8 mm + Ø 118 x 150 mm	0	fresh	0.048	0.35	0.25
2	EHC Ø 90 x 11 mm Ø 90 x 50,8 mm + Ø 118 x 150 mm	0-16 s 150 A	fresh	0.04	0.28	0.25
2	EHC Ø 90 x 11 mm Ø 90 x 50,8 mm + Ø 118 x 150 mm	0	aged	0.112	1.07	0.24
2	EHC Ø 90 x 11 mm Ø 90 x 50,8 mm + Ø 118 x 150 mm	0-16 s 150 A	aged	0.086	0.61	0.29
2 mod.	EHC Ø 90 x 11 mm Ø 90 x 50,8 mm + Ø 127 x 74,5 mm + 120 mm	0-16 s 150 A	aged	0.046	0.68	0.15
1 mod.	EHC Ø 70 x 10 mm Ø 70 x 50,8 mm + Ø 127 x 74,5 mm + 120 mm	0-16 s 100 A	aged	0.031	0.46	0.09

9 ON BOARD DIAGNOSIS (OBD II), SAFETY ASPECTS

In the USA, testing of emissions of relevant components such as catalytic converters, using procedures and limitations for on-board diagnosis laid down by authorities, has acquired great technical and economic significance owing to the longterm guarantees offered by automotive manufacturers. The legislation can be found in the respective Californian and US American CARB and EPA publications. As for the unheated catalyst, criteria have already been released for the EHC with regard to diagnosis. Relevant to ascertaining that the heated catalytic converter is heated, and the EHC has become active, as designated, is the monitoring of the electrical output e.g. current and voltage during the starting phase. Thus the EHC is the only ULEV system to be accepted and receive official approval for production. The catalytic function after chemical aging is being examined together with the main catalyst e.g. by means of the lambda sensor method. The objective of new developments include the reduction of tolerances determining effectiveness as well as the preparation of measurement methods and algorithms which allow the monitoring function during driving operation field „on-line“ [26]. These functions are monitored in that a faulty behaviour must be indicated by a warning lamp (MIL). Further, „in use“ tests by vehicles in the field are to provide observation with regard to adherence to test limits; the consequences could otherwise include official recall action.

One far-reaching aspect of legislation in the USA is the perpetration and responsibility principle in the industry with regard to the safety of products, which in order to avoid damages claims to require the utilization of the safest known techniques and components. Any risks to the consumer already known, therefore, are to be strictly avoided. For the EHC system, FMEA analyses have not determined any risks to drivers, either direct or indirect.

10 CONCLUSIONS

To increase the effectiveness of exhaust aftertreatment systems, various applications were examined with the Series 6 heated catalyst generation for LEV/ULEV and EC Stage III. The Series 6 heated catalyst offers the following advantages:

- a wide performance spectrum with current consumption between 50 and 300 A
- minimal mass with extremely short heating element heating-up time
- compliance with all electrical and mechanical durability requirements
- The application of a low power EHC shows that with supporting measures such as adaptation of engine control data in warmup, reduction of exhaust gas systems masses to be heated, catalyst position close to engine and optimization of secondary air injection, the effectiveness of the exhaust gas aftertreatment can be improved to such an extent that the ULEV values are fulfilled with both fresh and aged catalysts.
- To fulfil ULEV values with aged catalysts, the implementation of heated systems to ensure long term characteristics is necessary, particularly where unfavorable gas temperature and raw emission niveaux exists.
- Although the emission niveau in the EC Stage III test, with 11 s idle, was lower than in the FTP test, an EHC is also necessary for the aged condition to fulfil the requirements.
- It is possible to supply low power EHCs from the battery.

11 PROSPECTS

1. The emission results show that the EHC with a relatively low heating current can fulfil ULEV and EC III requirements even in an aged condition. Whether the EC results could also be achieved with a more economical, unheated catalyst system with regard to heat transfer has yet to be researched. Further, the development of accumulated HC emissions in the EC suggest that with a special heating strategy, substantially lower overall emissions can be expected.

2. The results lead to the conclusion that with favourable peripheral conditions, extremely low power, some 0.5 to 0.75 kW would be enough to fulfil the requirements in a ~ 2 l engine. With heating currents from about 100 A, it is possible to install a simple, inexpensive power switch or relay. Any particular requirements of the vehicle systems as far as the battery or alternator are concerned can probably be eliminated. The cross-section of heating current cables (and thus their weight) is greatly reduced when compared with the heating currents of 200 to 400 A.

3. In the Series 6 heated catalyst, a functionally stable component with a low energy consumption has been made available. The increasingly simplified and economical system is capable of reducing some 35 % of the Bag I emission (Fig. 14). In this respect, the costs of an EHC system for Bag I are comparable with those for the enlargement of a main catalyst, with which similar emission reductions can be achieved, however predominantly in Bags II and III.

ACKNOWLEDGEMENTS

Dr. Paul Küper of Dr. hc. Ferd. Porsche AG and members of the Emitec GmbH staff participated in the preparation of the paper and the provision of results. The authors wish to express their thanks for this support.

REFERENCES

1. Nonnemann, M.: „Metal Supports for Exhaust Gas Catalysts", SAE Paper 850131, 1985
2. Öser, P.: „Novel Autocatalyst Concepts and Strategies for the Future with Emphasis on Metal Supports, SAE Paper 880319, 1988
3. Hawker, P.N., C. Jaffrey, A.J.J. Wilkins, J. Alker: „Metal Supported Automotive Catalysts for Use in Europe", SAE Paper 880317, 1988
4. Maus, W., A. Reck: „Neuentwicklung metallischer Katalysatorträger für Automobile - Einfluß der Katalysatorträger auf Motor und Abgas", Technische Akademie Wuppertal Sept. 1989
5. Nishizawa, K., K. Matsuda, H. Horu, J. Hirohashi: „Development of Improved Metal-Supported Catalysts", SAE Paper 890188, 1989
6. Brüstle, C., F.-W. Kaiser, E. Rutschmann: „Optimierung der Vorgänge im Ein- und Auslaßsystem bei Porsche Motoren", 11. Intern. Wiener Motorensymposium, April 1990
7. Bode, H., W. Maus, H. Swars: „How Metal Substrates are Able to Have an Influence on Converter Efficiency", I. Mech. E. Conference 1991
8. Asano,M., K. Yoshimura, M. Yasukawa, T. Takada, A. Onishi: "Development of Metal Supported Catalytic Manifold Converter", Toyota Engineering Technology Vol. 40 No. 1
9. Takada, T., T. Tanaka: Development of a Highly Heat-Resistant Metal Supported Catalyst SAE Paper 910615, 1991
10. Pelters, St., F.-W. Kaiser: „Konkurrenz für den Keramik-Kat"; Automobilproduktion, Nov. 1988
11. Pelters, St., F.-W. Kaiser, W. Maus: „The Development and Application of a Metal Supported Catalyst for Porsche's 911 Carrera 4", SAE 890488, 1989
12. Vaneman, G.L.: „Performance Comparison of Automotive Catalytic Converters: Metal vs Ceramic Substrates": XXIII Fisita Congress, Report No. 905115, May 90
13. Nonnemann, M.: „New High-Performance Gas Equilizing Metal Support for Automotive Exhaust Gas Catalysts", SAE Paper 900270, 1990
14. Kaiser, F.-W., St. Pelters: Comparison of Metal-Supported Catalysts with Different Cell Geometries, SAE 910837, 1991
15. Held, W., M. Rohlfs, W. Maus, H. Swars, R. Brück, F.-W. Kaiser: "Improved Cell Design for Increased Catalytic Conversion Efficiency", SAE 940932, 1994
16. Piotrowski, G.K., „Evaluation of a Resistively Heated Metal Monolith Catalytic Converter on a Gasoline-Fueled Vehicle", U.S. EPA, OMS, EPA/AA/CTAB/88-12, Dec., 1988
17. Hellman, K.H., R.I. Bruetsch, G.K. Piotrowski, W.D. Tallent: „Resistive Materials Applied to Quick Light-off Catalysts", SAE Paper 890700, 1989
18. Hurley, R.G., L.A. Hansen, D.L. LaCourse, W.L.H. Watkins, H.S. Gandhi, W.A. Whittenberger: „Evaluation of Metallic and Electrically Heated Metallic Catalysts on a Gasolin Fueled Vehicle", SAE Paper 900504, 1990
19. Heimrich, M.J.: „Air Injection to an Electrically-Heated Catalyst for Reducing Cold-Start Benzene Emissions from Gasoline Vehicles" SAE Paper 902115, 1990

20. Hurley, R.G., D.L. Guttridge, L.A. Hansen, R.J. Pawlowicz, J.M. Smolinski, H.S. Gandhi: „Experiences with Electrically Heated Catalysts“; SAE Paper 912384, 1991

21. Heimrich, R.G., S. Albu, J. Osborne: „Electrically-Heated Catalyst System Conversions on Two Current-Technology Vehicles“, SAE Paper 910612, 1991

22. Gottberg, I., J.E. Rydquist, O. Backlund, S. Wallmann, W. Maus, R. Brück, H. Swars: „New Potential Exhaust Gas Aftertreatment Technologies for „Clean Car“ Legislation“, SAE Paper 910840, 1991

23. Kaiser F.-W., W. Maus, H. Swars, R. Brück: „Optimization of an Electrically-Heated Catalytic Converter System, Calculations and Applications“, SAE Paper 930384, 1993

24. Küper, P.F., W. Maus, H. Swars, R. Brück, F.-W. Kaiser: „Ultra-Low Power Electrically-Heated Catalyst System“, SAE Paper 940465, 1994

25. Brill, U., U. Heubner: „Werkstoffe für Metallträger von Automobil-Abgaskatalysatoren“ MTZ Motortechnische Zeitschrift 49 (1988) 9

26. Pelters, S., D. Schwarzenthal, W. Maus, H. Swars, R. Brück: “Alternative Technologies for Studying Catalyst Behaviour to Meet OBDII Requirements“, SAE Paper 932854, 1993

27. U.S. Patent: „Electrically Coductive Honeycomb Body, and Method of Manufacturing and Operating It As an Exhaust Gas Catalyst Carrier Body“, Patent No. 5,322,672, 1994

28. U.S. Patent: „Process for Accelerating The Response of an Exhaust Gas Catalyst, and a Apparatus and Electrically Heatable Carrier Bodies for Performing the Method“, Patent No. 5,146,743

29. Otto E., Dr. W. Held, A. Donnerstag, Dr. P. Küper, B. Pfalzgraf, A. Wirth: „Die Systementwicklung des elektrisch heizbaren Katalysators (EHC) für die LEV/ULEV und EU-III Gesetzgebung“, 16. Internationales Wiener Motorensymposium, Mai 1995

30. Donnerstag A., A. Degen, Dr. W. Held, K. Korbel: „Erfüllen der ULEV-Norm durch elektrisch beheizten Katalysator“, 16. Internationales Wiener Motorensymposium, Mai 1995

C498/34/135/95

The contribution of transmissions to vehicle fuel economy

D SIMNER BSc, MSc, AMIMechE
Rover Group Limited, UK

SYNOPSIS

This paper discusses the four ways that the fuel economy of a vehicle can be influenced by the transmission system. These are the parasitic losses, power proportional losses, mass of the system, and hardware aspects. The latter being the configuration of the transmission system and its control.

The need for improving the fuel consumption of vehicles and, in particular, the part played by the transmission is discussed along with how each of the above aspects of the driveline system influences it.

The use of performance prediction and simulation is suggested as the way forward. This is because of the importance of being able to predict and optimise the performance of the transmission, modifications to it, and the effect on the vehicle.

1 INTRODUCTION

A number of influences on the automotive industry are causing it to look increasingly seriously at reducing the fuel consumption of vehicles. These include the environmental 'movement', customer opinion or feedback, and the threat of taxation on fuel use of one form or another. While considerable contributions to fuel economy can be made by the vehicle as a whole, other pressures are having an increasingly negative effect on fuel usage. Examples of this are the emissions control on engine and exhaust treatment systems and the increasing weight of safety systems and impact absorbing structure.

The transmission systems of road vehicles have, in the past, rarely been looked upon as prime targets for fuel consumption improvements. This paper proposes that this will not always be the case in the future, and some of the reasons for this are discussed.

The scope of this paper is limited to the transmissions used in passenger cars (or light off road and commercial vehicles). However, a considerable amount of the discussion is fairly generic in nature so may be of relevance to heavier, commercial vehicles also.

Rather than looking in detail at an aspect of, for example, transmission efficiency or power train control, this paper deliberately takes a step back from such detail to look at the whole subject of transmissions and fuel economy. A number of excellent papers have been produced which discuss the processes involved in transmission losses, or the detail of the efficiency of a particular transmission. It is intended that this discussion takes a more holistic view of the subject and indicates that it is not just transmission efficiency that influences the fuel economy of a vehicle.

The trends that are occurring within the automotive transmission industry and externally to it are also discussed to highlight why the transmission will have an increasingly important role in vehicle fuel economy.

The final section of the paper looks at the need for a performance prediction capability. This is not only to model the efficiency of a transmission during a wide range of operating conditions but also to be able to predict the effect of changes to the transmission specification and the knock on effect to the vehicle. Such optimisation would involve both the mechanical parts of the system and the control algorithms for the whole power train.

2 THE NEED FOR IMPROVEMENTS

The reasons for any company to make engineering changes should always support the strategic direction. It is worth noting here, the need for fuel economy related improvements to transmissions and the strategic reasons why manufacturers may look at improving the fuel consumption of the vehicles they produce. These can be summarised as the following:

- Improving the cost position of the product,
- Improving the competitive position of the product,
- Ensuring the product complies with the legislative environment.

The situation with fuel economy on motor vehicles is that improvements may be required for any one of the three reasons listed above. It could be argued that the product cost example is somewhat tenuous. If, however, the scenario of, say, a CO_2 tax is considered then it can be seen that the fuel economy performance of a vehicle can have a direct bearing on the cost. Not all increases in product cost can always be passed onto the customer. Currently the legislative environment is fairly benign to fuel economy. The current emphasis is on emissions(1) other than CO_2, but this may not always be the case.

The major strategic reason for the improvement in fuel economy for any motor manufacturer is the improvement in their competitive position. This could manifest itself in a number of ways. The most obvious is simple comparisons in the fuel economy listed in the showroom on new cars. As well as this 'test cycle' fuel economy the actual figures achieved by the customer are equally important as the cost to the customer of running a vehicle can be significant.

In addition to looking at the external influences on the product, a number of trends or changes to the rest of the vehicle are significant to this subject. While a number of activities in the transmission field on vehicles maybe making it difficult to find opportunities for fuel economy improvements within the driveline, there is any number of trends on the engine and the rest of the vehicle which are actually having an adverse effect of fuel consumption.

Increasing weight of vehicles (1), especially in the structure due to the crash test requirements and the features fitted to improve the usability of the vehicle all helps to make things increasingly difficult for the manufacturer to reduce vehicle mass significantly enough to improve fuel consumption. It has to be noted that the electric "goodies" we all like on our cars and the increasing fitment of such things as air conditioning and power steering are all increasing the all up weight of the 'average family car'.

While concern for the environment has an influence on manufacturers, it is the legislation restricting vehicle emissions or fuel usage which obviously has a more significant bearing on product plans. The legislated emissions have been reduced by orders of magnitude over the last 20+ years, so the priority within engine development has been ever cleaner tail pipe emissions. The systems to ensure that these emissions are kept in check not only add weight to the car but some actually have a direct adverse effect on fuel economy.

As far as the transmission system is concerned it is the definition of the test cycles which is probably going to have the largest impact. The reduction of the start temperature on test cycles from fully warm to 'room temperature' has probably been the change which has meant the transmission system can have a part to play in fuel economy purely in terms of its efficiency. In the past the efficiency of most vehicle transmissions has been near enough to 100% during the test cycle to be able to ignore this aspect. As regulations change this is further from the truth. The lower the temperature that the test cycle is completed at the more the actual efficiency of the transmission will have an effect.

3 HOW TRANSMISSION SYSTEMS INFLUENCE FUEL ECONOMY.

It would be wrong to discuss just the efficiency of the transmission units as being the only or even the major influence that the systems have on the economy of the whole vehicle. For this reason four aspects have been identified that can be used to help visualise the total impact of the transmission and driveline system on the fuel economy of a vehicle. Using these concepts helps to formulate a range of actions to reduce fuel usage.

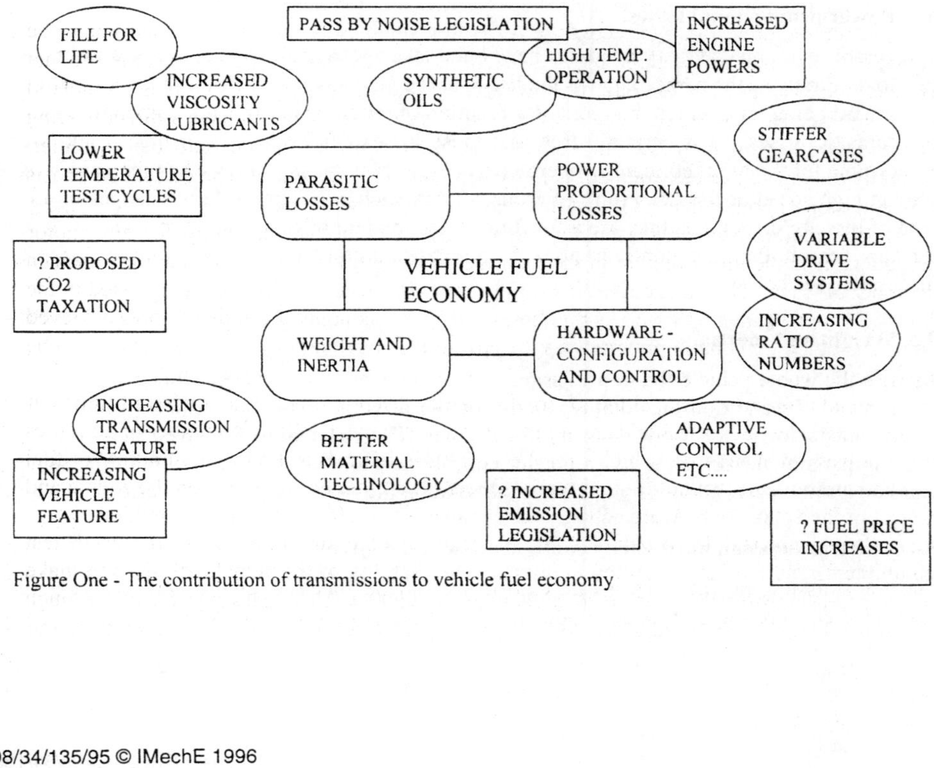

Figure One - The contribution of transmissions to vehicle fuel economy

A number of authors (inc. 2&3) have considered the two aspects of efficiency as separate entities but the other two have been added to complete the whole picture. The four aspects are:

- The parasitic losses in the transmission, i.e. oil churning, seal and bearing drag etc.
- The power proportional losses - at the gear, chain or belt mesh.
- The weight and rotational inertia of the transmission and driveline components.
- The configuration and control of the actual transmission hardware.

Figure One is an overview of how transmissions contribute to the fuel economy of a vehicle. Some of the major external influences are shown along with factors specific to transmissions. Obviously, the external influences are impact on other vehicle systems not just the transmissions.

3.1 Parasitic losses.

The operating temperature during a fuel economy test is the main factor in this respect. For passenger cars the tests used to be done with a warmed up vehicle, which meant that the oil churning and other drag torque's were comparatively small. Now these tests are completed on vehicles which are conditioned to temperatures of between 20°C and 25°C these losses are becoming more significant. This is occurring to the extent that the viscosity grade of a transmission oil can now have a noticeable impact on the fuel consumption of a vehicle. This is not just a feature of the test cycles. The average running temperature of a light passenger vehicle used for short journeys is probably below 50°C. If the low engine torque levels (which are discussed later) are also considered then it can be seen that the parasitic losses are important. The parasitic losses are considered to be a function of speed but it could be argued that lubricant temperature is more important.

3.2 Power proportional losses.

The reason why this category of losses have been taken separately to the parasitic losses is two fold. Firstly as the name suggests the losses at the gear, chain or belt mesh are a function of the load being transmitted, and less of a function of speed, viscosity and temperature than the parasitic losses. It is certainly true that these losses can be very high indeed and are responsible for the high lubricant temperatures achieved by transmissions when the vehicle is used at high speed or for some form of arduous work such as towing. Major changes in this type of loss are difficult as they are a function of the method of transmitting the power. It is for this reason that the configuration of the transmission is an important part of the discussion, see below.

3.3 Weight and inertia.

As with the whole vehicle the transmission system contributes to the fuel economy by virtue of its mass. In addition the rotating inertia of the driveline on some vehicles becomes a significant factor in the work done by the engine. The reduction of a given mass on the rotating parts of the driveline has a greater contribution to the fuel economy than the same mass on a static part. Obviously the faster the rotation the more significant the effect.

3.4 The transmission hardware - configuration and control.

The transmission can have its greatest influence on the fuel economy before the vehicle leaves the concept stage of its development. The 'configuration' of the transmission system refers to the choice of unit (4), the method of transmitting the drive, the ratios chosen and on more complex transmissions simply the layout of the whole system. Some or all of these can be varied if considered at an early enough stage in the design process, the duty cycle to which the vehicle is to be put can also be considered. The simulation and performance prediction of vehicles and their transmission systems is discussed later.

The optimisation of the control system of the whole power train is becoming a major development area for the reduction of fuel consumption. The increasing use of electronically controlled transmission units is facilitating this but further developments are required to take full advantage of the potential for complete optimisation. It could be argued that fully adaptive control of a variable transmission, engine and throttle would be required to reach this point.

4 TRENDS IN TRANSMISSION SYSTEMS - HELPING OR HINDERING FUEL CONSUMPTION?

There is a number of trends in the automotive transmission arena, some of which may help towards fuel economy, most probably will not. Examples of these are:

- higher viscosity oils for durability reasons.
- higher maximum operating temperatures.
- fill for life oils or longer drain intervals.
- move to synthetic lubricants.
- higher power due to higher performance requirements from the vehicle.
- smaller, more compact units.
- 'pass by' noise covers or refinement actions which can increase running temperatures.

The increasing average temperature should have a positive effect on efficiency but often this leads to use of higher viscosity fluids to provide protection for the gears bearings etc. Unless this is countered by the use of say synthetic fluids to retain the low temperature viscosity this trend can lead to higher parasitic losses.

There is an increasing use of variable drives, cvts, and an increase in the use of automatic gearboxes by customers in Europe. While these units have an inherently lower efficiency than conventional manual gearbox units for a whole variety of reasons, this may not necessarily increase the fuel consumption. A lot of the these types of unit have electronic control and the optimisation of this can negate at least some of the inefficiency on some installations although the customer will often prefer the performance potential instead. The use of electronic control indicates the potential for future developments, in that the customer will have a transmission system operating at its optimum condition for the current use but will also offer the choice of style of driving and hence the fuel economy the driver wants.

5 DELIVERY OF IMPROVED FUEL CONSUMPTION

It is important during a study of transmission efficiency that the overall objective of the exercise is the actual delivery of a vehicle to a customer who wants an improvement in its fuel consumption or a benefit from the improvement in fuel consumption. It is only possible to deliver one of these benefits with a holistic approach to the subject. The use of variable (and indeed conventional automatic) gearboxes is a good example. Both these types of unit can be optimised to deliver better fuel consumption than the equivalent manual transmission installation despite the poorer transmission efficiency. The simple expedient of putting the engine in a better operating condition at a particular time ensures lower fuel use despite higher losses in the transmission unit than say, an equivalent manual gearbox.

5.1 Consideration to duty cycles

Investigation of the torque levels during a drive cycle shows that very little time is spent at the higher torque levels. It is obvious that the transmission efficiency needs to be optimised throughout its operating range, especially at low torque levels where the parasitic losses are relatively high compared to the power proportional losses.

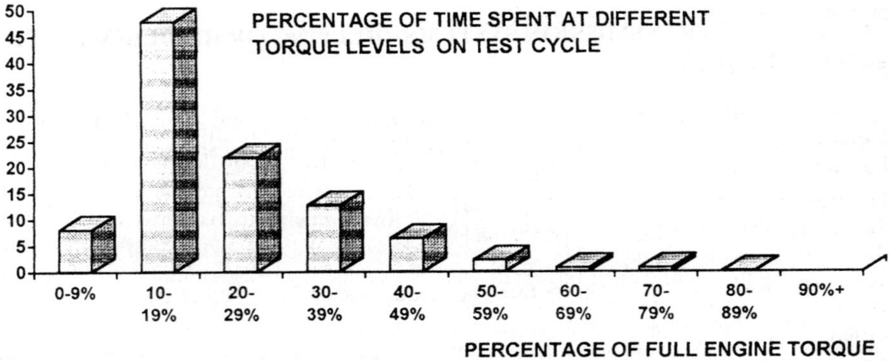

Figure Two - Typical transmission operating torque levels during a test cycle

The torque figures used to illustrate this trend are off an automatic vehicle on the FTP75 test cycle (taken at the engine crankshaft). The diagram, however, would look similar for different vehicles on different test cycles, it is the comparatively low average torque levels that are typical of both test cycles carried out on the rolling road and of 'average' everyday driving. The trend would move up the graph slightly if a more arduous cycle were chosen or down slightly if say a manual vehicle or a higher performance vehicle were chosen. What is interesting to note also, is the even more extreme graph that would be produced if the torque levels were considered at the final drive of a transmission unit or in the axle or transfer gearbox on a four wheel drive vehicle. The typical torque levels are then very low indeed, relative to the achievable maximum, and the parasitic losses in the units become much more crucial in determining the overall efficiency. Higher torque would only be seen in this part of the transmission due to arduous work in the lower gears so it is not surprising that this part of the transmission operates significantly below its maximum during typical/'normal' driving (on a test cycle or on the road).

As previously discussed the transmission can effect fuel economy by virtue of more than just its efficiency. The control strategy of transmissions not controlled by the driver is perhaps the area of most active development currently. This is for a number of reasons including the pace of automatic gearbox technology development and that of engine control, the increasing use of electronic control automatics and the relative ease which existing units can be tuned to suit the characteristics of different vehicles without effecting the hardware.

The weight and inertia of the unit can be reduced by the piecemeal engineering of individual components. This could be because of new materials or just as part of a weight saving exercise, but as with much of the engineering of the transmission system it would be best done at concept to maximise the effect and keep the costs to a minimum.

5.2 The marketing aspects

While considering the factors which can be optimised to improve fuel consumption on any particular vehicle, it is also important to appreciate that the customer requirements will always outweigh the pure engineering rationale. If this was not the case the product would be eclipsed by others in the market place. For example, executive class cars could possibly have their economy improved by fitting a manual gearbox with tall gearing but there would be a somewhat limited market demand for them. It is the job of the transmission engineer to provide the best optimum solution.

5.3 Lubricant choice

The transmission control system is able to put the power train in an optimum condition given the correct algorithms but cannot make up for poor mechanical efficiency. The choice of lubricants is very important to this aspect as it can effect both the parasitic and power proportional losses. Lubricants are always a trade off between a whole range of factors such as high temperature load carrying performance, low temperature viscosity, viscosity index, durability and cost etc. The increasing use of synthetic and semi synthetic fluids is perhaps complicating the picture even further as their performance has different emphasis and there is usually little opportunity during the development of a transmission to carry out exhaustive efficiency testing. A number of papers (5&6) clearly demonstrate the experience and care that is beneficial when efficiency testing. Therefore, the optimisation of the lubricant choice could be an expensive exercise if a large range were to be fully tested. Discussions within the CEC (7) have looked at this aspect of optimisation of lubricants. If laboratory tests were to be developed they would need to be able to compare the likely impact on transmission losses of different oils by understanding both the 'churning behaviour' of an oil and its likely performance in transmitting power, i.e. under load. The second of these, a comparative indication of the power proportional losses generated by different oils, would also have the added benefit of being an indication of the comparative performance of the oils in the high temperature running of a transmission which is of concern to so many manufacturers.

5.4 Consideration at the concept stage

It is certain that the best time to optimise the transmission system for a vehicle is at the concept design stage. It is here that the trade off between feature, derivability, performance, economy, etc. can be best achieved. Once an 'inefficient' unit is designed into a vehicle it is certainly costly perhaps impossible to reverse the situation. Some improvement of the transmission system can be achieved during the full production design phase or indeed after volume. An example of this may be changes to reduce the parasitic losses by alternative oil specification, but the power proportional losses are largely determined by the number of gear meshes in the system or say the selection of a chain or belt drive, in other words the transmission configuration.

6 OPTIMISATION REQUIRES PERFORMANCE PREDICTION

As can be seen from this discussion there a number of sometimes conflicting pressures on the design of the transmission system, and some equally conflicting solutions. One solution may prove better in one respect and another better in a different operating situation. It is suggested that the only way to obtain a completely optimum solution is the development of a transmission simulation technique. This should be able to model the transmission accurately at an early stage of a concept design and also help to optimise the mechanical and control systems later in the development cycle. This would have to include complete modelling of the variations in losses and hence efficiencies for the full range of torque, speed and temperature conditions encountered. Efficiency data for transmissions is often incomplete or approximate at lower input torque, or simply only gives data for torque conditions above 20 to 30%. It is in this lower region that much of the operation of a transmission occurs and also has the greatest variation. The details on losses during the transition from no load to light drive loads is often needed to 'complete the picture'. Figure three shows typical curves for transmission efficiency and why they may vary.

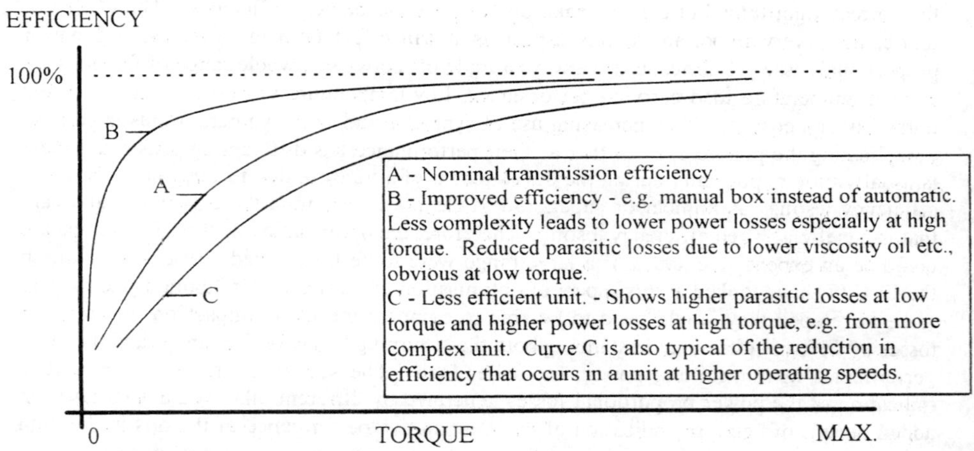

Figure Three - Typical transmission efficiency and effects of changing conditions

The curves shown could be very similar at higher torque where a lot of transmissions operate in the high 95%+ region. It can be seen, therefore, that it is in the lower torque region that simulation could be particularly useful to investigate the differences in efficiency due to different conditions. Following this performance prediction can establish the effect on the vehicle. Differences due to the lubricant choice are an important variable which need to be covered. This supports the work suggested by the CEC investigation group.

The use of engine modelling for performance, fuel economy and emissions, is becoming ever more sophisticated and transmission systems now require the same level of simulation. The development of simulation and particularly vehicle performance prediction systems has largely been carried out 'in house' by vehicle manufacturers in the past. The paper by Hedman (9) and recent developments on the Brite/Euram funded HYZEM (10) project probably show the way forward for transmission simulation. The modular nature of packages such as SIMULINK (11), used on HYZEM, are ideal for the development of a tool for simulating the performance, efficiency and control of transmission systems. This could in the long term reduce the need for detailed test rig development for transmission efficiency. Validation testing to correlate the predictions would still be needed.

In the near term experimental methods will be needed to determine transmission efficiency. This information on the mechanical aspects of the transmission can be used with data on the other vehicle systems to provide the optimum control strategy or indicate areas of weakness in the transmission. Examples include the choice of better ratios or the improved operation of a particular feature.

A number of specific issues can be tackled given the availability of accurate simulation techniques which today require some degree of vehicle or rig based development. These include:

- warm up - especially from cold start or in different ambient temperatures.
- variation in efficiency with different lubricants - i.e. synthetic and mineral based.
- different control algorithms or strategies.
- prediction of the effect of design changes:
  - oils.
  - configuration.
  - temperature.
  - ratio selection.
  - weight and inertia.

It is undoubtedly the case that the emphasis on these different aspects of transmission optimisation will vary from one application to another and from one manufacturer to another. This is one example where simulation may well lead to subtle variations in specification of say, lubricant on vehicles which even have the same transmission unit. The control of the units can also be varied to optimise the fuel consumption of different vehicles or vehicles with different duty cycles.

7 CONCLUSIONS

The transmission system of a vehicle effects the fuel consumption in a number of ways, not just by virtue of its efficiency. There is therefore the potential to improve the fuel consumption by optimising the whole of the transmission. This would need to be throughout the development process;

- the selection of the most appropriate configuration or method of transmitting the drive,
- detail design to minimise the parasitic losses and weight,
- optimisation of the control systems of the engine and transmission together.

This complete approach would ensure minimum fuel usage with acceptable emissions for the wide range of operating conditions encountered by the vehicle.

Trends in the legislation and the environment external to the manufacturer are changing the way we need to view transmissions. No longer can their impact on fuel economy be dismissed as insignificant as "the efficiency is close to 100%". The typical operating conditions during customer usage and current and future legislative test cycles do not support this view.

As the situation regarding transmissions is in many ways becoming more complex the overriding need is for accurate prediction methods. As is now the case with engines where the key performance parameters can be predicted at the drawing board the ability to predict the efficiency of transmissions at concept is long overdue. To be able to complete the picture this should be added to similar information on the engine, vehicle and control systems to enable complete performance prediction and fuel economy optimisation. It is recognised that some aspects such as the transient emissions prediction and the full optimisation of total power train controllers still require further work.

ACKNOWLEDGEMENTS

The author would like to acknowledge, with thanks, the permission granted by Rover Group to publish this paper. Also to colleagues for their valuable comments and input.

REFERENCES

(1) ACEA Newsletter No 24. Industry takes a dynamic approach to CO_2 reduction, June 1995.

(2) DONGEN, L A M van. Efficiency characteristics of manual and automatic passenger car transaxles, SAE Paper 820741, 1982, 12p

(3) DOREY, R E and McCANDLISH, D The modelling of losses in mechanical gear trains for the computer simulation of heavy vehicle transmission systems. IMechE C190/86, 1986, 14p

(4) GREENBAUM, J J et al. Manual transmission efficiency trends and characteristics, SAE Paper 942274, 1995, 4p

(5) KLUGER, M A et al. Proposed efficiency guidelines for manual transmissions for the year 2000, SAE Paper 950892, 1995, 8p

(6) KLUGER, M and GREENBAUM, J J. Automatic transmission efficiency characteristics and gearbox torque loss data regression techniques, SAE Paper 930907, 1993, 7p

(7) CEC is the Co-ordinating European Council for the Development of Performance Tests for Transportation Fuels, Lubricants and other Fluids.

(8) Transmission efficiency data supplied to Rover Group.

(9) HEDMAN, Dr A. Loss analysis of drivelines - application to vehicle performance simulation, VDI Berichte Nr.816, 1990, 5p

(10) FRIEDMAN, S et al. Hybrid technology development: Methods to develop new ways to efficient zero emission mobility, 1995, 12p

(11) SIMULINK is a high level simulation language running with MATLAB, a technical and numerical package for problem solving. Both are produced by The MathWorks, Inc.

C498/34/250/95

The development of a transient dilution tunnel to compare the transient performance of different piston features

A G L BLACKWOOD BEng, **D H TIDMARSH** BSc, PhD, CEng, FIMechE, MBIM, MIMgt, **H DAVIES** PhD, and **P CHURCH** BSc
University of Central England, UK
M WILLCOCK BEng, PhD
A E Goetze Limited, Bradford, UK

Synopsis

Particulate matter from diesel engines is collected either when the engine is operating under steady state or transient conditions. The assessment of particulate and soluble organic fraction (SOF) emissions from a transient test is only done when the engine has gone through a transient, from start to finish. This paper describes how a steady state dilution tunnel will be altered so that particulate matter can be collected, either under steady state or transient conditions. It will also enable particulate emissions to be collected during any part of a transient change, allowing transient particulate emissions to be analysed in greater detail.

1.0 INTRODUCTION

Many tests and procedures have been documented and published about the collection of particulates emitted from diesel engines (1,2,3)*, whether they are heavy duty or light duty high speed engines, indirect or direct injection. The particulates are collected with the engine operating under either transient or steady state conditions. This paper will describe the principle and planned operation of a tunnel that is part way through a conversion process to modify a steady state dilution tunnel so that it can be used to collect transient as well as steady state emissions.

The aim of the investigation is to compare the effects of different pistons by following a series of repeatable tests and determining which piston features contribute to the reduction of exhaust gas and particulate emissions. Undergoing two investigations simultaneously normally requires two test cells or, in this case, at least two dilution tunnels. Dilution tunnels, either for steady state or transient state work, are expensive items. For the comparison of piston performance it is hoped that this dilution tunnel configuration offers a method of collection that is reliable under either operating condition, but is also more cost effective.

* Numbers in parenthesis designate references at the end of the paper.

2. 0 THE ENGINE TEST CELL

The test cell is centred on a two litre direct injection Rover 'L' series diesel engine attached to a Schenk 230 Dynamometer. A Cadet 6 collects the data, and the engine fuel consumption is determined with either a Hurd Brown FM 3A or burette. Exhaust gas analysis will be determined using a Richard Oliver MultiGas analyser. The particulate emissions will be collected using a PSS-5 steady state particulate sampling system manufactured by SGRD LTD. See Figure 1.

2.1 The dilution tunnel

The tunnel consists of :-

1. 305 mm diameter x 3050 mm 316 stainless steel dilution tunnel and end cones.
2. Ambient air filters contained in a mild steel cabinet. The filters remove airborne particulates down to 0.0003 mm. It also has an activated charcoal filter to remove airborne hydrocarbons. Combined filter system is 99.997 % efficient with $17m^3$ / min passing through it.
3. Automatic control console. This unit features :
    a. A variable flow 0 to $17m^3$ / min centrifugal blower
    b. A refrigerated condensate unit
    c. Full system temperature monitoring
    d. Digital sample volume readout
    e. Digital pre-set sample volume
    f. Automatic sample flow control
4. Fully automated twin swing arm 47 mm filter holder unit with back up filter (4).

Its original operating configuration is shown in Figure 1 below. It is envisaged that one test cell, with one complement of equipment will be used to perform two independent investigations. One investigation is looking at the NOx / particulate trade off and how this relationship can be influenced by piston design. It is envisaged that different engine tunes will be tried at different steady state conditions. From the data collected the NOx / particulate trade off envelope can be plotted. Once this has been established, further experiments and tests will be carried out to try to simultaneously reduce both NOx and particulate emissions. As already stated, this investigation will require the dilution tunnel to operate under steady state conditions.

The second investigation is looking at cool starts and transient changes within the engine, again with the aim of reducing the emissions by altering the piston design. The aim is to take snap shots of the exhaust emissions as the engine goes through a transient. Ideally, 20-30 seconds worth of exhaust gases will be diverted into, and diluted by, the tunnel. The particulate is then collected on filters so that it can be analysed. It is conceived that this test can be repeated, this time with the next 20 - 30 seconds worth of gas being collected. See Figure 2. This will allow a picture to be pieced together showing how the particulate and, more importantly, the SOF changes as the engine reaches a state of equilibrium from a cold start. The investigation is directed at the first few minutes after engine start and before the oxidising catalyst lights.

Further investigations are also being considered, such as collecting and analysing the emissions from the engine as it is accelerated and decelerated. The aim is to see how the spectrum of emissions changes from a steady state mode to a transient mode.

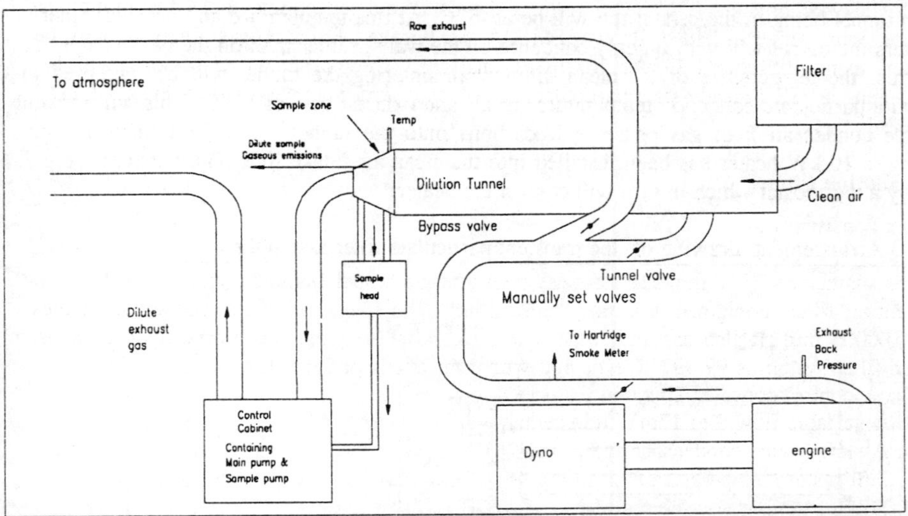

Figure 1 Schematic diagram of original operating configuration of the PSS-5 particulate tunnel.

The PSS-5 is designed to collect particulate under steady state conditions, and is therefore ideally suited for the first investigation, the NOx \ particulate trade off. It is not, however, suited for the capture of particulate under transient conditions.

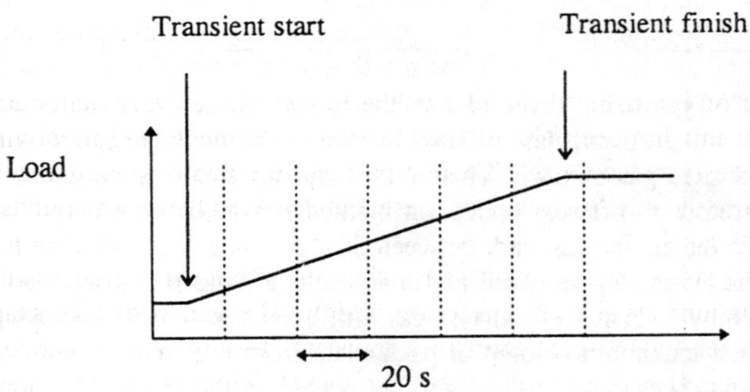

Figure 2 shows the transient broken down into 20 second intervals. Each 20 seconds will be collected separately.

3.0 MODIFICATION TO DILUTION TUNNEL TO ENABLE TRANSIENT AS WELL AS STEADY STATE PARTICULATES TO BE COLLECTED

When cool starts are investigated the temperature in the cell will be reduced to -5°C. The dilution tunnel being in the cell, it too will be at -5°C. At this temperature any organic material that enters the tunnel will very quickly condense on the walls, rather than on the particulates. To avoid this, the temperature of the clean dilution air entering the tunnel will be increased to maintain a particulate collection temperature, of air and exhaust gas, of 51°C. This will prevent excessive condensation of gas phase hydrocarbons onto the tunnel wall. To achieve the air warming, a 10 kW heater has been installed into the clean air intake line. The temperature will be set by a thermostat which in turn will control the heater.

Figure 3 Arrangement drawing of the transient particulate filter assembly

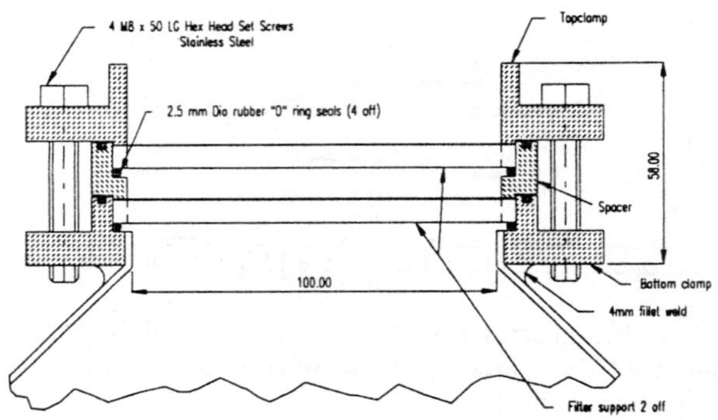

Note: Distance between filter papers is 15.50mm

With only a very small amount of gas being diverted into the tunnel, the existing particulate filtering method will not collect enough particulate to analyse. This is because the sample unit only takes a small part of the diluted gas mixture. This is adequate for steady state work, but not for the transient work. Therefore a different collection method has to be considered. The mass of exhaust gas emitted from the engine can vary between 26 Kg/hr at idle, to 475 Kg/hr at full load and speed. For example, if an engine speed and load emits a mass of exhaust gas of 225 Kg/hr, then in 20 seconds only 1.25 Kg of exhaust gas will enter the tunnel. Exposing a filter to all the gas will collect the maximum amount of particulate from the small sample. To achieve this a third filter holder has been designed and will be located on the end of the tunnel. Figure 3 shows its construction. This will allow the gas to be diluted and thoroughly mixed before it is collected. All the filter components will be made from 316 stainless steel. The bottom clamp will be welded to the nose of the tunnel. The top clamp will hold the two filter supports and spacer in place. The filter holder will clamp two 110 mm diameter filter papers, one behind the other, with a distance of 15.50 mm between the two. When steady state work is being carried out, the filter supports can be taken away so that they do not get contaminated with further particulates from different experiments. The system consists of two filters, primary and secondary. With such a small amount of particulate expected when sampling such a small amount of exhaust gas, the second filter may become redundant. This will be determined during the development of the test techniques.

3.1 Exhaust gas flow control

To be able to collect different time phases of the exhaust gas during a cool start it is essential that the flow of gas is controlled precisely. The aim of the tests, as already stated, is to compare

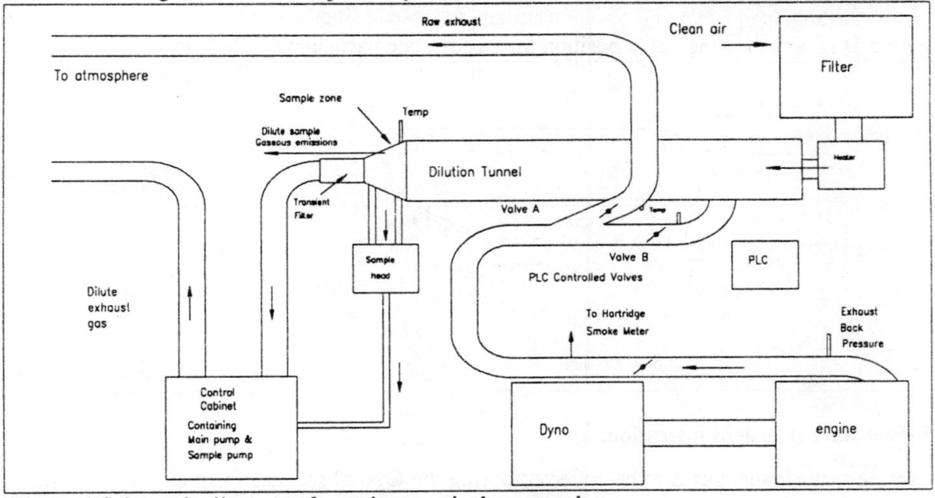

Figure 4 Schematic diagram of transient particulate tunnel.

the effect of different pistons on cool start emissions. It is vital that the tests are repeatable and reliable and that any difference can confidently be attributed to changes in piston design rather than variations in test equipment (5,6). To ensure precise gas flow to the tunnel two actuated high temperature valves will be installed into the exhaust line. See Figure 4. Valve A can be proportionally controlled. i.e. its position can be varied between open, closed and any position in between. Valve B can only be open or closed.

3.2 Steady state valve operation

Valve B is left open, allowing the passage of exhaust gas to the tunnel. When the engine is operating under different steady state conditions, different amounts of exhaust gas will be produced. Depending on the amount of exhaust gas produced, valve A will be set to allow enough gas into the tunnel so that it can be mixed and cooled, with particulate samples being collected at a maximum temperature of 52°C. When different pistons are tested, valve A can be set back to its original position determined from the base line tests. For example,

Valve settings

TEST	Valve position (% open)
Test 1	65%
Test 2	50%
Test 3	40%
Test 4	30%
Test 5	18%

Each piston will go through each test with the corresponding valve position. This assumes that the amount of gas generated with each different piston is similar.

Figure 5 Diagram showing valve position for steady state particulate collection

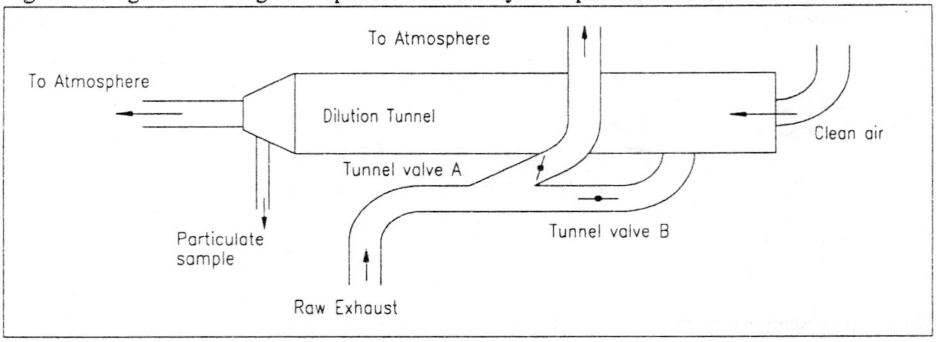

3.3 Cool start transient operation.

Assuming that exhaust gas is to be collected during the first 20 seconds from engine start, then valve A is closed, forcing the exhaust gas into the dilution tunnel. A temperature sensor detects that the gas has moved past Valve B. This activates a timer on a programmable logic controller (PLC). Once 20 seconds have elapsed, valve B closes and valve A opens. This allows the rest of the exhaust gas to escape to atmosphere unrestricted. At the same time, a short blast of clean, dry, compressed air is injected into the pipe, just past valve B. This keeps the exhaust gas flowing into the dilution tunnel. Once the multigas analyser has detected that the exhaust gas has passed the collection filter, the test has finished. The sequence of operations can be used to collect different sections of the starting engine's exhaust. The sequence of logic is shown below.

1. Enter program number.
2. Open valve B to allow gas into dilution tunnel.
3. Close valve A to force exhaust gas into dilution tunnel.
4. Set timer to zero.
5. Start engine.

6. Logic controller receives a signal from the temperature sensor and begins the sampling sequence.
7. Count 20 seconds.
8. Close valve B to stop gas flow into dilution tunnel.
9. Open valve A to allow gas to enter atmosphere.
10. Purge exhaust line immediately after valve A with dry compressed air for 5 seconds.
11. End.

Figure 6 Diagram showing tunnel configuration for transient particulate collection

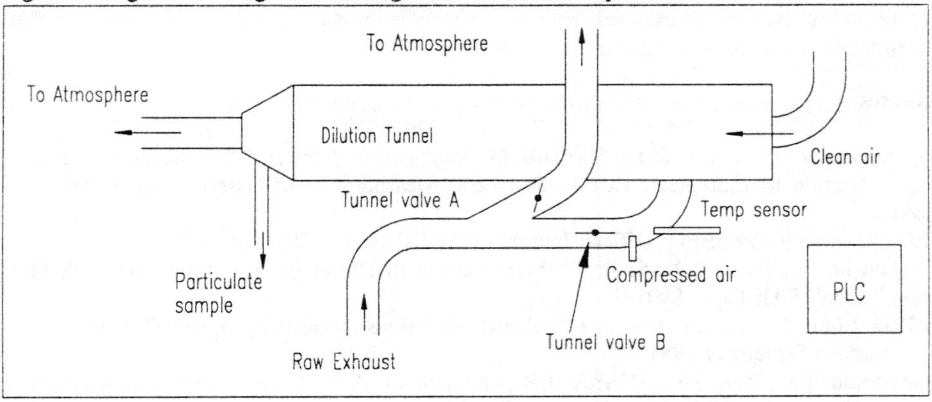

3.3 Hot start transient particulate operation.

The process and procedure for the collection of hot start transient emissions, in essence, will be the same as for cold start emissions. The engine will be run through a test cycle to warm the engine, then turned off. It will be left for a short period of time before being re-started. The collection method for the particulate will be the same as for the cold transient tests, twenty seconds of exhaust gas will be diverted into the tunnel and the particulate collected. The aim is to map out the changing exhaust emissions until the engine has reached a state of equilibrium. Hot transient particulate collection should be less time consuming than cool particulate transient collection because the engine will always be hot after the collection of each sample of exhaust, all that needs to be done for another collection of particulate is to re-condition the engine.

3.4 Change in engine conditions.

The third and final aim of the transient particulate tunnel is to collect the particulate from the engine as it goes through a process of increased load, attaining a new load and staying there for a period of time before returning to the original load. The area of interest, in this test, is to look at the change of the SOF as the engine undergoes a change of state when it is running.

4.0 CONCLUSION

To conclude, the modifications highlighted in this paper will allow a steady state dilution tunnel to collect timed intervals of exhaust gas from a diesel engine whilst in a transient mode. It does this without compromising the ability of the tunnel to collect steady state emissions if so required. The design allows the transient emissions to be separated into small intervals each of which can be analysed individually, allowing detailed analysis of the complete transient event.

Another benefit of this modification is that it also allows two sets of experiments to be carried out in one engine cell on one engine. This offers a substantial saving of resources in terms of cost, time and space.

References

1. BSi Standards 93/714176 Draft Standard for Reciprocating Internal Combustion Engines Exhaust Emission Measurement. Part 1. Test Bead Measurement of Gaseous and Particulate Emissions

2. EEC Directive 91/542/EEC. Official Journal of the EC 1991, L295 Vol. 34.

3. Hirakouchi, N., Fukano, I., Shoji, T. Measurement of Diesel Exhaust Emissions with Mini Dilution Tunnel. SAE Paper 890181.

4. SGRD Limited company brochure "Exhaust Emissions Measuring Systems" Page 17. Brochure issued September 1984

5. MacDonald, J.S., Plee, S.L., D'Arcy, J.B., and Schreck, R.M. Experimental Measurements of the Independent Effects of Dilution Ratio and Filter Temperature on Diesel Exhaust Particulate Samples. SAE Paper 800185.

6. Plee, S.L., MacDonald, J.S. Some Mechanisms Affecting the Mass of Diesel Exhaust Particulate Collected Following a Dilution Process. SAE paper 800186.

Mechanical Developments

C498/39/189/95

Engine instrumentation for piston system development

A M ABBEY
T & N Technology Limited, UK
A J WOODWARD
T & N Piston Products Group, UK
P JÖHREN
A E Goetze GmbH, Germany
E HARTLEY
A E Goetze Automotive Limited, UK
T WALTER
A E Goetze North America, USA

SYNOPSIS

Market and legislation pressures demand increasingly rapid and detailed analysis of piston assembly performance, requiring sophisticated measurement and analysis techniques which are often not available commercially. This paper describes the more recent developments both in routine testing and more specialised experimental testing within the T&N Piston Products Group, including details of the methodology and techniques used and typical application examples.

1 INTRODUCTION

In developing a piston system which successfully meets both customer expectations and increasing legislation, the technical capability of the piston assembly has necessarily increased to the extent that it is more refined in terms of noise vibration & harshness (NVH), emissions and fuel economy, whilst operating in an environment which is thermally and mechanically more hostile. In addition, the market demands increasingly more rapid determination of the optimal piston and ring designs for each specific engine. A quicker, yet more detailed, quantification of piston assembly performance is therefore required during the development phase, which has resulted in the development of new measurement techniques as well as the refinement of those existing.

Testing of new piston assemblies can be divided broadly into two categories, namely routine quantification of component performance and the more advanced engine testing techniques. General component performance evaluation includes measurement of oil consumption, emissions, blow-by, wear, and durability. The advanced measurement techniques allow quantification of piston noise, and evaluation of piston assembly friction and its influence on specific fuel consumption. Additionally, measurement of particular component performance features such as piston temperatures, piston and ring motion and inter ring gas pressures can be made using dedicated telemetry systems.

This paper describes the more recent developments in both of these areas within T&N Technology and AE Goetze, together with the relevant methodology and typical examples of their use.

2 GENERAL TESTING TECHNIQUES

Development of a complete piston system to attain maximum performance demands the ability to fully assess all requirements of the system, including oil consumption, gas pressure sealing, emissions, noise, friction and durability. These functions must all be optimised to achieve the best total system. In addition to the specialised techniques detailed later in this paper, AE Goetze uses a range of general evaluation techniques during product development.

Blow-by is routinely measured and used as a primary indicator of ring/bore performance. As a first step, cylinder bore geometry and distortion may be mapped using an Incometer, a sophisticated instrument originally developed by AE Goetze. Bore surface finish is examined and quantified in terms of various surface characteristics, highlighting areas of deficiency and courses of correction.

Hot and cold scuff resistance is evaluated using procedures which provide engine coolant at temperature extremes from +140°C to -35°C. Oil consumption measurements are taken using a number of techniques including gravimetric and exhaust tracer methods. Advanced emissions equipment is now in use at T&N Technology and several AE Goetze companies.

2.1 Oil Consumption

The variation of oil quantity in the sump occurs as a result of three different mechanisms, denoted by the three arrows in Figure 1. This oil consumption is defined in this paper as 'sump oil consumption' and includes lubricant lost in the formation of deposits within the engine and gains due to the products of combustion which can, under some conditions, be found in the oil sump. These are particularly important in gasoline engines, where up to 20% of fuel derivatives can be found in the lubricant.

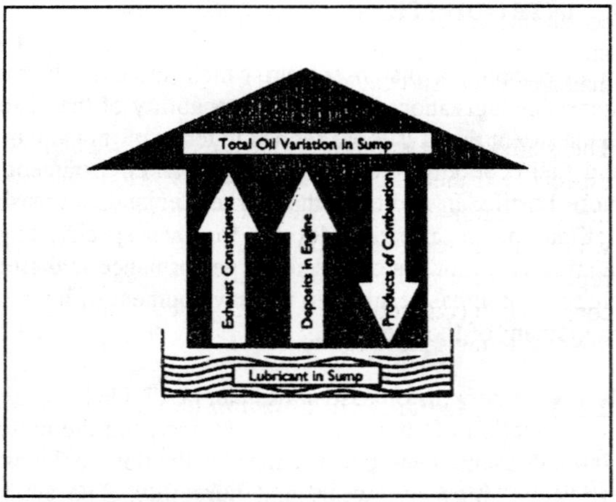

Fig 1 Contributions to total oil consumption

The oil lost through the exhaust system only is defined as 'exhaust oil consumption' and is related, though not equal, to the sump oil consumption because of the other two components of the total variation, depending on the running conditions of the engine. Volumetric and gravimetric measurement systems are used to measure sump oil consumption, whereas exhaust oil consumption can be measured using some form of tracer. The two techniques can give different results which may be interpreted in different ways.

An on-line continuously measuring gravimetric system for the measurement of sump oil consumption, and a sulphur trace system for the measurement of exhaust oil consumption are both used within AE Goetze Companies.

2.1.1 Sump oil consumption measurement

Oil consumption is measured routinely within AE Goetze using the MOVAN system, an on- line system developed by AE Goetze and shown schematically in Figure 2. During operation, oil is pumped continuously around the external circuit with the level in the engine maintained by the stack pipe in the sump. As oil is consumed, the quantity of oil in the external reservoir falls and the mass change is measured by a load cell in the supporting reservoir. A standard conditioning and test procedure is employed to ensure that engine operating conditions are always the same at the start of a test, thereby maintaining consistency of results.

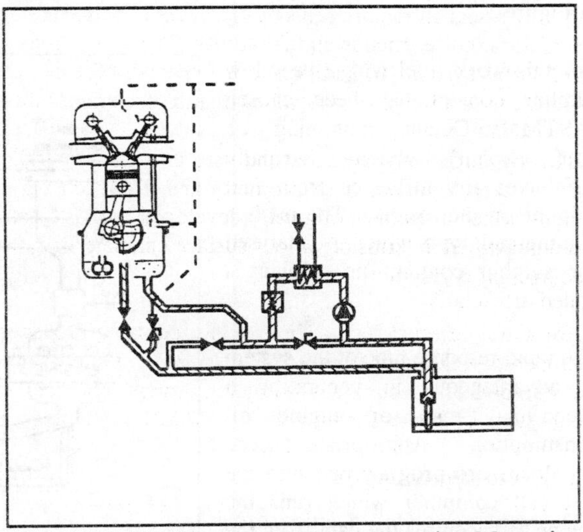

Fig 2 MOVAN Gravimetric Oil Consumption Measuring System

The statistical calculation is an important element in evaluating the pan oil consumption, especially if the oil circuit is not optimised, as in this case. This increases the complexity of the evaluation and increases the time needed to obtain the data for each test, as a result of which an automatic drain and weigh system is employed which includes automatic replenishment of fresh oil. In this way it is possible to control the measurement of pan oil consumption continuously. A more detailed description of oil consumption measurement using this technique can be found in (1).

2.1.2 Exhaust Oil Consumption

AE Goetze North America have been using SO_2 real-time oil consumption measurement since 1990 to provide fast, accurate comparisons of piston and ring design influence on engine oil consumption. This method uses the fact that oil consumed by an engine forms sulphur compounds in the exhaust, the amounts of which are proportional to the quantity of oil being consumed. Efforts have been focused on spark ignition engines, where continual improvements in oil consumption rates have made accurate drain and weigh methods more difficult.

The equipment used is the Horiba MEXA 1120 SUL analyser, specifically tailored for SO_2 oil consumption measurement. Engine exhaust passes through an oxidising furnace at 1000°C to ensure conversion of all sulphur compounds to SO_2. Heated lines are used to keep the sample well above the dew point, preventing absorption of SO_2 by water. The instrument uses NDIR detection of SO_2 and an on-board computing unit, combined with engine air flow measurement and manually input oil sulphur content, to measure in real time oil used by the engine in grams/hour. A time integrated output is also provided. The system is schematically illustrated in Figure 3.

For this work, fuel with a very low sulphur content has been chosen (ASTM Iso-Octane, containing <1 ppm sulphur), which virtually eliminates any influence from fuel sulphur on the results. The oil is a commercial type which is analysed for sulphur content; no sulphur is added artificially.

The rapid response time of the system is advantageous in generating a speed/load map of engine oil consumption. Appropriate speeds and loads are programmed into the test cell computer which runs the engine and acquires the data from the sulphur analyser. It has been found that operation for two minutes at a given condition is adequate to obtain a fully stabilised reading. After the map is run, a three-dimensional plot is generated (Figure 4)

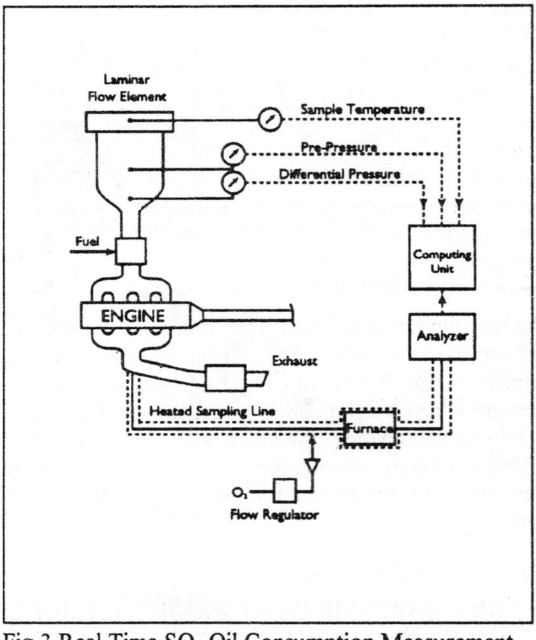

Fig 3 Real Time SO_2 Oil Consumption Measurement

The time-integrated output can also be used as an accurate, short-run equivalent to the 'drain and weigh' procedure. The engine is operated on a simulated driving schedule for 1 hour and the total oil used during this time is measured by the 1120 SUL. The processed data is then used to predict the service performance of piston and ring designs.

This technique could also potentially be used for bank-to-bank or cylinder-to-cylinder measurements, by taking exhaust gas samples directly from the exhaust manifold. The accuracy of the system may well improve in the future through enhanced, alternative SO_2

detection methods.

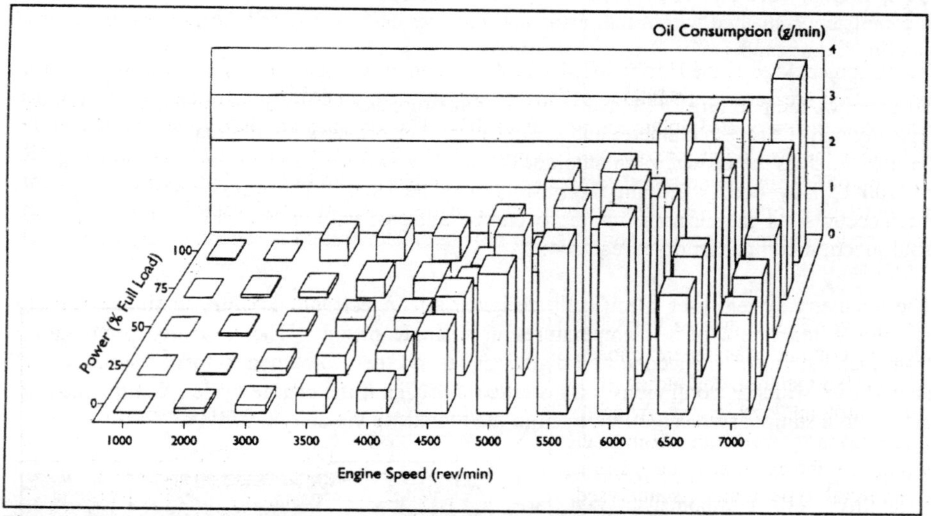

Fig 4 Real Time Oil Consumption as a Function of Engine Speed and Load

2.2 Exhaust Emissions

The most important task in the future is to reduce the contribution made by the piston assembly to both gaseous and particulates emissions, whilst maintaining the durability of the system. Improvements in the emissions performance of the engine must be verified directly by measuring the emission components. Over the past few years AE Goetze has sponsored work at several universities aimed at a better understanding of the piston assembly contribution to engine emissions. Gaseous and particulate emissions measurements equipment is now being installed at a number of AE Goetze companies to supplement that already in place at T&N Technology.

3 ADVANCED MEASUREMENT TECHNIQUES

Within the T&N group of companies, a number of specialised measurement techniques have been developed to allow characterisation of component performance. Among these are systems to quantify bore shape following tightening of the cylinder head, transducers to measure head to block movement in operating engines and dedicated telemetry systems for piston assembly performance measurement. More recent developments in this area are described below.

4.1 Wireless Telemetry For Piston Temperatures.

Successful monitoring of piston temperatures within the piston assembly of an engine can be crucial to the understanding of the piston behaviour and performance. The two beam mechanical linkage developed by T&N Technology(3) remains a highly versatile and successful technique for monitoring of a range of piston parameters. A more rapid technique is required, however, so T&N Technology have developed a wireless telemetry system which is completely contained within the piston and requires no modifications to the connecting rod and big end assembly.

To monitor temperatures within the piston, miniature bead thermistors are used which are typically only 1.6mm in diameter and can measure temperatures up to a maximum of 325°C; for areas such as the crown of an automotive diesel or two stroke piston, where temperatures of 400°C may occur, 2.5mm thermistors may be used. Thermistors are preferred to thermocouples as greater noise immunity can be achieved with correct circuit design, and no cold junction compensation is required.

The telemetry system uses specifically designed microelectronic circuitry to simultaneously monitor a maximum of 13 temperatures in a single piston. The measured temperature from each thermistor is encoded as a digital signal and then combined to form a serial digital data stream which is continuously transmitted throughout the engine cycle from the moving piston to a stationary receiver mounted in the crankcase via an optical link.

To provide the maximum performance and reliability, all components used are designed to military specifications and are capable of operating at local temperatures of 150°C. Power for the system is taken from a small battery which is also mounted in the piston. Figure 5 shows a cross-section of a typical instrumented piston, illustrating the location of the main components on the inside of the piston skirt. The mass of the microelectronic circuit, the battery, and the optical components is only 11g and therefore has minimal effect on the performance of the piston assembly.

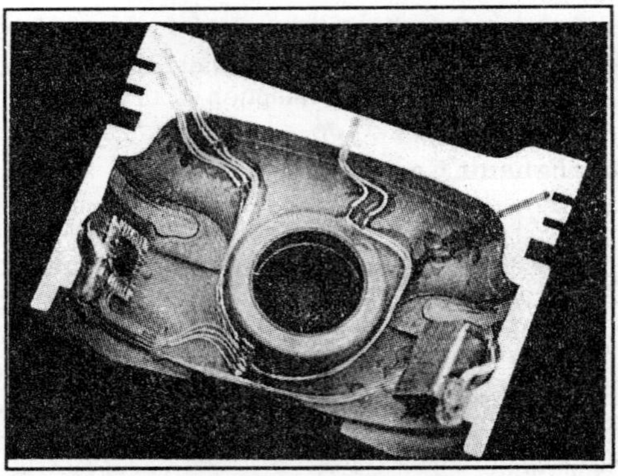

Fig 5 Wireless Telemetry Equipment Installed in a Gasoline Piston

Because of the inherent variation in the separation of the transmitter and receiver diodes, the signal output from the stationary receiver varies greatly throughout the cycle of the engine. Severe degradation of the signal also occurs due to the effects of the oil mist within the crankcase and the high ambient temperatures in which the optical link must operate. A high performance amplifier circuit is therefore needed to recover the data signal, before passing it to a microprocessor based system which synchronises to the data stream and locates each data channel. The recovered data is then converted to RS232 format and transferred to a personal computer, where custom software is used to display the measured temperatures in real time. If required, the data can be stored on disk or magnetic tape for more detailed analysis at a later date. Figure 6 shows a functional block diagram of the complete telemetry system.

During development of the system tests were performed in a wide variety of engines, including a 2.5 litre DI diesel, a 1.8 litre 16V gasoline, a 500cm^3 single cylinder motorcycle 2 stroke and a 3.8 litre North American V6. The electronic circuits are designed to operate to speeds in excess of 6000 rev/min, and have been tested at a maximum speed of 6500 rev/min. Because of present limitations on battery technology, battery life above 5000 rev/min is extremely short due to the massive reversals of acceleration which are experienced. However, tests at up to 5000 rev/min have achieved several hours endurance with durability above 6000 rev/min long enough to collect some temperature data. Typical results recorded during testing of the 3.8 litre V6 are shown in Figure 7 for the full load condition, up to the maximum engine speed of 4800 rev/min. During these tests, comparison data was recorded using a conventional two beam linkage, and these results are shown for comparison. General agreement between the two techniques can be seen to be excellent across a wide range of conditions, particularly in the medium to high speed ranges which are of most interest.

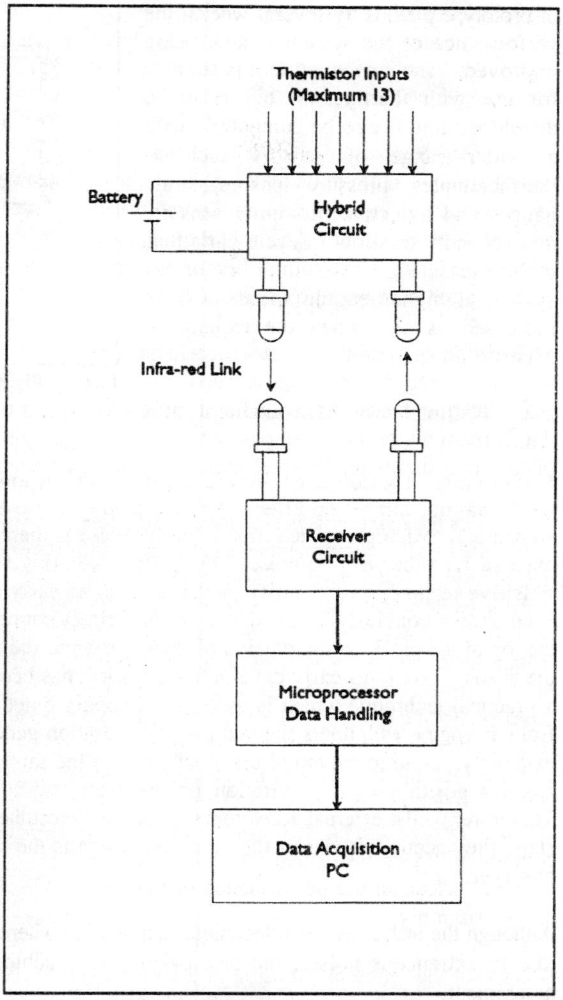

Fig 6 Schematic of Wireless Telemetry System

Using the wireless telemetry system described, considerable cost and time savings can now be achieved over alternative measuring technologies. Design layout for a new system within a piston and crankcase, instrumentation of the engine, and performance of a series of tests now typically takes only eight weeks. Cost savings are considerable, especially over a number of tests where the majority of the hardware and all of the software can be transferred from one engine to another. Transfer between different engines only requires a new piston to be instrumented and relocation of the crankcase receiver.

Having developed a working telemetry system which can speed up the evaluation and testing of prototype pistons by several weeks, the performance of the system is now being improved. The current system is suitable for use with thermistors, but is to be developed initially to be compatible with a wider range of sensors such as thermocouples, pressure sensors, strain gauges and inductive proximity sensors, and secondly to allow in-cycle variations to be measured. The durability of the system at higher engine speeds will be increased as and when the technology exists to do so.

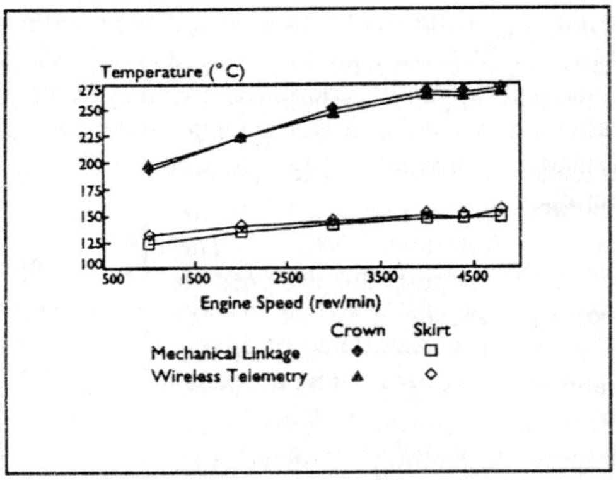

Fig 7 Piston Crown and Skirt Temperatures Measured at Full Load

3.2 Piston Noise Measurement and Analysis

Piston noise in gasoline engines is a short duration low frequency impulsive sound which, while having almost no effect upon the overall noise level, can be readily detected by the human ear. Microphone detection of piston noise is therefore very difficult as it can easily be masked by other engine noises. As piston assembly suppliers, AE Goetze require a more sensitive technique to quantify piston noise at an early stage of engine development. It has been shown conclusively(3) that there is a direct connection between vibration measured at the top of the block on the outside of the engine and the impact magnitude when measured in the piston. From this early development work it has been possible for AE Goetze to develop a practical technique which is used in developing quiet pistons. Recently acquired test data from an engine with liners showed that the vibration generated by piston slap when measured externally, close to the top of the block, was of the same frequency and occurred at the same angular position as the vibration on the liner itself but was attenuated by about 7dB. Therefore, whilst external accelerometers do not detect the true magnitude of the piston to bore slap, they accurately detect the vibration which is the source of the noise transmitted to the observer.

Although the instrumentation technique can be used where subjective assessment is impossible, due to extraneous noises, the best evaluation is achieved when combined with subjective assessment.

3.2.1 Measurement technique

For the detection of piston noise, piezoelectric accelerometers are attached to the outside of the block (on the bore centre line), on the thrust and non thrust axes near to the top of the block. Where physically possible this is carried out simultaneously on all cylinders. Pin noise is detected from accelerometers mounted on the main bearing caps. The accelerometer outputs are amplified and fed directly to the inputs of a multi-channel tape recorder.

A peg is inserted into either the front pulley or flywheel so that it is co-incident with a magnetic pick up when the engine is at #1 piston Top Dead Centre (TDC). The magnetic pick up pulse is conditioned to give a square wave signal, the positive going edge of which represents Top Dead Centre Firing (TDCF) #1 cylinder. Conventional ignition engines are fitted with an optical transducer in #1 cylinder plug lead to indicate the firing event. For wasted spark engines either the engine management signal from the cam sensor is used, if it is suitable, or a separate peg is attached to the cam shaft which is used to activate a magnetic transducer.

The engine coolant circuit is modified, with the thermostat being omitted or fixed open to accept external inflow and outflow from a mobile cooling module. Antifreeze is normally used either 100% concentrated or diluted 50% with water. The module is capable of cooling an engine from normal running temperature to -15 °C in an hour. Its mobility enables use on any bed as well as being suitable for cooling down engines in vehicles which are driven to the module. In order to prolong the cold state on the test bed once the engine is running, the cooling module remains in the coolant circuit permitting extra time to evaluate the cold noise profile. As the engine warms up the effect of increased piston and coolant temperatures (vastly different coefficient of expansions can give rise to the worse noise condition being not at the lowest temperature) can be subjectively and objectively assessed.

Since all the electronic equipment is portable it can be installed in the passenger compartment of a car where it is powered by an uninterruptible power supply. The car engine coolant circuit is modified to accept the cooling module input and output. The engine is then cooled to -15°C, the cooling module disconnected and the engine cooling circuit re-made. The car can then be driven on the test track or road where the required road load conditions can be generated. Data recorded on the multi-channel tape recorder is taken to the laboratory for analysis.

3.2.2 Analysis

The analogue output from the tape recorder is digitised for computer analysis by T&N Technology developed software. A time domain windowing technique is used where up to four windows can be selected of variable angular position and width. Typically a window of 50° is used for each location, these being pre and post TDC firing and non-firing. The programme then filters the data to select the frequency range of interest, determines the vibration level in each window for every engine cycle, averages the values for a wide range of engine speeds and accelerations, and produces a data table.

It has been found that for currently manufactured gasoline pistons, that the frequency of the vibration caused when various parts of the piston impact the bore fall into two frequency bands. The stiffer parts of the piston, ie the lands and the top of the skirt on a solid skirt piston, generate vibration of frequency around 4 kHz whilst skirt slap is at around 1.5 kHz frequency. Accordingly two filters are available, of frequency bandwidth 0.7 to 3 kHz and 3kHz to 7kHz.

Output from the programme can be varied according to the requirements but is generally tabular or graphical data of vibration levels (in dB), or in the form of a noise map.

3.2.3 Typical applications

As part of gasoline piston development programmes, noise analysis is routinely carried out. Traditionally one of the cures for piston noise is to reduce the piston to bore clearance. This solution may however, lead to additional skirt collapse during engine use, which may, in the longer term increase piston noise.

A 4 cylinder engine was run with standard piston to bore clearance for 100 hours on the test bed. Noise measurements were taken after break in, and after 100 hours of cycling. These are shown in Figure 8.

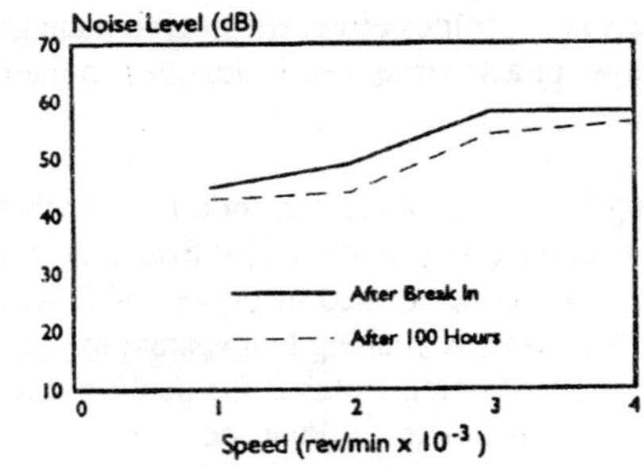

Fig 8 Piston Noise Results before and after 100 hours running

3.3 Piston Assembly Friction Measurement

Piston assembly friction is the single most important factor affecting mechanical losses within an engine, contributing as much as 25%(5), whilst the contribution of the piston assembly to the total engine friction has been estimated between 40% and 80%(6,7,8). Typically, a 10% reduction in engine friction could lead to a 3% improvement in fuel economy(5). The reduction of piston assembly friction can therefore clearly have a significant influence on fuel consumption and emissions performance of the engine, and the ability to quantify the frictional benefits of different piston system designs is necessary.

The ability to quantify directly the frictional force in a single cylinder test engine using the floating liner method has been adopted for sometime at T&N Technology, and has been reported elsewhere(9). This technique allows the direct measurement of the influence of generic design features such as piston skirt form, but is not representative of modern, high speed gasoline engines. The system now being adopted uses a "switch over" type test to measure accurately both fired and motored torque and fuel consumption in a standard production engine. The principle of operation of this system is to run an engine with carefully controlled inlet air temperature, pressure, steady state speed and manifold depression conditions. Switch over motoring tests are also carried out at each speed condition. Here the engine is switched over rapidly from firing to motoring with wide open throttle, and motoring torque measured over a short period immediately following stabilisation of the load. Thus motoring loads at realistic operating temperatures can be obtained.

The engine chosen in the development of the system was an in line 4 cylinder 2.0 litre gasoline engine with 2 valves per cylinder and fuel injection. This was mounted on a Mawdsley 186kW regenerative brake dynamometer. Inlet air is supplied to the engine inlet via a 200 litre reservoir in which the temperature and pressure are automatically maintained using a heater and inlet throttle. Hence tests are run at above ambient temperature and below ambient pressure. A small heater installed in the air filter box compensates for temperature gradients in the inlet piping. The dynamometer itself has been retuned to damp its response and a more accurate, temperature corrected load cell fitted. The engine is typically run at a series of steady state speed and manifold depression conditions, derived from points on the ECE emissions test cycle. At each condition, a fuel consumption measurement is taken and the load sampled within a series of time intervals over the total fuel time. The data is stored on a PC along with the engine running conditions which are downloaded from the test bed control computer.

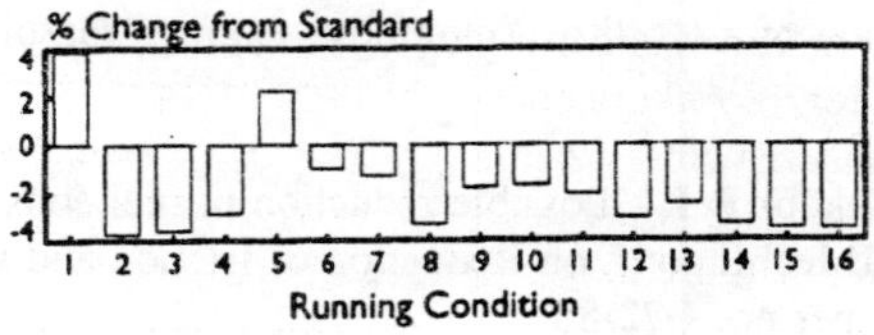

Fig 9 Change in Specific Fuel Consumption with Modified Piston Skirt Finish

Typical data output from the system is shown in Figure 9. This shows the percentage improvement in specific fuel economy which can be achieved by modifying the piston skirt surface finish to give a smoother profile.

4 CONCLUSIONS

In response to increased market demands, in particular, tightening emissions legislation, the technical requirements of the piston assembly have correspondingly become greater. In addition to routine measurement of durability, wear, oil consumption and blow-by, the ability to quantify performance in respect of emissions, NVH and specific fuel consumption, and directly measured performance parameters such as piston temperature, is of fundamental importance. With much shorter leadtimes to produce new piston assembly prototypes now possible, performance testing although more sophisticated, must be carried out in a shorter timescale.

AE Goetze companies are fully equipped with the resources necessary for performance evaluation in piston system development for customers worldwide.

REFERENCES

1. Johren P-W and Newman, B. A. "Evaluating the oil consumption behaviour of reciprocating engines in transient operation". SAE Technical paper 880098.

2. Gazzard et al. Emissions paper , T&N Technical Symposium 1995

3. Woodward AJ and Poyant R. "The contribution of instrumentation to piston assembly development" T&N Technical Symposium 1990.

4. Mechanical Engineers Institute paper by John Richmond in 1987

5. Long O.R. and Kernen I.R., 'Reibungsverluste in Verbrennungsmotoren', Schmiertechnik.

6. Ball et al. The friction of a 1.6 litre Automotive engine - Gasoline and diesel' SAE 860418.

7. Kulmann P.E. and Jakobs R.J. 'Possible reduction in fuel consumption by modifying piston rings' IMechE conf. on Reduction of friction and wear London, 1985, March paper no. C72/85.

8. Masaali Sai, Chikashi Asahi, Shisei Kai, Teturou Murata. A study of Piston Friction loss reduction for an automotive engine. SAE? 911055.

9. Parker, D.A., Adams, D.R. and Donnison, G.M. 'The measurement and reduction of piston assembly friction'. Proc. IMechE. 1989. Paper No. C375/017.

C498/39/190/95

Predictive and measurement techniques in cylinder head gasket development

G P WHITHAM BSc, MSc
T & N Technology Limited, UK
T TRONEL
Curty Payen SA, France

SYNOPSIS

The automotive sealing gasket is a high technology product that interacts with its complex engine operating environment.

Understanding this environment is necessary to design quality products and to fulfil the requirements of the customer. Thus, the Automotive Sealing Group of T&N has developed both measurement and predictive design techniques.

In the paper investigations involving both aspects are described. Particular attention is paid to the need to integrate thermal cycling of the engine with the resulting deformation and its effect on gasket life. It is shown that the design of cylinder head gaskets can be optimised to benefit the whole engine and particularly to reduce cylinder bore distortion.

The described approach to development is shown to provide innovative, high quality products in reduced lead time.

1. INTRODUCTION

The design of cylinder head gaskets has become increasingly complex during recent years. The reasons for this include the requirement for engines to have increased durability, higher performance, reduced emissions and reduced weight. Simultaneous satisfaction of these sometimes conflicting requirements has led to reduced structural stiffness and higher temperatures.

This paper describes the development process for cylinder head gaskets used to achieve good sealability, reduced lead times and full integration within the engine system.

The development process involves an iterative application of finite element (FE) techniques, experimental methods and many years experience of gasket design. A flow chart outlining the process is shown in Figure 1; the following Sections of the paper discuss the several stages in chronological order.

2. DESIGN STUDY

2.1 Collecting engine information from the customer

The first step in the process of designing a gasket is to gather information about its operating environment. For this purpose a questionnaire is addressed to the customer requesting information to characterise the engine including; type, operating data, dimensions and clamping conditions.

In recent years the use of FE analysis has helped to reduce the lead time required for gasket development. The technology available means that FE analyses of the engine structure can be performed to examine the gasket within its operating environment. To minimise lead time it is desirable that an FE model of the engine be supplied by the customer.

2.2 Initial gasket design

Based on the collected engine information, an initial design is made using the experience of the applications engineer and in-house software that predicts the load balance between the bore eyelet and the gasket body material. The software must be provided with the relative stiffnesses of the bore eyelet and body region of the gasket, together with the overstand of the eyelet above the body and the bolt load. It then calculates the load balance between the eyelet and the body of the gasket. It is therefore a useful tool in assessing the sealing load on the gasket and hence in the design of an initial gasket layout.

Figure 2 shows a schematic representation of the load balance software. Two non-linear springs represent the eyelet and body materials, and a gap element represents the eyelet overstand above the gasket body. Figure 3 shows the load-deflection curves of representative gasket body and eyelet materials used at this stage of the design process. The overall gasket load deflection curve is determined by combining the body and eyelet curves. In this example of a fibre gasket, it can be seen that 90% of the load is applied around the bore eyelet and only 10% is applied to the body material.

Table 1 gives a summary of the load balances predicted for two gasket technologies, before, during and after engine operation.

Table 1 Proportion of Bolt Load Transferred to Each Area of the Gasket

	Fibre Gasket		Steel Gasket	
	Eyelet	Body	Eyelet	Body
Cold Clamped	90%	10%	50%	50%
During Engine Operation	75%	25%	40%	60%
Residual Clamp Load	65%	35%		

The software is quick to use and gives the applications engineer an initial estimate of the load on each region of the gasket. However, it assumes that the load is applied evenly around the eyelet or over the body. This is not so, since distortions of the head and block occur due to localised thermal and bolt loads. It is therefore more appropriate to make more accurate predictions of the loading by FE simulations of the engine.

3. THE FINITE ELEMENT METHOD FOR GASKET PREDICTION

To predict accurately the performance of the gasket design, including its dependence on the behaviour of the engine during operation, FE methods are used. Such predictions can aid the applications engineers by highlighting weaknesses in the initial design and quantifying improvements due to design changes. FE methods can reduce lead times and improve the number of prototype gaskets that are 'right first time'.

3.1 Engine structure modelling

The flexibility of the head and block are important aspects to consider when designing gaskets for modern lightweight engines. Continuing increases in computing power mean that these deformations can be predicted more and more accurately using FE techniques and large 3D models of the full engine structure, such as that depicted in Figure 4. The benefits include detailed information about contact pressures upon the gasket and deformations of the engine during various cycles of engine operation. Any weaknesses in the gasket design can be highlighted and changes to the design can be effected rapidly.

T&N have developed a flexible method of handling the complexity of the entire clamping system (1). The analyses predict deformations of the block, head, bolts and gasket. The gasket itself is modelled using complex non-linear descriptions of material behaviour including temperature dependency. Programs have been developed to apply temperatures rapidly to the head-gasket-block models, enabling the gasket performance to be predicted under operating conditions including thermal transients. Further programs apply the boundary conditions, thus enabling customers' models to be incorporated and analysed rapidly. The analyses have been accelerated by superelements within the ABAQUS FE code to represent block and head stiffnesses. Quick turn round time of the analyses means that parametric studies of the gasket design can be undertaken and suggestions for improvements to the applications engineers' initial designs can be made.

As an example, Figure 5 shows the pressure predicted by the analyses of a fibre gasket in static cold clamped conditions. The largest pressures are predicted around the bore eyelet as suggested by the simpler load balance software, Table 1. However, the FE predictions show variations in pressure around the eyelet caused by deformations of the head and block due to the localised clamping force of the bolts. This is further illustrated in Figure 6 that shows the cold clamping pressure around the eyelet with peaks at positions closest to the bolts. Minimum pressures are seen between bolts at around 90° and 270°, the bore bridge regions. It is these regions of minimum pressure that must be examined to ensure that gas sealing will be effective.

When an initial gasket layout has been agreed, the tooling required to produce the gasket is designed. Determination of the precise shape of the tooling can also involve FE analyses, often in conjunction with more detailed investigations of the gasket design.

3.2 Manufacturing process modelling

The function and durability of metal bore eyelets and steel embossments is dependent upon the processes employed in their manufacture. The stresses and strains within the material are not visible and cannot be easily measured. Therefore the processes of forming eyelets and embossments have been studied using FE analyses to gain an understanding of the stress-strain histories of the materials, and to optimise the forming processes (2,3).

3.2.1 Eyelet manufacture

The process of forming and sizing of eyelets in composite gaskets has been examined to gain an understanding of the three dimensional aspects of variable width eyelet manufacture. The variable height of the eyelet after punching implies a non-axisymmetric shape, which cannot be modelled accurately using two dimensional FE models.

Instead, a three dimensional shell element model of the forming process of a quarter of an eyelet has been analysed using the software package ABAQUS Explicit. Figure 7 shows the variable height eyelet after the punching stage. The manufacturing process is then continued by an operation to throw over the eyelet and finally completed by a compression operation. The 3D FE prediction of eyelet formation has shown good correlation with experimental measurements of eyelet roundness, Figure 8. The point of minimum eyelet diameter corresponds to the largest eyelet throwover and results from complex 3D circumferential stress and strain fields.

Such analyses are used to develop innovative toolings to minimise the non-uniformity of the eyelet roundness. They can also ensure that the eyelets do not suffer amounts of strain during manufacture that could lead to reduced fatigue life and the possibility of eyelet cracking in service.

3.2.2 Embossment manufacture

The manufacture of steel embossments in the expanding field of metal gaskets has been investigated. Using FE techniques, a design database for both coated and uncoated steel has been established. Full embossments for gas seals and half embossments for fluid seals have been extensively modelled. The models allow optimisation of the tooling, to achieve the required embossment design for a specific application. The manufacturing optimisation includes design criteria based on the space available, safe levels of plastic strain during manufacture and the contact pressures that will be available in service to seal the embossment.

An embossment showing the non-recoverable plastic strain remaining after the forming process is shown in Figure 9. Such strains are kept within safe limits to maximise the fatigue resistance of the embossment. In this example with open tooling, the maximum strains occur where the steel is bent around the corners of the tooling. Upon release from the tooling the steel recovers slightly but retains much of its embossed profile due to the non-recoverable plastic strains.

3.3 In-service performance modelling

After the manufacturing process of the gasket components has been simulated, a further analysis stage consists of predicting the behaviour of the gasket under operating conditions. This typically includes simulation of the compression of the gasket between head and block. The results of the analyses include internal stress and strain, load-deflection characteristics, contact pressure at the interfaces with the head and the block and the deformed profile of the component. The detailed information provided by these analyses allows the engineer to determine whether the gasket design will seal adequately and whether there will be sufficient resistance to fatigue failure throughout the engine's life.

3.3.1 Eyelet component

Results of simulations of eyelet manufacture have already been described in (2) and

(3). However, work has continued to improve the accuracy of the analysis techniques and the understanding of the results they provide. These methods are now in everyday use throughout the Group.

3.3.2 Embossment component

After analysis of the manufacture of embossments as described in Section 3.2.2, the embossments are further analysed with respect to performance in service. The analyses consist of compressing and releasing the embossment between two rigid surfaces. Compression and recovery curves of both coated and uncoated embossments have been predicted and the results have correlated well with experimental data. Figure 10 shows measured and predicted load and unload curves for a full embossment of rubber coated stainless steel. Using facilities within the ABAQUS FE code, the analysis compensates for the rubber coating without the difficulties of modelling the thin and highly deformable layer itself.

The predicted curves can be used to ensure an adequate load on the embossment to prevent leakage. Additionally, the results show whether the recovery after compression will be sufficient to conform with the vertical movements between the head and block during the combustion cycle, although not so large that a dangerous fatigue cycle is initiated.

Figure 11 shows a compressed embossment along with the internal permanent plastic strains and contact pressures against the upper rigid platen. The maximum plastic strain increases during compression and again must be kept within safe limits. Those limits are determined through an experimental fatigue program. To assess local sealing capability (micro-sealing), the contact pressure between the gasket surface and the upper and lower plates are monitored. The pressure profiles provide information about the pressure peaks that create barriers against leakage. In this example, two peaks are seen near to the centre of the compressed embossment and contribute to micro-sealing. The maximum contact pressures must be kept below the yield point of the block and head material to avoid marking and damage.

The results of many in-service performance analyses have been included within a design database for steel embossments that is currently used by applications engineers.

3.3.3 Elastomeric component

Several different technologies can be used to provide fluid sealing in metallic gaskets. These include half embossments in steel layers and elastomeric beads attached to the metallic gasket body.

There are important design aspects to consider when specifying a gasket with an elastomeric bead. The elastomeric material is characterised by its softness, high recovery, incompressibility, and the ability to withstand large strains before rupture. However, although such high conformability makes the elastomeric component an excellent sealant, it cannot withstand large loads or high temperatures such as those reached around the bore.

Consequently, a design optimisation program has been developed using FE methods. The aim has been to limit internal shear stress and compressive stress during clamping between head and block by redesigning the profile of the bead. Figure 12 shows an optimised elastomeric bead in a compressed state. The stress within the bead has been greatly reduced from the unoptimised design. The main reductions are obtained along the interface between the bead and the gasket where the optimised bead shows stresses that are only 10% of those in the unoptimised design. This stress reduction means that one of the main causes of failure in elastomeric beads, that of debonding, is eliminated.

4. EXPERIMENTAL METHODS FOR GASKET VALIDATION

When the design of the prototype gasket has been determined by the applications engineer's experience and FE modelling, the acceptability of the design must be validated by a series of experimental tests.

The experimental processes involve both static and dynamic testing techniques, some of which are described in the following sections. Any deficiencies in the gasket design are modified by design iteration, as shown in Figure 1. Further FE simulations may not be necessary if only minor changes are made to complete the gasket design.

The measured results are also essential for validation and modification of the predictive FE techniques.

4.1 Static tests

4.1.1 Pressure sensitive paper

To record detailed information about the pressure distribution in the plane of the gasket, a pressure indicating paper can be used between the gasket and cylinder head or block plate. The pressure evaluation is made by measuring the absorption of incident light and a processor is used to calculate and record the pressure values in MPa.

This method can be used to detect areas of low sealing pressure and the effectiveness of any necessary changes to the gasket design. Because similar results will have already been predicted by the FE engine structure modelling techniques, correlation with the FE analyses is made to help validate the method.

Figure 13 shows the measured and predicted pressure distributions around one cylinder sealed by a multi-layer steel (MLS) diesel engine gasket. High pressures are concentrated on the stopper around the bore and on the areas of the body around the bolts. Additional high pressure areas are measured around the bore where the full embossments are in contact with the head and block. These are seen as the narrow rings of high pressure and provide sealing in addition to the stopper during the engine firing cycles. Similar high resistance paths are recorded around the fluid holes and the edge of the gasket where half embossments are used to provide fluid sealing.

The FE pressure prediction shows only one element layer of high pressure around the bore. This is because the mesh refinement of this model allows the bore seal of stopper and full embossment to be represented by one element only. Even so, the predicted pressure distribution is reasonably accurate, with the highest pressures around the bore and the outside of the bolts. Both the measured and predicted results show lower pressures around the bore in the region of the prechamber and over the left hand side bore bridge. The FE methods therefore correlate well with measurements made using the universally accepted pressure sensitive paper.

Validation of the FE methods is carried out whenever possible to help develop the confidence of the FE engineers, the applications engineers and the customers in the predictive methods being used.

4.1.2 Gasket thickness measurement

To determine the installed thickness of the gasket, and to examine the deformations of the head and block under bolt load clamping, lead pellets are incorporated into the gasket that is then clamped within the engine by applying the bolt loads. Again, the information collected can be compared with the results obtained using the FE methods for further validation. Additionally, it helps to determine whether the customer's requirements concerning combustion compression volume are met.

4.1.3 Pressure tests

Static pressure tests of the combustion area using nitrogen gas are universally used and give information about the micro-sealing capability of the gasket under cold clamping conditions. The micro-sealing is dependant upon the surface finish of the gasket, head and block. Leakage due to the surface roughness of any of these faces is usually prevented by the application of suitable coatings to the gasket. The coolant system can also be pressurised to check for internal and external fluid leakage.

4.1.4 Ultrasonic bolt load measurement

To determine the tensile stresses in the bolts once the gasket has been fitted into the engine, ultrasonic measurements of the bolt elongation are performed. These measurements can be repeated after various stages of an engine test program. Ultrasonic measurements give a more accurate indication of the residual bolt load than the retorque-to-position method. Figure 14 shows the initial bolt load and the loss of bolt load after different stages of an engine test. It can be seen that almost all the bolt load reduction occurs during the first 12 hours of engine operation. During this period the new gasket settles under the bolt and thermal loading, causing bolt load loss. The bolt load reduction is about 30% and must have been anticipated during the prototype design stage for the gasket to seal successfully.

4.2 Dynamic tests

To test all the load situations that a cylinder head gasket may face during an engine's lifetime, a Modular Test Program has been developed (4). This includes an initial leakage test, a full load hot test, an ice start and a thermal shock program. The structure of the Modular Test Program allows the test sequence to be varied according to the requirements. Used in the development of prototype gaskets, this test gives the maximum information within the limited time span available.

The decisive dynamic test is the validation according to the customers own specification and no gasket design is complete until this test has been successfully concluded.

Dynamic testing in the Automotive Sealing Group also includes continuous measurement of the combustion gas leakage rate and sealing gap height. These methods are summarised in the following sections.

4.2.1 Gas leakage measurement

The leakage of combustion gases during engine operation can be continuously detected and quantified (5). The principle of the method is that of separating the leakage gas from the coolant and recording the collected volume. This is an improvement over the conventional static method that requires the engine operation to be stopped, thereby introducing the possibility for error. From a plot of the volume of leakage gas versus time, the rate of gas leakage can be determined.

Measurements of the combustion gas leakage rate during an engine test and the corresponding static pressure test results are shown in Figure 15. The static test results show a continuous decrease in the static sealing pressure to a value, after 100 hours of testing, that is well below the customer's requirements. The leakage rate measurements show increases during the tests, but reductions as the switches between tests are made. The leakage rate increases are most rapid during the hot (full power) tests with a leakage rate after 100 hours of 7 litres per hour, far more than a customer would tolerate.

Gasket redesign would be necessary in this example, whether static or dynamic testing methods were used as the arbiter. However, the dynamic method not only avoids reliance on single readings, but produces continuous information about changes in leakage rate during each of the tests.

4.2.2 Sealing gap height measurement

A system to enable the continuous measurement of sealing gap height during engine operation has been developed at T&N Technology (6). The system involves capacitive transducers that are built into the gasket and connected to a data collection unit.

The capacitive transducers and the gap height between head and block surfaces recorded from three transducers during the heating up of an engine are shown in Figure 16. It can be seen that several transducers, and adjacent thermocouples, can be located within a gasket so that a large amount of information can be gained about the gap height during transient conditions of engine operation. The system is useful in helping to understand any leakage problems caused by vertical deformations of the block and head during thermal cycling.

5. SUPPLEMENTARY PREDICTION & MEASUREMENT TECHNIQUES

5.1 Bore distortion prediction and measurement

Bore distortion is caused by localised application of pressure due to bolt loading and thermal expansions arising from a non-uniform temperature distribution. The consequences of a non circular bore profiles are a risk of increased oil consumption and therefore raised emissions levels. Consequently, the gasket designer must consider the effects on bore distortion of the sealing system designs.

FE models of the engine system can be used to predict bore distortion. Figure 17 shows finite element predictions of the bore shapes under conditions of cold clamping and engine operation. The cold clamped model shows the classic clover leaf profile caused by the arrangement of the four bolts around the bore. During engine operation the bore distortion changes due to the thermal expansion of the components, particularly the aluminium cylinder head and, in the case of IDI diesel engines, the prechamber. In this particular case, additional forces due to the prechamber expansion superimpose an oval bore distortion and cause the apparent buckling of the cylinder bore wall. FE methods allow an insight into bore distortion during engine operation, something that cannot be easily achieved experimentally.

Measurement of static bore distortion with the INCOMETER system is a world-wide standard and is described in several publications (7,8). The elastic bore distortion caused by the clamping forces is calculated by subtracting any initial distortion of the unclamped cylinder from the distortion after clamping. A Fourier analysis procedure divides the distortions into different harmonic orders. Of these orders, the second order distortion represents an oval profile, while the fourth order distortion represents a clover leaf profile. The fourth order profile is the most critical because the piston rings can experience difficulties conforming to the clover leaf shape. The experimental measurements again allow correlation with the FE analyses.

Figure 18 shows the bore distortion induced by a prototype gasket, which was reduced by minor changes to the gasket design. These changes involved extending the eyelet material around the bolt holes, thus transferring some of the eyelet load onto the material around the bolts with beneficial results.

5.2 Temperature measurement in the gasket plane

To study the temperature variations over their structure, cylinder head gaskets can be equipped with thermocouples, as shown in Figure 19. In composite gaskets, thermocouples of 0.5 mm diameter can easily be fitted into the facing material.

Typical locations for thermocouples include the bore eyelet, the bridge area between the cylinders, the prechamber of an IDI diesel engine and the gasket body close to coolant or oil passages. Most of these locations are included in the sixteen examples of Figure 19.

A knowledge of the temperatures around the gasket is essential for the choice of the appropriate materials such as elastomers for surface coatings, or for inserted rubber grommets. This essential thermal information is gathered as a function of engine speed and load.

The temperatures around the cylinder bore for engine operation at maximum power output from thermocouples located as in Figure 19 are shown in Figure 20. The highest recorded temperatures in this IDI diesel engine are found in the prechamber area of the aluminium head and reach a maximum of 280°C. Temperatures in the head and block regions away from the prechamber were much lower, in the range 150°C to 175°C.

6. CONCLUSIONS

This paper has described the design process, and some of the development techniques, that are used to design modern cylinder head gaskets.

The use of Finite Element (FE) analyses enable predictions of gasket and engine behaviour to be made before engine hardware is available for experimental testing. FE analyses allow an insight into the internal stresses and strains of the gasket during manufacturing and in-service. Such knowledge is used to design gaskets that provide better sealability and increased durability. Several gasket design options can be analysed rapidly using FE methods. Thus, lead times can be reduced because gasket designs can be made that are increasing 'right first time'.

The measurement techniques described allow gasket designs to be evaluated for sealability and durability within the engine. Modular engine test programs have been designed to test gaskets under extreme conditions in relatively short periods of time. In a similar way to FE analyses, these tests are aimed at reducing lead times for new gasket designs. The measurement techniques are also used to provide information for the validation of the FE analyses, an essential provision to ensure confidence in the analytical results.

Using a combination of predictive and measurement techniques, gaskets are being designed to seal modern, high performance, reduced emissions engines in continuously reduced lead times.

REFERENCES

(1) **Voit, K., Knowland, C. and Gordon C.,** 'Cylinder Head Gaskets with System Capability'. AutoTech 94.

(2) **Bono, A., Percival, P. and Harland, C.,** 'Gasket Technology - Development of Design Techniques'. T&N Technical Symposium, 1990 Paper No. 7.

(3) **Harland, C., Novaria, P. and Robinson, M.,** 'Process and Performance Modelling of Gasket Components'. SAE Technical Paper 930118.

(4) **Jost, W.,** 'The Goetze Modular Test'. Goetze Dichtungen Brochure.

(5) **Jost, W. and Ehrenpreis, R.,** 'The Dynamic Sealing Behaviour of Cylinder Head Gaskets Under Thermal Shock Conditions'. Goetze AG Technical Paper D14.

(6) **Woodward, A.,** 'A System for the Static and Dynamic Measurement of Cylinder Head to Block Separation in Operating Engines', SAE Technical Paper 940290.

(7) **Loenne, K.,** 'The Effects of Engine Operating Conditions on Static Cylinder Distortions', Goetze AG Technical Paper D7.

(8) **Loenne, K. and Ziemba, R.,** 'The Goetze Cylinder Distortion Measurement System and the Possibilities of Reducing Cylinder Distortions'. SAE Technical Paper 880142.

ACKNOWLEDGEMENTS

The authors would like to extend their thanks to Dr K. Voit and Mr C. Winter of AE Goetze, Germany, for their invaluable contribution to the engine structure modelling work reported in this paper.

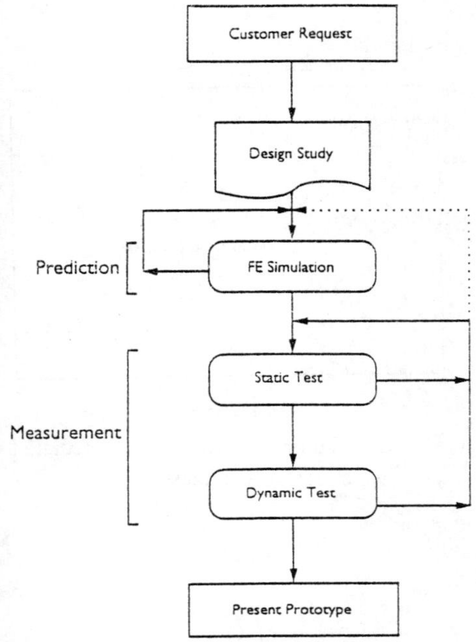

Figure 1 Gasket design process flowchart

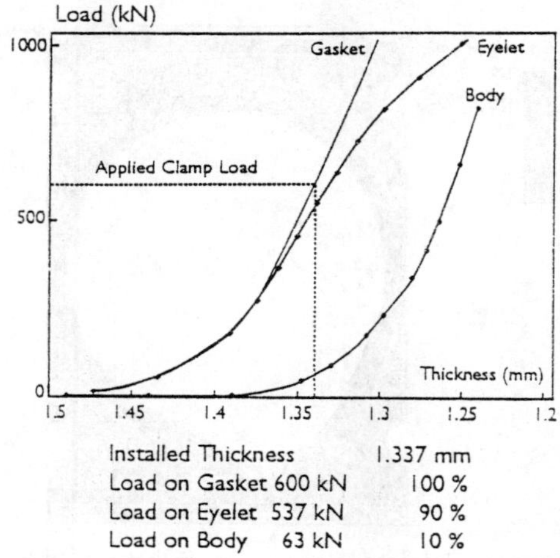

Figure 3 Load deflection curves of gasket body and eyelet materials

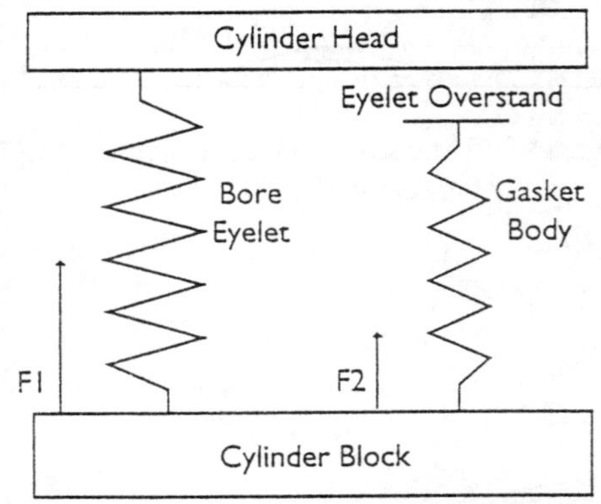

Figure 2 Representation of load balance software

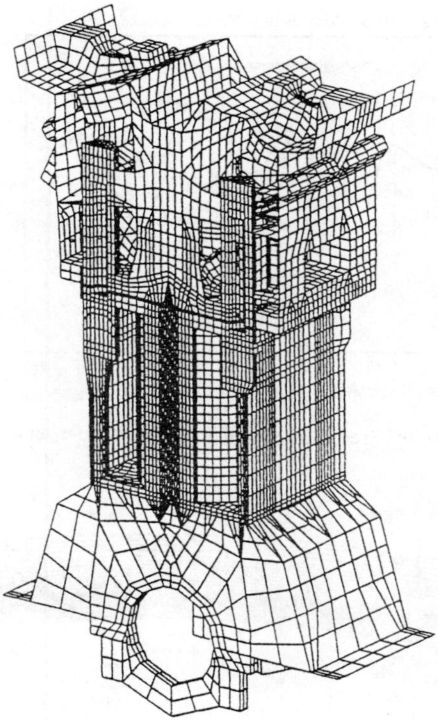

Figure 4 3D FE engine model

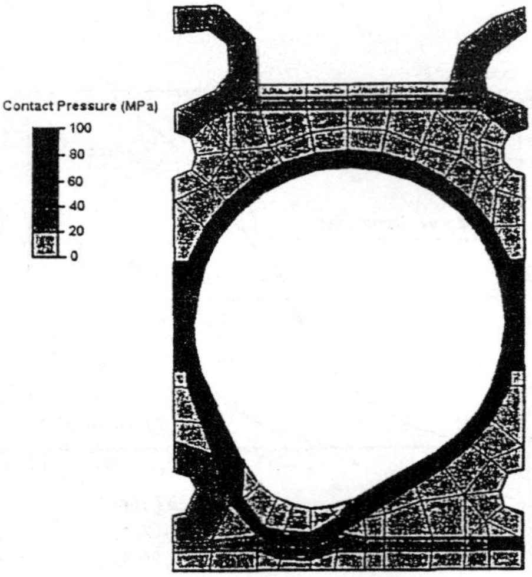

Figure 5 Predicted pressure on gasket

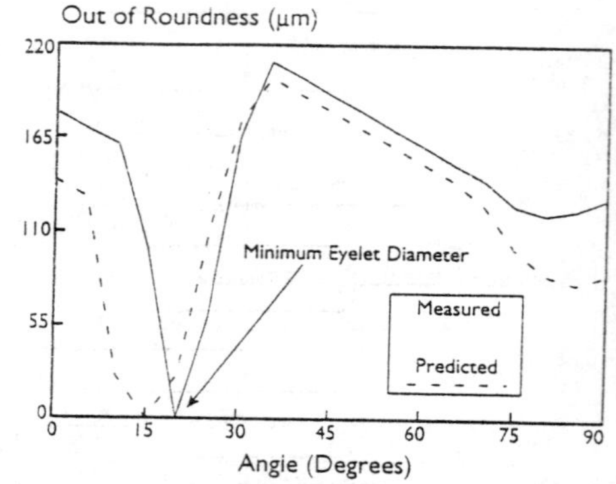

Figure 8 Correlation of measured and predicted eyelet out of roundness

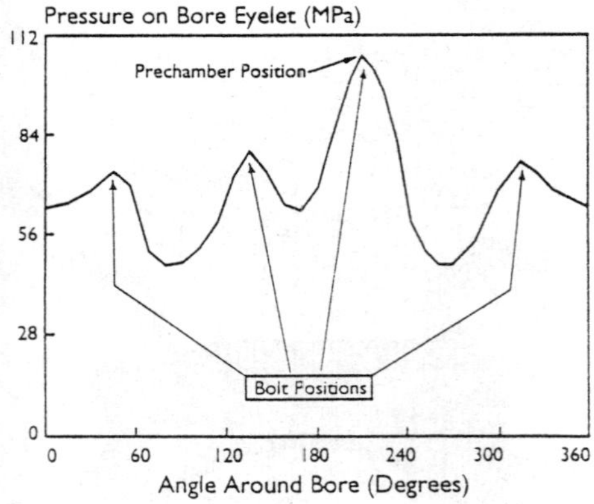

Figure 6 Predicted pressure around bore eyelet

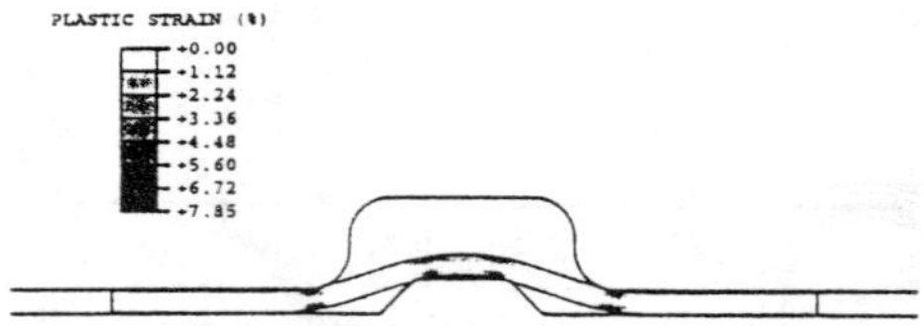

Figure 9 Plastic strain in MLS embossment after forming

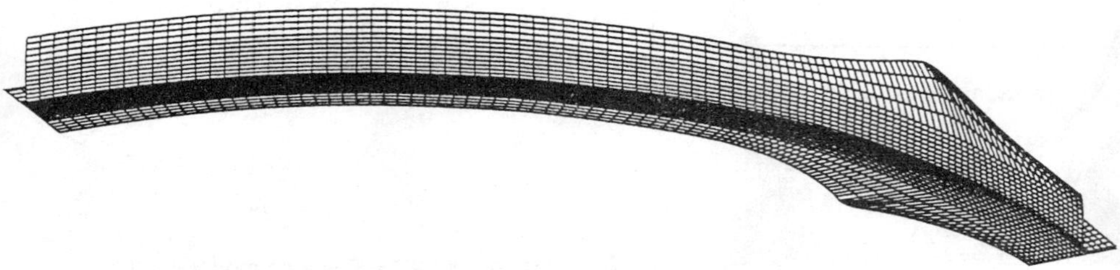

Figure 7 3D FE eyelet model used for manufacturing simulation

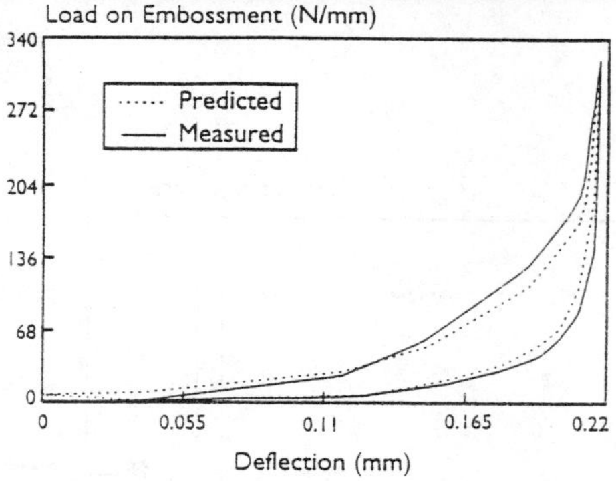

Figure 10 Correlation of measured and predicted embossment load/unload curves

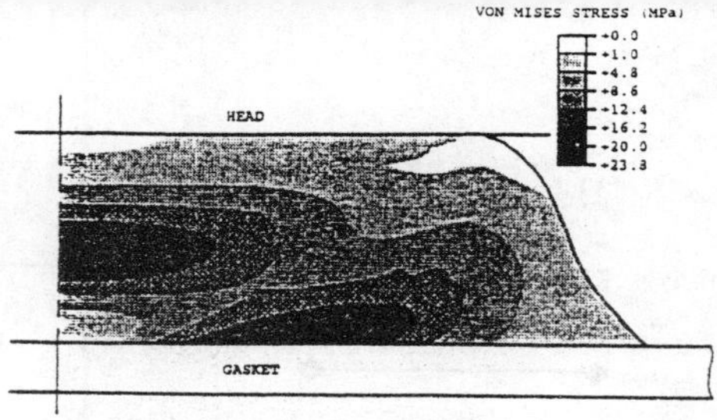

Figure 12 Stress in compressed elastomeric bead

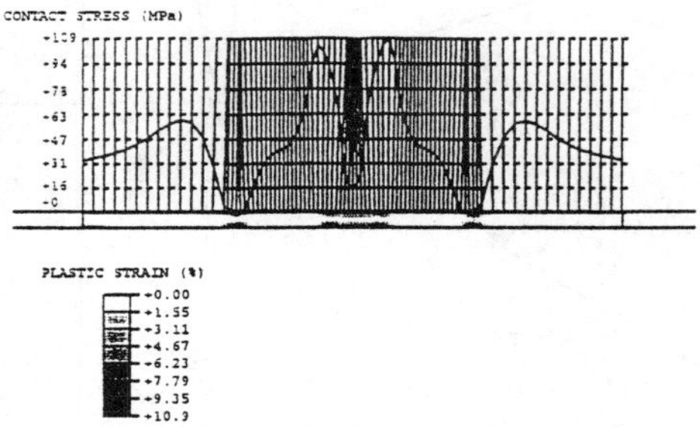

Figure 11 Plastic strain and contact pressure of compressed embossment

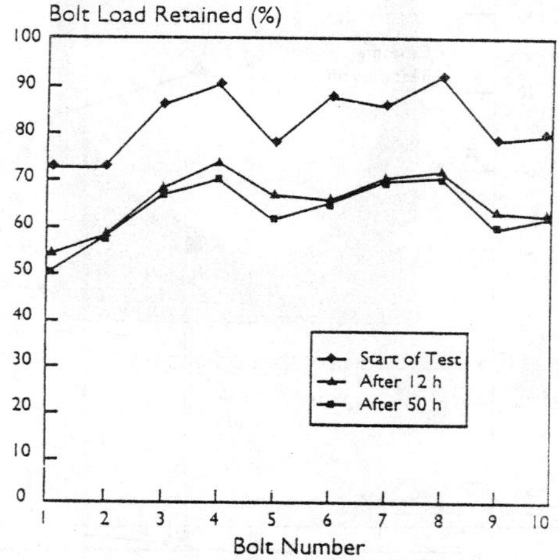

Figure 14 Ultrasonic bolt load measurements during engine test

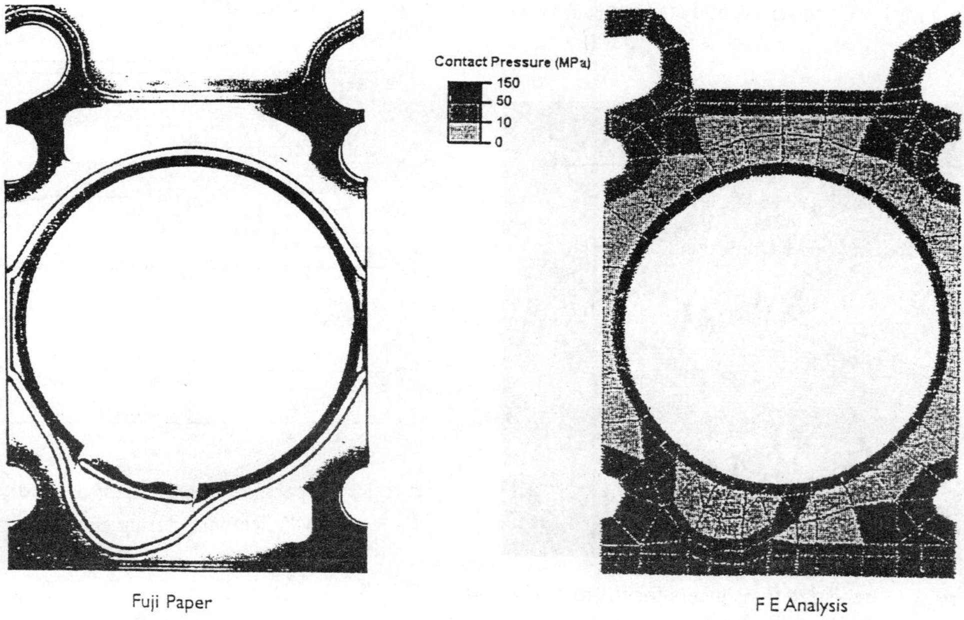

'igure 13 Comparison of contact pressures measured by Fuji paper and predicted by FE analysis

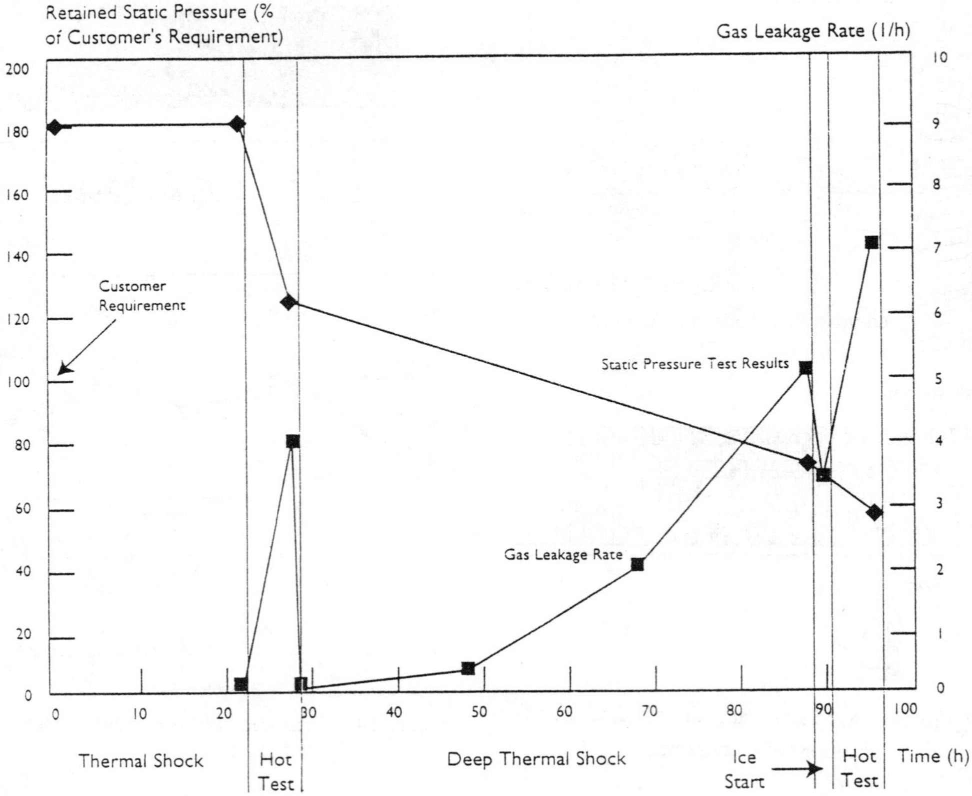

Figure 15 Comparison of dynamic gas leakage rates and retained static pressures

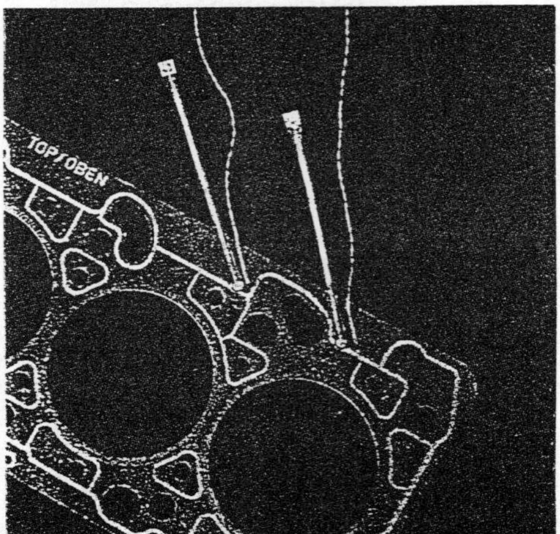

Figure 16a Capacitive transducers for the measurement of head to block clearance

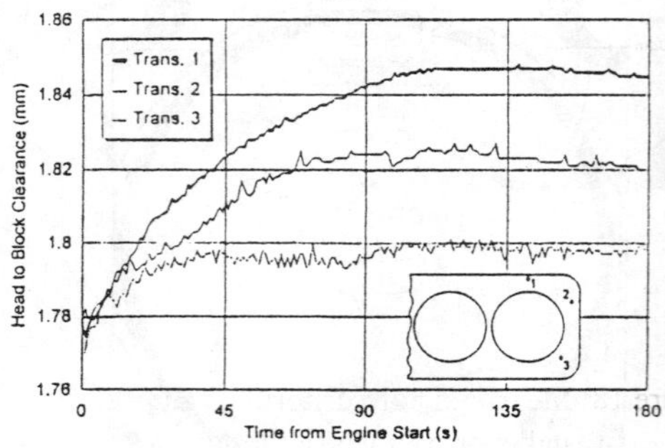

Figure 16b Capacitive measurement of head to block clearance during an engine start

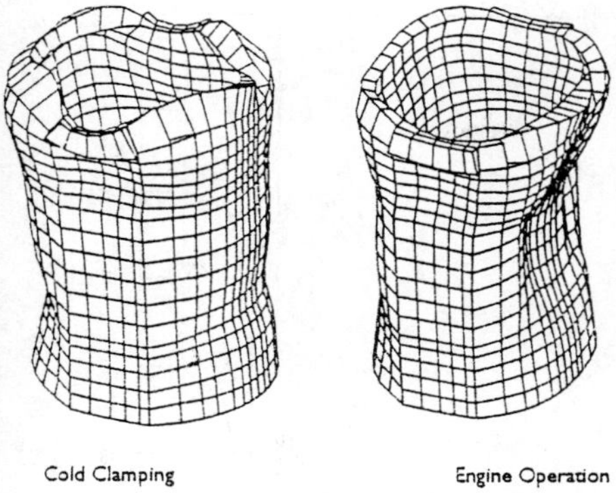

gure 17 Predicted bore shapes for cold clamping and engine operation

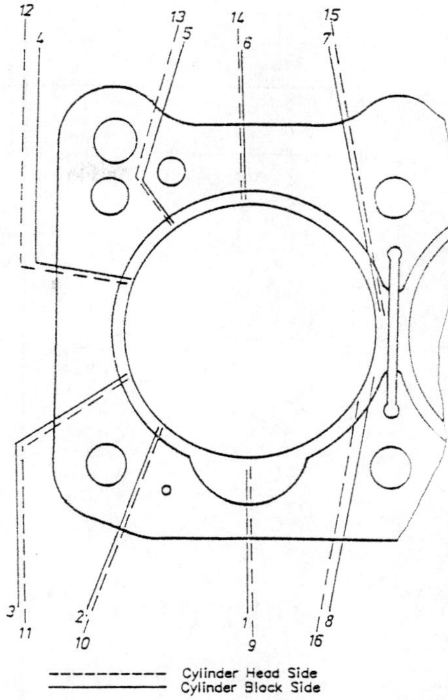

Figure 19 Thermocouple locations around bore

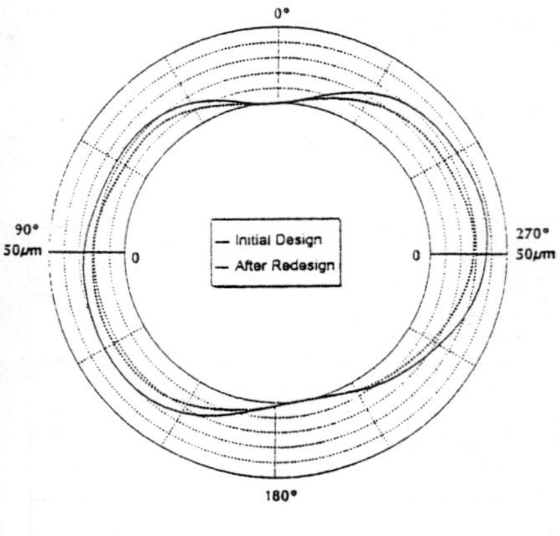

gure 18 Measured bore distortions induced by initial and improved gasket designs

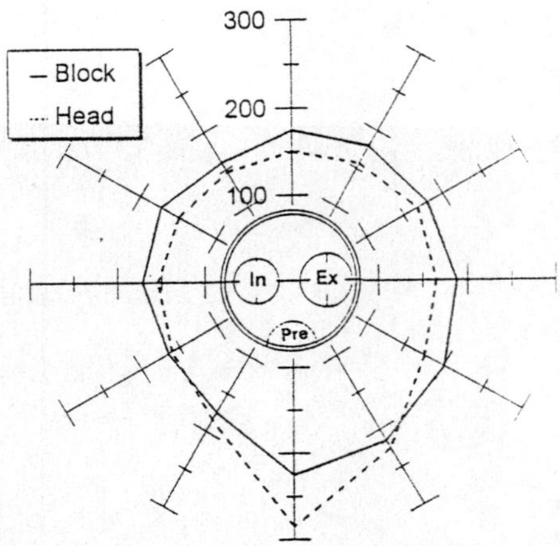

Figure 20 Measured temperatures around bore (°C

C498/39/206/95

The influence of crankshaft surface characteristics on the security of the bearing system

C S CROOKS and **D D PARKER**
Glacier Vandervell, UK

SYNOPSIS

A fluid film bearing is a load transmitting system which has three components: two surfaces in relative motion and a separating fluid. In the internal combustion engine this translates to a crankshaft journal, a bearing and lubricating oil. System failure is usually characterised by damage to the bearing but it is often strongly influenced by the other two components, especially the crankshaft journal. Today's higher demands place greater emphasis on the interaction of all components. Understanding these interactions can significantly help in the development of new or improved bearing materials and also point the direction of overall system improvements.

INTRODUCTION

The development of bearing materials requires the measurement of bearing material properties whilst closely simulating engine conditions. Most important are fatigue strength and surface properties, the latter usually referred to as seizure resistance.

Fatigue strength is perhaps the simplest of these properties to measure although in the context of bearings by no means being fully understood. Most manufacturers achieve this under test rig conditions not dissimilar to an engine. The Sapphire and Underwood machines are widely accepted in this context and have been described elsewhere(1).

'Seizure' resistance is a complex property which generates much discussion amongst tribologists. It comprises many elements: compatibility, conformability and embeddability to name but three. Only an engine test can embody all elements and then in no known proportions from test to test or engine to engine. By comparison, a test rig is generally used to isolate particular characteristics of a bearing material which may lead to insufficient similarity to an engine's complexity. In the authors' company several procedures have been developed, used and sometimes discarded because they failed to rank materials as engines appeared to do.

It has been apparent for some time that engine developments require not only bearings with increased fatigue strength but also increased seizure resistance. Methods adopted by the industry to date to measure seizure resistance have been unreliable and were not able to quantify accurately either conformability or compatibility. Conformability may be defined as the system's ability to accommodate minor geometrical inaccuracies of the surfaces whilst compatibility is the ability of the two surfaces to be in contact without system failure. A test has been developed which has been applied not only to the study of these elements with regard to bearing materials, but also to investigate the bearing system as a whole, particularly shaft surface characteristics.

DEVELOPMENT OF THE NEW SEIZURE TEST

The new seizure test is based on the Sapphire test machine which dynamically loads the test bearing thus simulating the combustion load. Many manufacturers have used seizure tests based on the application of a steady load to the test bearing. In the authors' experience such tests can be dramatically influenced by minute errors in concentricity of the test journal. A few microns eccentricity can be introduced when such test shafts are ground. The resultant small cyclic element to the test load substantially thickens the oil film by inducing squeeze effects and an order of magnitude effect on load carrying capacity has been observed. Where a dynamic load is applied via a deliberately eccentric shaft the effect is controlled.

A previous development of a seizure test using the Sapphire machine, whilst providing sensible discrimination of materials relative to field behaviour generally resulted in extensive plastic deformation of the test bearing lining prior to actual seizure. Specific loads on the test bearings were extremely high and could exceed 200 MPa. The new test was developed to eliminate this distortion. Severe deformation of the lining is generally not a field characteristic of scuffing or seizure events, although extensive deformation usually follows seizure.

The major prerequisite of the new test was to operate under much thinner oil films in order to induce asperity contact between shaft and test bearing. This leads to seizure at loads more typical of those found in the internal combustion engine. It was achieved by the introduction of a narrow axial groove in the lining near to the bearing crown which disrupts the oil film and severely reduces its load carrying capacity. Experience has shown that increasing the load stepwise produces uncontrollable, almost random, thermal instabilities. Therefore an important additional feature of the test is that the peak value of the dynamic load is continuously ramped up under computer control, producing controlled instability.

The principle of the seizure test machine, now widely used in the authors' company, is illustrated in Fig. 1 together with the new test schedule.

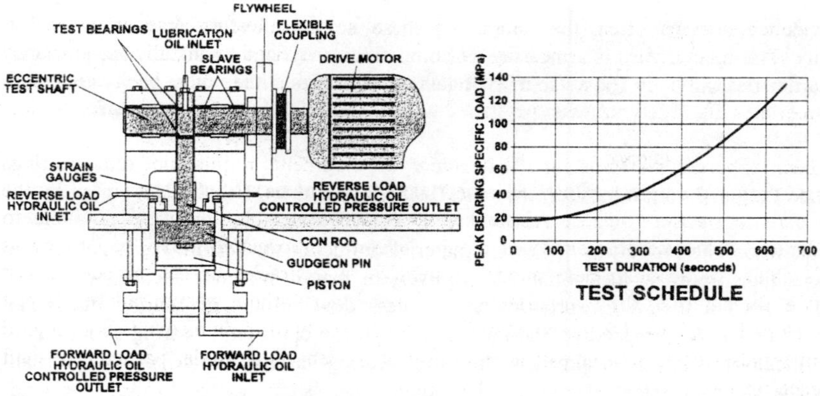

Fig. 1 Sapphire test rig principle and test conditions

Examples of the test results can be seen in Fig. 2. Plotted, using a common y axis - load and temperature, are the back of bearing temperature measured in the rod half, oil inlet temperature and the peak specific load.

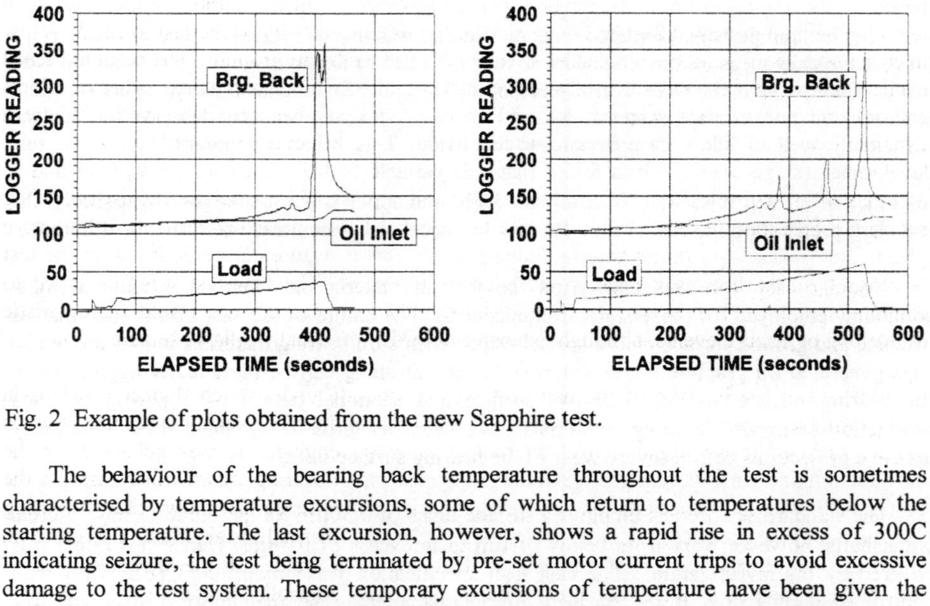

Fig. 2 Example of plots obtained from the new Sapphire test.

The behaviour of the bearing back temperature throughout the test is sometimes characterised by temperature excursions, some of which return to temperatures below the starting temperature. The last excursion, however, shows a rapid rise in excess of 300C indicating seizure, the test being terminated by pre-set motor current trips to avoid excessive damage to the test system. These temporary excursions of temperature have been given the name "scuff" events, the final event being called a "seizure". Examination of test bearings after a scuff event gives indications of localised modifications to the bearing surface and geometry, frequently barely visible and quite restricted in area. Some change in the bearing surface composition can also occur, possibly because of a local loss or redistribution of a soft phase constituent of the bearing material such as tin or lead.

This evidence suggests that the nature of these scuffing events is a measure of conformability. The final seizure is a measure of compatibility. Characteristically the stronger and hence harder bearing alloy shows scuff type behaviour whereas the softer alloys generally do not.

STUDIES OF THE BEARING SYSTEM - COUNTERFACE ROUGHNESS

It is clear that, whilst the properties of bearing materials influence the seizure resistance of the bearing system, they are by no means the only, or even the major, influence. Hence, whilst the primary use of such a test is for product and process development of bearing materials, considerable effort has been placed on the investigation of the crankshaft bearing system as a whole. Specifically this has been aimed at the other bearing surface, the crankshaft journal. Traditionally the technical effort of crankshaft producers has been directed towards the load carrying characteristics of the component. However, as demands on the bearing system increase, e.g. thinner oil films and higher loads, the characteristics and condition of the mating journal surface has to assume greater significance. Over the last few years the authors have encountered many instances of bearing damage being directly related to journal geometry or surface finish. These factors have largely stimulated the search for an improved understanding of the factors affecting system security.

The first and perhaps simplest aspect considered was that of journal surface finish. Surface finish is typically measured by engineers as Ra or the centre line average roughness and though many other parameters were considered, both 2D and 3D, obtained using stylus or quite sophisticated non contact systems, Ra was chosen. Ra appeared to describe the surface sufficiently well to allow comparisons to be made. This became apparent during the test development stage where it was found that this variable had to be closely controlled and a tolerance of $\pm 0.02\mu m$ Ra was ultimately achieved and applied. It is understood that should the test shaft preparation method change Ra may not necessarily accurately describe the difference.

Nodular cast iron was chosen as the journal material this being the material most commonly employed for passenger car crankshafts. The nature of the cast iron is such that its machine and ground surface places special requirements on bearing materials in that the ferrite caps generated on graphite nodules during surface finishing can be particularly aggressive to the bearing surface[2]. Fig. 3 shows surface and sectional details of a poorly polished production crankpin. Bearings materials have been designed to eliminate these potentially abrasive projections before severe wear of the bearing surface occurs[3].

The standard seizure test employs a surface finish of 0.30μm Ra measured axially. Actual crankshafts, however, vary from below 0.1μm to in excess of 0.4μm. Hence this range was covered by the investigation. The cast iron specification was grade 600/3 (BS.2789) with control of nodule size at the finished journal diameter. The metallurgical structure was consistent with that seen on customer examples. Final grinding of the test shaft was carried out in the favourable direction, i.e. with the burrs generated over the graphite nodules lying 'with' the relative direction of bearing rotation. This notation is illustrated in Fig. 4

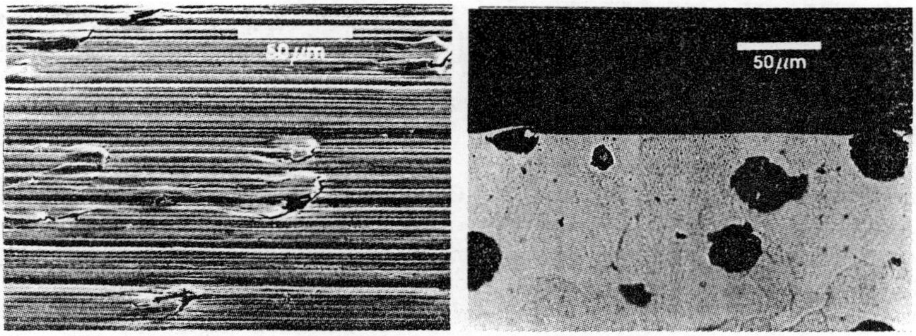

Fig. 3 Surface and section through a poorly finished crankpin.

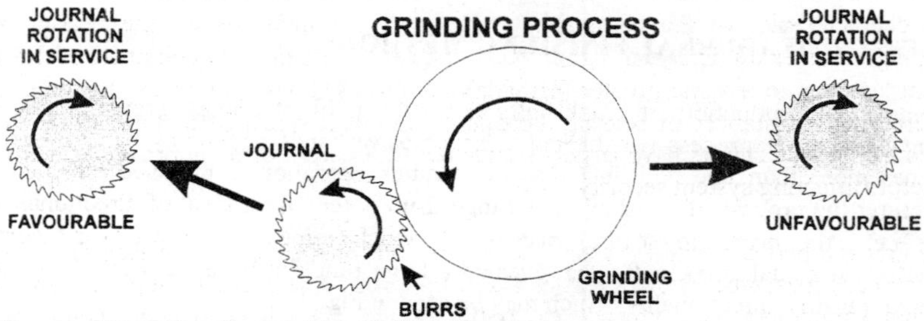

Fig. 4 Favourable, and unfavourable grinding directions.

An aluminium tin-silicon-copper alloy was selected as an appropriate test bearing material to quantify the effects of journal surface finish. Different aluminium bearings materials show a range of behaviour on the new test, particularly with regard to number or existence of scuffing events. This alloy was also developed with satisfactory operation against nodular cast iron as a major objective. Typically under standard test conditions it shows several 'scuff' events prior to seizure.

The load at seizure clearly indicated a very strong dependence on surface finish (Fig. 5). Plotting this data on a log log basis gave a straight line, slope - 1. Hence, in this test, shaft surface finish was inversely proportional to the seizure resistance of the system; halving the surface finish doubled the load at seizure. This is significant in that it would appear that appreciable improvements to the seizure resistance of a bearing system may be obtained by improved journal characteristics, perhaps even more than bearing material modification itself.

Interestingly the effect on scuff resistance was much less significant. This suggests that scuff is a much more complex characteristic, perhaps more dependant on specific or bulk bearing characteristics than operating conditions and counterface surface topography. This is consistent with the interpretation that the scuff rating is a measure of conformability. Conformability involves elastic and also possibly plastic deformation, both of which are dependant on bulk material properties. This argument is discussed in greater detail elsewhere along with its relevance to bearing material development[4].

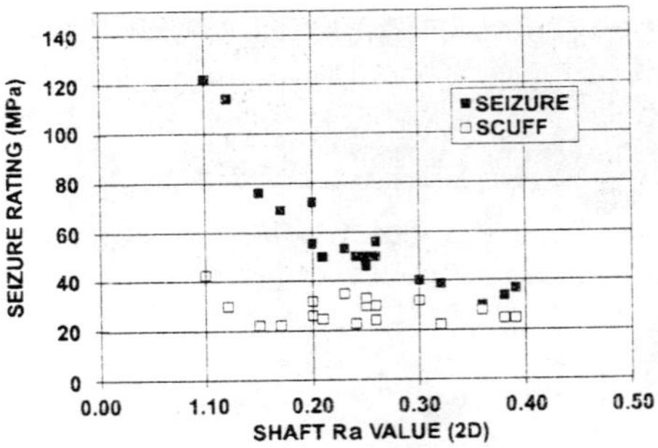

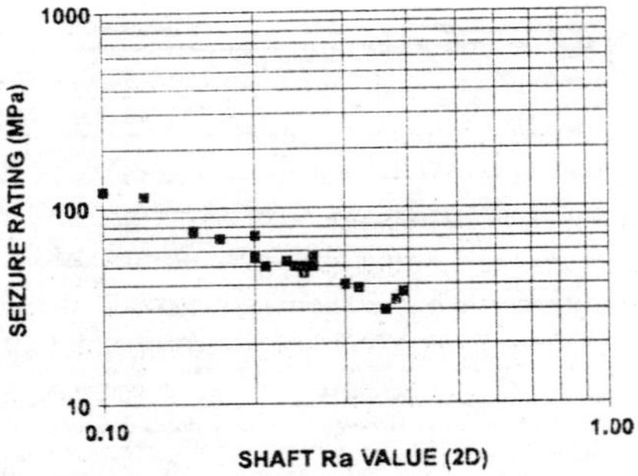

Fig. 5 The effects of journal surface finish on bearing surface properties.

THE EFFECT OF JOURNAL FINISHING METHODS

The commercial production of crankshafts normally involves plunge grinding followed by polishing. The latter process is achieved using abrasive bands backed by hard or soft shoes. Sometimes more than one polishing direction and/or operation is involved. Depending on manufacturer the process of polishing may range from potential removal of directionality in the surface, i.e. the burrs, to several microns of stock removal and geometry correction. Examination of actual crankshafts has revealed a large range in roughness and in some cases detrimental residual directionality. Such may be seen in Fig. 3

It is often stated that grinding direction relative to journal rotation is an important consideration to system performance[2]. In the favourable direction the lay of the finish and therefore the direction of burrs above graphite nodules is with the relative bearing rotation or counter the direction of journal rotation. Hence it was felt important to verify this statement and the new test appeared to offer an opportunity to quantify the performance differences.

Accepting the limitations on our ability to reproduce commercial finishing processes an investigation was undertaken to examine and compare four variants.

Treatment T1	Ground in the favourable direction (60 grit)
Treatment T2	Ground in the unfavourable direction (60 grit)
Treatment T3	Ground in the unfavourable direction (60 grit) and polished in the favourable direction (440 grit paper)
Treatment T4	Ground in the unfavourable direction (50 grit). Tufftrided and polished in the favourable direction (400 grit paper).
Normal surface finish range 0.1 to 0.4 μm Ra	

Table 1 Journal finishing variants

Once again an aluminium-tin-silicon-copper alloy was the chosen material against which to quantify the finishing routes and the same grade of cast iron was also employed. Prior to use the test journal surfaces were characterised using 2 dimensional stylus equipment. Again Ra was selected as that most commonly used in the engineering industry. This is not to refute however that in some situations specific relationships between surface properties and such functions as bearing area may dominate[5]. Results of the load at seizure from the standard T1 condition gave excellent agreement with the previous data reported earlier in this paper, indicating the test to have good reproducibility, Fig. 6.

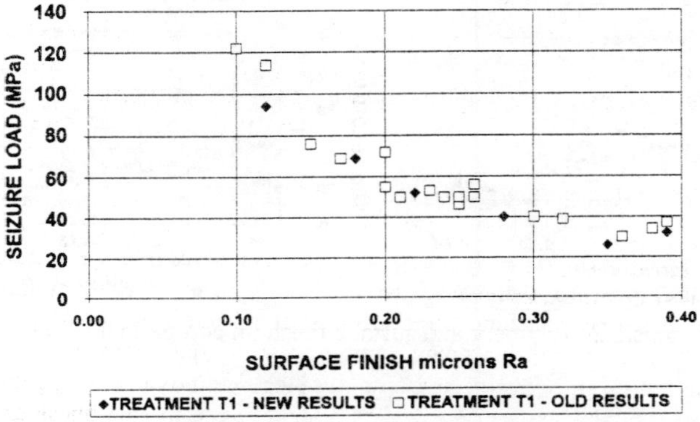

Fig 6 Test reproducibility for favourably ground journal surfaces.

A strong relation of seizure resistance to surface finish was noted for all treatments, similar to those obtained in the previous tests. Results for each treatment are shown in Fig. 7 which also has a summary of results against mean surface finish. Conformability results obtained showed a similar weak relationship with surface finish as in the previous test and are not reported in this paper.

The rankings of the preparation routes were the same for both surface properties but more clearly defined in the case of seizure. The data provide strong evidence that not only the surface finish but also the means by which it is achieved can significantly influence the security of the bearing system. Confirmed are the effects of grinding directionality on a cast iron crankshaft and also the fact that a reverse and favourable polishing operation may not necessarily eliminate such effects. This is also observed in actual engine production situations.

Tufftriding, which is occasionally employed to surface harden crankshafts, is a low temperature cyanate based nitriding process. It produces a thin (10-20μm) hard, coherent layer. Because the process causes slight roughening of the surface, subsequent polishing is recommended for bearing applications. Condition T4 ground in the unfavourable direction, Tufftrided, polished in the favourable direction, produced a significant improvement compared with T3 which was untreated. The implication of this is that either the burrs are more effectively removed from the hardened surface, or the increased surface hardness compensates for the adverse grinding direction. Possibly both mechanisms are operative.

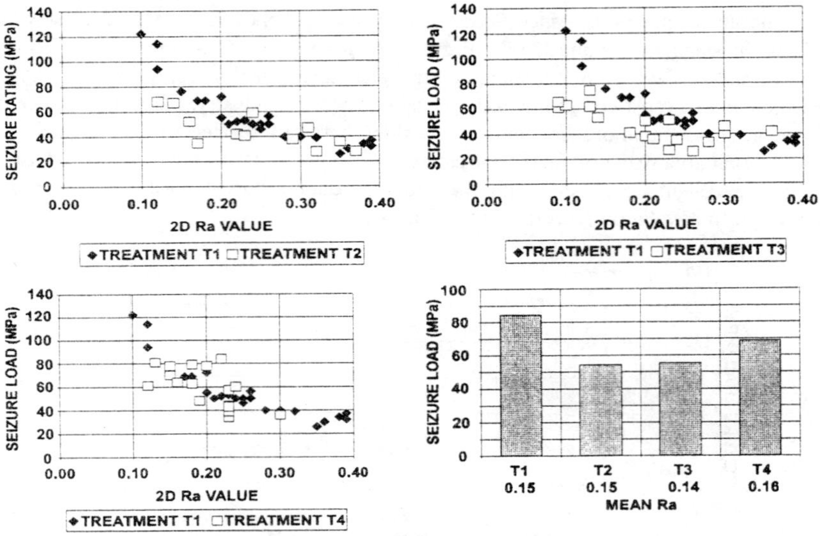

Fig. 7 The effect of finishing treatment and surface finish on compatibility (seizure rating)

CONCLUSIONS

A dynamically loaded bearing test has been developed which can evaluate and hence compare surface interactions. The test has been successfully supplied to the development of improved crankshaft bearing materials and to the investigation of the bearing system as a whole.

It has been shown that journal surface finish can have a profound effect on system security, smoother surfaces considerably reducing the risk of seizure. Methods of journal finishing i.e. grinding and polishing, can also influence performance. Grinding direction relative to journal rotation is highly significant for nodular cast iron shafts and it cannot be assumed that final polishing will necessarily remove adverse directionality. Surface hardening may promote additional security.

REFERENCES

1. Pratt, G.C., 'Materials for Plain Bearings'. Institute of Metals, Interanational Metallurgical Review 1973.
2. Davies F.A. 'Crankshaft Grinding and Plain Bearing Performance'. Automotive Engineer, IMechE 1976.
3 Davies, G.J., Eastham, D.R. and Eastwood, B.J., 'The Development of New Bearing Materials and Overlays'. AE Technical Symposium 1986, Paper 10.
4. Brown W.P.,Galand H.,et al, 'Aluminium Bi-metal Bearing Developments for Automotive and Medium Speed Diesel Applications'. T&N Technical Symposium 1995, Paper 1
5 Conway-Jones, J.M. and Eastham, D.R., 'Parameters for Control of Roughness of Surfaces Operating with Thin Oil Films'. Int. J. Mach Tools Manufact., Vol. 35 No. 2

